THE HISTORY OF THE ROYAL AND INDIAN ARTILLERY IN THE MUTINY OF 1857

Lieut. Colonel Henry Tombs V.C., C.B.
Bengal Horse Artillery
after a photograph taken in 1858

THE HISTORY OF THE ROYAL AND INDIAN ARTILLERY
IN THE MUTINY OF 1857

BY COLONEL JULIAN R. J. JOCELYN

WITH MAPS AND ILLUSTRATIONS

The Naval & Military Press Ltd

Published by
The Naval & Military Press Ltd
Unit 10 Ridgewood Industrial Park,
Uckfield, East Sussex,
TN22 5QE England
Tel: +44 (0) 1825 749494
Fax: +44 (0) 1825 765701
www.naval-military-press.com

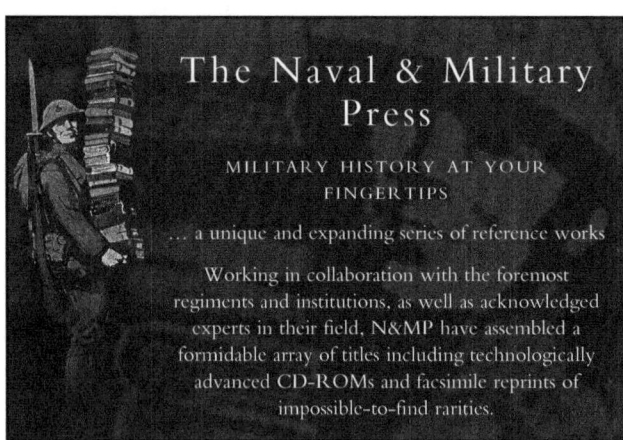

The Naval & Military Press

MILITARY HISTORY AT YOUR FINGERTIPS

... a unique and expanding series of reference works

Working in collaboration with the foremost regiments and institutions, as well as acknowledged experts in their field, N&MP have assembled a formidable array of titles including technologically advanced CD-ROMs and facsimile reprints of impossible-to-find rarities.

In reprinting in facsimile from the original, any imperfections are inevitably reproduced and the quality may fall short of modern type and cartographic standards.

PREFACE

WHEN I began to trace the services of the Royal Artillery in 1857-8 I found them to be inextricably involved with those of the Bengal Artillery and in a less degree with those of the Madras and Bombay Artillery, and that the amalgamation of the four regiments, promulgated officially in 1861, took place practically at an earlier period on the battlefields of India.

With the consent of the Committee of the Royal Artillery Institution I have related the services of all four, and the present volume thereby becomes a military history of the Indian Mutiny. In telling the story of the campaign, which has sometimes been regarded as a series of local episodes, it has been my object to narrate events in their proper sequence, and to show that each skirmish, siege, and battle possesses its own place in a connected history.

The narrative is based almost entirely on official despatches and on the books, diaries, letters, and conversations of military eye-witnesses whose names will be found in the List of Authorities.

For the illustrations I am indebted to Colonel

G. A. Crawford, late R.A., Lieutenant-Colonel E. D. Swinton, D.S.O., R.E., and Major H. G. Lloyd, D.S.O., R.A.

I wish again to acknowledge the kindly encouragement of the Committee of the Royal Artillery Institution and my gratitude to Lieutenant-Colonel H. W. L. Hime, (late) R.A., with regard to whom I wish to say, as I said on a former occasion, that whatever merit the book possesses is due, in no small degree, to his never-failing and untiring assistance.

JULIAN R. J. JOCELYN.

LONDON,
June 1915.

LIST OF AUTHORITIES, Etc., REFERRED TO IN THE TEXT

When an abbreviated title is used it is added in italics

OFFICIAL PUBLICATIONS

The London Gazette, 1857, 1858, 1859.
Medal Rolls and other Documents in India Office.
Official Army Lists, Hart's, Bengal, Madras, and Bombay.
Selections from the Letters, Despatches, and other State Papers preserved in the Military Department of the Government of India. Edited by G. W. Forrest, C.I.E. Vols. i., ii., iii., and iv., 1902–12. (*State Papers.*)
Despatches of R.A. Officers published in Occasional Papers, R.A. Institution, Woolwich. Vol. i., 1860.
Reports on the Engineering Operations at the Siege of Lucknow, March 1858, by Major-General Sir R. Napier, K.C.B., Colonel Harness, C.B., and Lieutenant-Colonel Lennox, R.E. Professional Papers of the Corps of Royal Engineers, vol. x. (New Series), p. 67, etc.
Report on the Defences of Lucknow, by Brigadier-General Sir R. Napier, R.E., and other Officers, R.E., March 1858. *Ibid.*, vol. ix. (New Series), p. 17, etc.
Journal and Reports of the Operations of the Royal Engineers at the Siege of Jhansi, by the late Colonel T. Fenwick and Major J. B. Edwards, R.E., April 1858. *Ibid.*, p. 33.
Narrative of the Siege of Delhi, by Lieutenant H. W. Norman, D.A.A. and Q.M.G. (*State Papers*, vol. i. p. 429, etc.)
Madras Artillery Records: Regimental Intelligence—Miscellaneous, vols. 12 and 15 (Royal Artillery Institution Library, Woolwich).

GENERAL PUBLICATIONS, Etc.

Author, etc.	Publication, etc.
Adye, Gen. Sir J. M., G.C.B.	The Defence of Cawnpore by Major-Gen. Windham. London, 1858.
Alison, Gen. Sir A., Bt., G.C.B.	Lord Clyde's Campaign in India, by Lieut.-Col. A. Alison. Blackwood's Magazine, October 1858.
Anderson, Capt. T. C., late 12th Ben.N.I.	Ubique. War Services of all the Officers of H.M. Bengal Army, Calcutta (undated).

LIST OF AUTHORITIES, Etc.

Author, etc.	Publication, etc.
Anonymous	The Defence of Lucknow, by Staff Officer.
Arbuthnot, Major G., late R.A.	Letters.
Arbuthnot, Major-Gen. H. T., C.B., late R.A.	Letters and Conversations.
Ball, Mr. C.	The History of the Indian Mutiny, by Charles Ball (undated). The London Printing Company. (*Ball.*)
Bonham, Col. J., C.B., late R.A. (formerly Ben.A.).	Letters and Conversations.
Bourchier, Col. Sir G., K.C.B., late R.A. (formerly Ben.A.)	Eight Months' Campaign against the Bengal Sepoy Army, by Col. G. Bourchier, C.B., Ben.H.A. 1859.
Burne, Major-Gen. Sir O. T., K.C.S.I.	Rulers of India : Clyde and Strathnairne, by Major-Gen. Sir O. T. Burne, K.C.S.I. 1891. (*Clyde and Strathnairne.*)
Coghill, Col. K. J. W., late 19th Hussars and formerly 2nd European Bengal Fusiliers.	Letters.
Crowe, Major-Gen. T. C., late R.A. (formerly Bom.H.A.).	Letters and Conversations.
Ford, Col. A., C.B., late R.A.	Diary and Letters.
Forrest, Sir G., C.I.E.	A History of the Indian Mutiny, by G. W. Forrest, C.I.E. Vols. i., ii., iii., 1904 ; vol. iv., 1913. (*Forrest.*)
Geary, Lieut.-Gen. Sir H. Le G., K.C.B.	Diary.
Grant, Gen. Sir J. H., G.C.B.	See Knollys.
Gubbins, Mr. M. R., Bengal Civil Service.	An Account of the Mutiny in Oudh and the Defence of the Lucknow Residency, by M. R. Gubbins. 1858. (*Mutinies in Oudh.*)
Hills-Johnes, Lieut.-Gen. Sir James, V.C., G.C.B. (formerly Ben.H.A.).	Letters and Conversations.
McLeod Innes, Lieut.-Gen. J. J., V.C., R.E.	Lucknow and Oudh in the Mutiny, by Lieut.-Gen. McLeod Innes, V.C., R.E. 1896. (*Lucknow and Oudh.*)
,, ,, ,,	The Sepoy Revolt, a Critical Narrative by Lieut.-Gen. McLeod Innes, V.C., R.E. 1897. (*Sepoy Revolt.*)
Jones, Capt. Oliver J., R.N.	Recollections of a Winter Campaign in India, by Capt. Oliver J. Jones, R.N. 1859.

LIST OF AUTHORITIES, Etc.

Author, etc.	Publication, etc.
Kane, John, Lieut. and Adj. Invalid Artillery.	List of Officers of the Royal Regiment of Artillery. 4th ed., 1900. (*Kane's List.*)
Kaye, John William, F.R.S.	A History of the Sepoy War in India. Vol. i. (2nd ed.), 1865; vol. ii., 1865; vol. iii., 1870. (*Kaye.*)
Knollys, Col. Sir H., K.C.V.O., late R.A.	Incidents in the Sepoy War of 1857-8. Compiled from the private journal of Gen. Sir Hope Grant, G.C.B., with Explanatory Chapters by Capt. H. Knollys, R.A. 1873. (*Knollys.*)
Kyle, Col. S. C., late R.A.	Diary.
Leslie, Major J. H., late R.A.	A List of Officers who have served in the Madras Artillery. 1900. (*Leslie's List.*)
Majendie, Col. Sir, V. D., K.C.B., late R.A.	Up among the Pandies, by Lieut. V. D. Majendie, R.A. 1859.
Malleson, Col. G. B., C.S.I.	History of the Indian Mutiny, 1857-8. Vol. i., 1878; vol. ii., 1879; vol. iii., 1880. (*Malleson.*)
Maude, Col. F. C., V.C., C.B., late R.A.	Memories of the Mutiny, by Francis Cornwallis Maude, V.C., C.B., late Col. R.A. 2 vols. 1894. (*Memories of the Mutiny.*)
Maunsell, Gen. Sir F. R., K.C.B. (formerly Ben.E.).	Various Papers and Memoranda and Conversations.
Rideout, Maj.-Gen. A. K., C.B., late R.A.	Letters and Conversations.
Roberts, Field-Marshal Earl, V.C., K.G., K.P., G.C.B., O.M., G.C.S.I., G.C.I.E.	Forty-One Years in India, by Field-Marshal Lord Roberts of Kandahar, V.C., G.C.B., G.C.S.I., G.C.I.E. 2 vols., 1897. (*Forty-One Years in India.*)
Smith, Maj.-Gen. W., late R.A.	Letters.
Spring, Col. F. W. M., late R.A. (formerly Bom.A.).	The Bombay Artillery List of Officers. (*Spring's List.*)
Strange, Maj.-Gen. T. Bland, (late R.A.).	Letters, Memoranda, and Conversations.
Stubbs, Maj.-Gen. F. W., late R.A. (formerly Ben.A.)	List of Officers who have served in the Regiment of the Bengal Artillery. 1892. (*Stubbs' List.*)

Author, etc.	Publication, etc.
Stubbs, Maj.-Gen. F. W., late R.A. (formerly Ben.A.)	History of the Organization, Equipment, and War Services of the Regiment of Bengal Artillery, by the same. Vol. iii., 1895. (*Stubbs.*)
Taylor, Miss	Life of Gen. Sir Alexander Taylor, K.C.B., R.E., by his daughter. 1913.
Thomson, Capt. Mowbray, late 53rd Ben.N.I.	The Story of Cawnpore, by Capt. Mowbray Thomson, 53rd Ben.N.I. 1859. (*Mowbray Thomson.*)
Verney, Capt. Sir E. H., Bt., R.N.	The Shannon Naval Brigade in India, by Lieut. E. H. Verney, R.N. 1858.
Wood, Field-Marshal Sir E., V.C., G.C.B., G.C.M.G.	The Revolt in Hindustan, by Field-Marshal Sir E. Wood, V.C., G.C.B. G.C.M.G. 1908.

CONTENTS

PREFACE . . . . pp. v, vi

LIST OF AUTHORITIES, ETC. . . . . pp. vii–x

CHAPTER I

INTRODUCTION . . . . pp. 1-6

CHAPTER II

THE OUTBREAK OF THE MUTINY

I. *Bengal:* Mutinies at Berhampore and Barrackpore — The First Reinforcements. II. *Meerut:* Inactivity of the Garrison—Escape of the Mutineers. III. *Delhi:* The Moghuls—Arrival of the Meerut Mutineers — Loss of the City — The Blowing-up of the Magazine—The Flight from Cantonments. IV. *Umballa:* Concentration of British Troops — Subsequent Movements. V. *The Punjaub:* Lahore—Lawrence at Peshawar—Sepoy Regiments Disarmed—Mooltan—Jullundur—Jhelum—Sialkote—The Movable Column. VI. Rajputana—The Native States—Nusseerabad—Neemuch—Reinforcements from Deesa. VII. *Central India:* The Native States—Jhansi—Jalaon and Hamirpore—Gwalior—Indore—Hungerford at Mhow—Saugor. VIII. *Rohilkhund, Agra, and Northern Doab:* Loss of Rohilkhund—Mutinies round Agra—Action at Sassiah, July 5—Futtehghur—Hansi, Hissar, and Sirsa. IX. *Oudh and Cawnpore:* Oudh in 1857—Sir Henry Lawrence—Garrison of Oudh—Mutinies in the Province—Bonham's Escape—Cawnpore—The Nana Sahib—The Entrenchment—Mutiny of the Native Troops. X. *Allahabad, Benares, and Lower Doab:* Allahabad—Benares—The Disarming Parade—Mutiny at Allahabad—Defence of the Fort—Relief of Allahabad. XI. *Benares District, Behar,* and *Chota Nagpore:* Mutinies in the District. XII. *Madras.* XIII. *Bombay:* Energetic Action of the Governor—Risings in the Presidency. XIV. *Hyderabad:* Shorapore and Chicumba, etc. XV. *The Task of the British.* Appendix to Chapter II, *Native Fidelity* . . . . . pp. 7-67

xii   CONTENTS

CHAPTER III

The Delhi Campaign

I. *The Occupation of the Ridge :* Brigadier Wilson on the Hindun, May 30 and 31—Barnard's Advance on Delhi—Action of Badli-ki-Serai, June 8—Seizure of the Ridge. II. *Strength of the Rebels in Delhi and the British on the Ridge :* The Defences of Delhi—Strength of the Rebels and their Reinforcements—The British Position—British Force and Reinforcements. III. *The Defence of the Ridge :* Occupation of the Ridge—Rebel Attacks, June 9, 10, and 11—Rebel Attacks, June 12 and 15—Constant Artillery Fire—Kishangunge, June 17—Ochterlony Gardens, June 19—The Racecourse Picket—Subzi Mundi, June 21 —New British Pickets—Rebel Attacks, June 22–30—British and Rebels Reinforced—Affair of July 4—Brigadier Wilson succeeds to the Command—Cavalry Raid, July 9—Subzi Mundi, July 9—Subzi Mundi, July 14—Subzi Mundi, July 18—Improvements in the Defences of the Ridge—Metcalfe's House, July 23—Rebel Attacks, August 2— Rebel Pickets driven in, August 12—The Movable Column—Battle of Najafghur—Rebel Attack, August 25—The Last British Reinforcement. IV. *The Capture of Delhi :* The Question of the Assault of Delhi—Brigadier Wilson's Views—Baird-Smith and Alexander Taylor—The Proposed Plan of Attack—The Plan Accepted by Wilson —The Samee House and No. 1 Siege Battery—No. 2 Siege Battery— No. 3 Siege Battery—No. 4 Siege Battery—Action taken by the Rebels —The Assault—No. 1 Column—No. 2 Column—No. 3 Column—No. 4 Column—Defeat of No. 4 Column—The Cavalry Brigade—The Reserve —The Work of September 14—Capture of Delhi—A Notable Siege. V. *The Pursuit of the Rebels :* Greathed's Column—Action at Bulandshahr, September 28—Panic at Agra—Action at Agra, October 10— Grant replaces Greathed in Command—General Van Courtland and Brigadier Showers—Gerrard's Column—Action at Nurnaul, November 16

pp. 68–144

CHAPTER IV

The Lucknow Campaign

I. *Cawnpore :* Anxiety about Cawnpore—Defence of the Entrenchment—Treachery of the Nana Sahib—The Massacre. II. *The Defence of the Residency :* Activity of the Rebels—The Battle of Chinhut, June 30—The Residency invested—The Defences of the Residency— Repulse of Rebel Attacks—The Relief. III. *Havelock's Efforts to relieve Lucknow :* Lines of Communication—Formation of Relief

Columns at Allahabad—Havelock at Futtehpore—Action at Futtehpore, July 12—Action at Aong, July 15—Action at the Pandoo River, July 15—Battle of Cawnpore, July 16—Occupation of Cawnpore—Havelock crosses Ganges—Action at Unao, July 29—Capture of Bassiratgunge, July 29—Retirement to Mangalwar—2nd Action at Bassiratgunge, August 5—Action at Boorhya-ki-Chowki, August 12—Action at Bithur, August 16—Havelock's Dangerous Position. IV. *The Lines of Communication*: Mutiny at Dinapore—Kunwar Sing—Mutiny at Segowlie—Mr. Vicars Boyle—The Siege of Arrah—Defeat of Dunbar's Column—Vincent Eyre—Action at Bibigunge, August 2—Relief of Arrah—Capture of Jugdispore—Help from Nepaul—Trouble on the Grand Trunk Road—Action of the Government—Sir James Outram. V. *The Advance on Lucknow*: Outram's Self-negation—Havelock's Force—Capture of the Alam Bagh, September 23—Capture of the Char Bridge, September 25—Advance to the Moti Mahal—Action of the Rear Guard—The Advance to the Bailey Guard—Final Movements—The First Relief of Lucknow . . . . pp. 145–205

CHAPTER V

THE LUCKNOW CAMPAIGN (*Continued*)

I. *Arrival of Sir Colin Campbell*: The New Commander-in-Chief—Preparations for an Advance on Lucknow—Menace to the Grand Trunk Road—Action at Khujwah, November 1—Hopes of the Mahrattas—Sir Colin goes to the Front. II. *The British Forces in Oudh and at Cawnpore*. III. *The Advance on Lucknow*: Sir Colin's Plans—The Advance on the Dilkhusha—Halt at the Dilkhusha—Preparations in the Residency—November 16—The Sekundra Bagh—Capture of the Barracks—Capture of the Sekundra Bagh—The Advance continued—Capture of the Shah Najaf—Casualties on November 16—Action taken by the Residency Garrison—November 17—Capture of Khurshid Manzil and Moti Mahal—Skirmishes on the Left and Centre—Bombardment continued, November 17 and 18—The Exodus of November 19—Bombardment continued, November 21–24—Relief of the Residency—Evacuation of the Residency. IV. *The Defeat of Tantia Topi*: Sir Colin's Plans—Death of Havelock—Sir Colin marches to Cawnpore—Windham at Cawnpore—His Plans—Combats round Cawnpore, November 26—Tantia Topi's Force—November 27—November 28—Sir Colin crosses the Ganges—His Available Force—Battle of Cawnpore, December 6—Affair of Serai Ghat, December 9
pp. 206–41

CONTENTS

CHAPTER VI

THE MILITARY SITUATION AT THE END OF 1857

I. *Rajputana and Central India:* The Jodhpore Legion—Kotah—Prince Feroz Shah—The Malwa Field Force—Action at Dhar, October 22—Capture of Dhar, October 31—Skirmish at Rawal, November 8—Relief of Neemuch—The Nagpore Column. II. *Events North of the Jumna:* Benares District—Allahabad—Futtehpore—North-West Provinces and the Punjaub. III. *Plans of the Government*   pp. 242–250

CHAPTER VII

THE LUCKNOW CAMPAIGN (*Continued*)

I. *The Advance into Oudh:* Seaton at Gangaree, December 15—Action at Patiala, December 18—Seaton's Junction with Walpole—The Commander-in-Chief's Movements—Action on the Kali River, January 2—Concentration at Futtehghur—Walpole on the Ramgunga—Action near Shamshabad, January 26—The Camp at Bantara—Capture of Miangunge, February 23. II. *The Reinforcements from the East and South-East:* The Gurkhas—The Jaunpore Field Force—The "Nazim" of Sultanpore—Affair near Sekundra, January 22—Action at Chanda, February 19—Action at Sultanpore, February 20—Action on Fort Dhowrara, March 4. III. *Outram's Defence of the Alam Bagh*—The Attacks on his Position. IV. *The Defences of Lucknow:* The Inner Line of Defence—Second Line—First Line—Defences in the City. V. *The Army of Oudh.* VI. *The Siege and Capture of Lucknow:* The Advance to the Dilkhusha, March 2—Occupation of the South of Lucknow, March 3 and 4—Sir Colin's Plan of Attack—The Siege Batteries begun—Passage of the Gumti—March 6—March 7—March 8, Right Attack—March 8, Left Attack—March 9, Right Attack—March 9, Left Attack—March 10, Right Attack—March 10, Left Attack—March 11, Left Attack—March 11, Right Attack—March 12, Right Attack—March 12, Left Attack—March 13, Left Attack—March 13, Right Attack—March 14, Left Attack—March 14, Right Attack—March 15, Left Attack—March 16—March 17–18—March 19—March 21—Action at Kursi, March 23 . . . . . pp. 251–96

CHAPTER VIII

CENTRAL INDIA

I. *The Central India Field Force:* Sir Hugh Rose at Sehore—Siege of Rahatghur—Action at Barodia, January 31—Relief of Saugor—Movements of the 1st Brigade—Siege of Chanderi, March 5–17—

CONTENTS

Jhansi—The Siege of Jhansi, March 23 to April 3—Tantia Topi's Attempt to relieve Jhansi—Battle of the Betwa, April 1—Storming of Jhansi, April 3. II. *The Rajputana Field Force :* Capture of Awah—General Roberts' Force—Siege of Kotah, March 24–30. III. *The Advance on Kunch :* Occupation of Poonch—Affair of Lohari, May 2—Reappearance of the Begum of Jhansi—Battle of Kunch, May 5—The Nawab of Banda. IV. *The Saugor Field Force :* Whitlock's Advance on Banda—Battle of Banda, April 18—Flight of the Nawab to Kalpi. V. *Capture of Kalpi :* Sir Hugh Rose's Movements—Skirmishes outside Kalpi, May 15-17—Concentration of the Field Force—Plans of the Rebels—Battle of Gulauli, May 22—Capture of Kalpi, May 23—Affair of Sher Ghat, May 16—Dispersal of the Troops. VI. *The Gwalior Campaign :* The Seizure of Gwalior—Rose's Plans—Battle of Morar, June 16—Brigadier Smith at Kotah-ki-Serai—Action at Kotah-ki-Serai, June 17—Battle of Gwalior, June 19—Capture of the Fort, June 20—Action at Jaora-Alipore, June 22. VII. *The Saugor Field Force :* Movements of the Saugor Field Force. VIII. *Beyt Island :* Attack on Beyt Island—Capture of Beyt Island . . pp. 297-351

CHAPTER IX

THE EXTINCTION OF THE MUTINY

I. *The Situation in Upper India, March* 1858 : The Forces of the Enemy—The Plans of the Commander-in-Chief. II. *Rohilkhund, March to June*, 1858 : Action at Kankar, April 7—Walpole at Ruiya—Affair at Sirsa, April 22—Sir Colin's Advance on Bareilly—The Meerut Column—The Rurkhi Column—The Battle of Bareilly—Combats round Shahjehanpore—Pacification of Rohilkhund. III. *Benares District and Behar*: Mahndi Husain and Kunwar Sing—Combats round Azimghur, March 22 to April 6—Lugard's Column—Kunwar Sing's Retreat on Jugdispore—Defeat of Le Grand—Clearing the Jugdispore Jungle—Mahndi Husain driven across the Rapti. IV. *The Summer Campaign in Oudh*, 1858 : Sir Colin's Plans—The Defences of Lucknow—Action at Bari, April 12—Affair of Simri, May 10—Action at Newabgunge-Barabanki, June 12—Relief of Maun Sing—Operations at Sultanpore in August—Affairs at Salimpore, September 22 and 23—Skirmishes about Derriabad—Brigadier Berkeley's Operations. V. *Operations in Oudh during October :* Sir Colin's Plans—Skirmishes at Miangunge and Kantur, October 5 and 7.—Affair at Doadpore, October 20—Skirmish at the Kandu River, October 27—Operations on the Western Frontier of Oudh—Affair at Jamo, October 8—Capture of Birwah, October 21—Capture of Ruiya, October 28—Affair at Bunkagaon, October 8—Brigadier Troup's Operations—The Eastern Frontier of Oudh. VI. *The Winter Cam-*

xvi   CONTENTS

paign in Oudh, 1858-9: The Force at the Disposal of the Commander-in-Chief—Capture of Rampore Kussia, November 3—Capture of Amithi and Shunkerpore, November 10 and 16—Brigadier Evelegh's Movements—Affair at Dundeakhera, November 24—Gordon and Carmichael's Columns—Brigadier Horsford's Operations—Completion of the First Part of Sir Colin's Plans—Passage of the Gogra—Lord Clyde's March from Lucknow to Fyzabad—Grant's Movements—Lord Clyde's Movements—Grant's Movements—Action at Tolsipore, December 23—Action at Burgeedia, December 26—Capture of Musjeediah, December 27—The Rebels enclosed—Action at Banki, December 30—Subsequent Movements. VII. *Pursuit of Tantia Topi*, 1858-9: Position of British Troops after Capture of Gwalior—Rebel Hopes—Tantia Topi's March to Tonk—His Flight along the Banas—Action on the Banas, August 13—Tantia Topi's Flight to Jhalawar Territory—His Attempts to reach Indore—Action at Beora, September 15—Action at Mangaoli, October 10—Action at Sindwaho, October 19—Action at Kurai, October 25—Action at Bagrode, October 26—Tantia Topi crosses the Nurbudda—Skirmish at Rajpore, November 19—Action at Chota Udepur, December 1—Affairs at Pertabghur and Zirapore, December 25 and 29—Affair at Chubbra, December 31—Meeting of Tantia Topi with Maun Sing and Feroz Shah—Maun Sing's Outbreak—Siege of Paori—Affairs of Bijepur and Kundri—Reappearance of Feroz Shah, Ranode, December 17—Affair at Sarpore, December 22—Tantia Topi's Dash for Jeypore—Action at Dosa, January 14, 1859—Surprise at Sikar, January 21—Capture of Tantia Topi. VIII. *The Final Operations of the Saugor Field Force*: Jhansi District—Jalaon District—Action at Sahao, September 5—Hamirpore District—Kirwi District—Affair at Nyagaon, August 13—Carpenter's March to Kothi—Attack on Kirwi Palace, December 25—Action at Panwari, December 29—Affair on the Tons, March 4, 1859—Chanderi District. IX. *Reorganisation of the Indian Army*: Changes in the Bengal, Madras, and Bombay Armies—The Record of the Native Artillery—The Amalgamation of the Royal and Indian Artillery   .   .   .   .   . pp. 352-431

APPENDICES

	PAGE
APPENDIX I. THE ORGANISATION OF THE BENGAL, MADRAS, AND BOMBAY ARTILLERY	433
APPENDIX II. THE GARRISON OF INDIA, 1857-9:	
I. EUROPEAN FORCE IN INDIA, APRIL 1857 . . .	435
II. THE NATIVE ARMY	438
III. THE REINFORCEMENTS	439
IV. NEW REGIMENTS, ETC.	443

CONTENTS

APPENDIX III. CASUALTIES:

		PAGE
I.	DELHI, MAY 30–SEPTEMBER 20, 1857	444
II.	CHINHUT, JUNE 30, 1857	445
III.	LUCKNOW GARRISON, JUNE 30–SEPTEMBER 26, 1857 .	445
IV.	HAVELOCK'S FORCE, July 11 TO AUGUST 16, 1857 .	446
V.	OUDH FIELD FORCE, SEPTEMBER 25, 26, 1857 . .	446
VI.	FIELD FORCE UNDER GENERAL SIR COLIN CAMPBELL, NOVEMBER 12–22, 1857	447
VII.	ARMY UNDER SIR COLIN CAMPBELL, MARCH 2–21, 1858 .	448
VIII.	SIR HUGH ROSE'S JHANSI AND KALPI CAMPAIGN, 1858 .	449
IX.	SIR HUGH ROSE'S GWALIOR CAMPAIGN, 1858 . .	449

APPENDIX IV. SERVICES OF THE ROYAL AND INDIAN ARTILLERY, 1857–9:

I.	OFFICERS OF THE ROYAL ARTILLERY HOLDING COMMANDS OR STAFF APPOINTMENTS	450
II.	DETAIL OF TROOPS R.H.A. AND COMPANIES R.A. .	452
III.	OFFICERS OF BENGAL ARTILLERY HOLDING COMMANDS AND IN STAFF EMPLOYMENT	458
IV.	OFFICERS OF BENGAL ARTILLERY UNPOSTED OR ABSENT FROM THEIR TROOPS OR COMPANIES AT OUTBREAK OF MUTINY	460
V.	OFFICERS OF TROOPS BEN.H.A., COMPANIES BEN.A., ETC., WHICH MUTINIED	461
VI.	OFFICERS OF BENGAL ARTILLERY IN THE ORDNANCE DEPARTMENT AND VARIOUS MILITARY APPOINTMENTS	462
VII.	OFFICERS OF BENGAL ARTILLERY IN CIVIL EMPLOYMENT PRESENT DURING THE MUTINY	463
VIII.	DETAIL OF TROOPS OF BEN.H.A. AND COMPANIES BEN.A.	464
IX.	OFFICERS OF MADRAS ARTILLERY HOLDING COMMANDS, APPOINTMENTS, ETC.	474
X.	DETAIL OF TROOPS OF MAD.H.A. AND COMPANIES OF MAD.A.	475
XI.	OFFICERS OF BOMBAY ARTILLERY HOLDING COMMANDS, APPOINTMENTS, ETC.	478
XII.	DETAIL OF TROOPS OF BOM.H.A. AND COMPANIES OF BOM.A.	479
XIII.	OFFICERS OF CAVALRY AND INFANTRY ATTACHED TO ARTILLERY	482

CONTENTS

	PAGE
XIV. KEY TO SERVICES OF EXISTING BATTERIES R.H.A., R.F.A., AND COMPANIES R.G.A.	482
XV. LIST OF SURGEONS, ASSISTANT SURGEONS, AND VETERINARY SURGEONS, ROYAL ARTILLERY	484
XVI. LIST OF SURGEONS, ASSISTANT SURGEONS, AND VETERINARY SURGEONS OF THE INDIAN ESTABLISHMENT SERVING WITH ARTILLERY	485
XVII. LIST OF BATTLES, SIEGES, ACTIONS, ETC.	487
NOTES TO APPENDIX IV	491
APPENDIX V. THE INDIAN MUTINY MEDAL, BY MAJOR J. H. LESLIE, R.A. (Retired List)	493
INDEX	495

LIST OF ILLUSTRATIONS

MAJOR-GENERAL SIR HENRY TOMBS, K.C.B., V.C. . *Frontispiece*
When Lieut.-Colonel H. Tombs, V.C., C.B., etc.[1]

HALF-TONE ENGRAVINGS

	FACING PAGE
THE FLAG-STAFF TOWER, DELHI	18

From a contemporary photograph (Major H. G. Lloyd, D.S.O., R.A.).

A "FORT" IN OUDH 46
From Ball's "History of the Mutiny."

HINDU RAO'S HOUSE, DELHI 80
From a contemporary photograph (Major H. G. Lloyd, D.S.O., R.A.).

THE LANE WHERE NICHOLSON WAS WOUNDED . . . 120
From a contemporary photograph (Major H. G. Lloyd, D.S.O., R.A.).

THE KASHMIR BASTION, DELHI, SEPTEMBER 14, 1857 . . 128
From a contemporary photograph Major H. G. Lloyd, D.S.O., R.A.).

ARTILLERY CROSSING A NULLAH 248
From "Recollections of a Winter Campaign in India," by Captain Oliver J. Jones, R.N

THE NAVAL BRIGADE, H.M.S. "SHANNON," AT KHUDAGUNGE . 254
From "Recollections of a Winter Campaign in India," by Captain Oliver J. Jones R.N.

THE FORT AND CITY OF JHANSI, APRIL 3, 1858 . . . 308
From a War Office print.

---

[1] This picture was prepared from a photograph taken in the summer of 1858 by Colonel C. Hogge, C.B., Ben.A., a copy of which has been preserved by Lieut.-General Sir J. Hills-Johnes, V.C., G.C.B., and kindly placed by him at the disposal of the author. The date of the picture is fixed by the following:

Capt. and Brev.-Major H. Tombs, Ben.H.A., received his brevet as a Lieut. Colonel on Jan. 19, 1858; his C.B. on Jan. 22, 1858; his V.C., April 4, 1858; his brevet as Colonel, July 20, 1858; and the first G.G.O. concerning the Indian Mutiny Medal was issued on Aug. 15, 1858. In the picture he has the badges of a Lieut.-Colonel and is wearing the V.C., the C.B., the Punniar Star, and two Sikh Medals each with two clasps.

LIST OF ILLUSTRATIONS

ENGRAVINGS IN TEXT, ETC.

THE SEKUNDRA BAGH, NOVEMBER 16, 1857 . . *Title Page*
From a photograph in "State Papers," vol. ii. (Lieut.-Col. E. D. Swinton, D.S.O., R.E.).

THE BAILEY GUARD GATE OF THE RESIDENCY, MARCH 1858 .

Facing 1
From a photograph in "State Papers," vol. ii. (Colonel G. A. Crawford, late R.A.).

THE KASHMIR GATE OF DELHI, SEPTEMBER 14, 1857 . . 144

DHOOLY SQUARE, LUCKNOW, SEPTEMBER 25, 1857 . . . 205
From a photograph in "State Papers," vol. ii. (Lieut.-Col. E. D. Swinton, D.S.O., R.E.).

THE SHAH NAJAF, LUCKNOW, NOVEMBER 16, 1857 . . . 241
From a photograph in "State Papers," vol. ii. (Lieut.-Col. E. D. Swinton, D.S.O., R.E.).

THE KHURSHID MANZIL (MESS HOUSE), LUCKNOW . . . 428
From a photograph in "State Papers," vol. ii. (Lieut.-Col. E. D. Swinton, D.S.O., R.E.).

PLANS IN TEXT

BADLI-KI-SERAI, JUNE 8, 1857 73
CAVALRY RAID, DELHI, JULY 9, 1857 89
DEFENCES OF THE RIDGE, DELHI 97
NAJAFGHUR, AUGUST 25, 1857 101
SIEGE WORKS AT DELHI 131
BULANDSHAHR, SEPTEMBER 28, 1857 134
AGRA, OCTOBER 10, 1857 139
WHEELER'S ENTRENCHMENT, CAWNPORE, JUNE 1857 . . 148
CHINHUT, JUNE 30, 1857 156
THE DEFENCES OF THE RESIDENCY, LUCKNOW, JULY 1857 . 161
CAWNPORE, JULY 16, 1857 174
BASSIRATGUNGE, JULY 29, AUGUST 5, 1857 177
BITHUR, AUGUST 16, 1857 180
BIBIGUNGE AND ARRAH, AUGUST 1857 190
THE CHAR BRIDGE, LUCKNOW, SEPTEMBER 25, 1857 . . 197
APPROACHES TO THE BAILEY GUARD, LUCKNOW RESIDENCY,
    SEPTEMBER 25, 1857 201
    After a photograph from a model in "Lucknow and Oudh."
APPROACHES TO THE BAILEY GUARD (PLAN) 203
THE KALI RIVER, JANUARY 2, 1858 255
    From "Recollections of a Winter Campaign in India," by Captain Oliver J. Jones, R.N.

LIST OF ILLUSTRATIONS

	PAGE
SHAMSHABAD, JANUARY 26, 1858	258
From "Recollections of a Winter Campaign in India," by Captain Oliver J. Jones, R.N.	
ALAM BAGH DEFENCES, NOVEMBER 28, 1857 TO MARCH 1, 1858	271
RAHATGHUR, JANUARY 28, 1858	299
CHANDERI, MARCH 17, 1858	305
JHANSI, MARCH 23 TO APRIL 3, 1858	307
THE BETWA, APRIL 1	310
SIEGE OF KOTAH, MARCH 24–30, 1858	317
KUNCH, MAY 7, 1858	321
KALPI AND GULAULI, MAY 22 AND 23, 1858	333
BAREILLY, MAY 5, 1858	361

MAPS, ETC.

MAP
- I. INDIA, 1857
- II. THE RIDGE AND THE CITY OF DELHI
- III. LUCKNOW CITY (1ST AND 2ND RELIEF)
- IV. CAWNPORE CITY
- V. LUCKNOW CITY (SIEGE AND CAPTURE)
- VI. COUNTRY ROUND GWALIOR
- VII. BEHAR, BENGAL, ETC.
- VIII. OUDH (THE DOAB AND SOUTHERN ROHILKHUND)
- IX. PUNJAUB AND DELHI
- X. RAJPUTANA, CENTRAL INDIA, SAUGOR, AND NURBUDDA TERRITORIES
- XI. LUCKNOW AND ITS ENVIRONS

N.B.—The reader may find it useful to consider the above list of maps before reading any particular chapter.

N.B.—The coloured maps in this reprint are placed at the end of the volume

GLOSSARY OF INDIAN WORDS USED IN TEXT

TITLES OF FEUDATORY PRINCES

Begum (feminine)	Title of ruler of Bhopal; also applied to queen consorts and queen mothers.
Maharajah	Title of the rulers of Gwalior, Indore, Kashmir, and Rewah.
Maharao	Title of the ruler of Kotah.
Nana	Title used by representative of the deposed Peshwa.
Nawab	Title of the ruler of Banda.
Nizam	Title of the ruler of Hyderabad.
Peshwa	The ruler of the Maharattas (extinct in 1852).
Rajah	A title in general use.
Rao	Title used by the rulers of Kirwi. Also used as an honourable addition to a name, *e.g.* Dinkar Rao, Bala Rao, etc.
Thakhur	Title used by ruler of Awah. Also in general use in the same manner as Rajah.

MILITARY RANKS, ETC.

Soubadar Major	
Soubadar	Native officers.
Jemadar	
Havildar	Native N.C. officers.
Naik	
Sowar	Private soldier, Cavalry.
Golandauz[1]	„ Artillery.
Sepoy[2]	„ Infantry.
Lascar (Artillery)	An enlisted native employed to clean guns, carriages, etc.
Syce (Artillery and Cavalry)	A stable-helper and groom.
Grasscutter (Artillery and Cavalry)	A native employed to hoe and collect short tufts of grass for forage.
Poorbeah (Eastern)	A term used by Sikhs and others for the down-country or Hindustani soldier.

[1] Literally "a ball thrower."
[2] The mutineers generally, of whatever arm of the service, were spoken of as the Sepoys.

GLOSSARY OF INDIAN WORDS USED IN TEXT

GENERAL TERMS

Badmash	A rogue, village bully, or " corner boy."
Bagh	A garden or small park.
Bearer	A body-servant.
Chupatti	A thin round cake of unleavened flour baked in a pan.
Chuprassi	A government messenger or porter.
Culassi	A native employed in the Customs Department.
Dâk	The post.
" Deen ! Deen ! "	" For the Faith ! "
Doab (two waters)	The country between two rivers. When used without qualification it refers to that between the Ganges and the Jumna.
Eedghur	A house of prayer or temple.
Fakir	A travelling Hindu priest, mendicant, or religious recluse.
Feringhi	A foreigner or European.
Gharri	A carriage or post-chaise.
Ghazi	A fanatic Mohammedan.
Jheel	A swamp or bog.
Koti	A house or establishment.
Kotwali	The court-house or Police Office.
Lakh	A hundred thousand rupees.
Maidan	An open plain.
Moulvie	A Mohammedan priest.
Musjid	A mosque or temple.
Nuddee	A river.
Nullah	A ravine, sometimes dry, sometimes flooded.
Rajput	A strict Hindu sect.
Rupee	A silver coin of nominal value two shillings.
Serai	A large building with interior courtyard, used as a rest house for travellers.
Talukdar	A land tenant under Government—also a term employed for the " feudal barons " of Oudh.
Tope	A grove of trees usually planted in rows.
Zemindar	A farmer or yeoman.

ABBREVIATIONS

A., Artillery; Ben., Bengal; Bom., Bombay; C. Cavalry; C. of O. Commissary of Ordnance; D. Gs., Dragoon Guards; E., Engineers; E.F., European Fusiliers; E.R., European Regiment; F.B., Field Battery; Ft., Foot; H.A., Horse Artillery; H.F.B., Heavy Field Battery; I., Infantry; Irr., Irregular; K.R.R., King's Royal Rifles; L.C., Light Cavalry; L.Ds., Light Dragoons; L.I., Light Infantry; Mad., Madras; N., Native; O., Oudh; P., Punjaub.

ARTILLERY DESIGNATIONS

1/1 R.A.	No. 1 Company 1st Battalion Royal Artillery.
1/2 R.A.	No. 1 Company 2nd Battalion Royal Artillery.
1/1 Ben.H.A.	1st Troop 1st Brigade Bengal Horse Artillery.
1/1 Ben.A.	1st Company 1st Battalion Bengal Artillery.
A/1 Mad.A.	A Company 1st Battalion Madras Artillery.
1/1 Bom.A.	1st Company 1st Battalion Bombay Artillery.

Designations such as:
No. 3 F.B. 2/3 Ben.A. would mean the 2nd Company 3rd Battalion Bengal Artillery with No. 3 Field Battery; No. 7 H.F.B. 3/4 Ben.A. would mean the 3rd Company 4th Battalion Bengal Artillery with No. 7 Heavy Field Battery.

A Horse Field Battery means a Field Battery with horse draught in contradistinction to a Bullock Field Battery.

The reader is also reminded that that portion of a battery now termed a *section* was known in 1857-9 as a *division*.

NOTE.—*The full designation of regiments, troops, and companies as they existed in 1857-9 will be found in Appendix II, p. 435, etc.*

THE BAILEY GUARD GATE, LUCKNOW, AFTER THE SIEGE

THE HISTORY OF THE ROYAL AND INDIAN ARTILLERY IN THE MUTINY OF 1857

CHAPTER I

INTRODUCTION

THE wide spread of British power which took place in India during the half-century preceding the Mutiny of 1857 was not accomplished without provoking suspicion and dislike, which were aggravated by subsequent legislation. Sedition, often unsuspected and always disregarded, grew apace. Crushed ambitions, hardships real or imaginary, misunderstandings, jarring creeds, the eternal friction between Eastern and Western ideals, all tended to nourish a horde of agitators who had not far to seek for willing listeners in the old territories and new provinces of the Honourable East India Company. When Lord Dalhousie's governorship ended in 1855 official minutes were full of self-congratulation, and Britons generally regarded with complacency the splendid empire they had gained; but there was hardly a bazaar from Calcutta to Peshawar where treason was not preached, or a sepoy cantonment where mutiny was not advocated. Many eyes were turned upon the native army, especially upon the Bengal army, which growing with the growth of our empire far outnumbered the British garrison,[1] and those who privily pulled the wires of revolt hoped to find in this army an instrument that should free India from the hated rule of a Christian Government.

In the past the sepoys had displayed excellent fighting qualities coupled with the most devoted loyalty. British officers of the native army had a very steadfast belief in their own men; but mutiny and discontent, though only of a sporadic kind, were not altogether unknown, and in the years succeeding the Sikh Wars the bonds of discipline were allowed to slacken, and a kind of lethargy followed in the track of habitual success. Officers whose age or infirmities forbade the possession of energy or initiative were left in responsible positions; many of the juniors sought posts in civil employment;[2] while

---

[1] The strength of the army in India in May, 1857, was as follows:

	H.M. British and Indian European Regiments.	Native Regiments.
Bengal	24,366	137,757
Madras	10,726	51,244
Bombay	10,430	45,213
TOTAL	45,522	234,214

[2] The following is from the Bengal Army List of 1856:

8th Battalion Ben.A.: No. of Officers, 30.		9th Battalion Ben.A.: No. of Officers, 33.	
Staff and Civil Employ	11	Staff and Civil Employ	13
Furlough, etc.	4	Furlough, etc.	12
Effective	15	Effective	8
	30		33

the sepoys were sometimes pampered and sometimes irritated by pettifogging regulations. Finally the General Service Enlistment Act of 1856—by which all recruits were required to enlist for oversea service —brought about a dangerous state of affairs in the Bengal Army. At first sight such a regulation appeared not only necessary, but unobjectionable. For years past a similar act was in force in the Bombay Army, and in the Bengal Army itself this act was already applicable to six battalions and to all the artillery, but under existing circumstances such an order could not " pass into general circulation without being ignorantly misunderstood by some and designedly misinterpreted by others." [1] The warlike province of Oudh, both before and after annexation, was the principal recruiting-ground of the Bengal Army. The sepoys as a rule were Hindu Rajputs, high-caste men of good fighting stock, in many of whose families military service was a tradition; but to cross " the black water " was to lose caste, and the new-comers who might accept the new conditions would be men unworthy of companionship. The old soldier, back in his village on pension, never again would see a son succeed him in the old regiment; things would be changed, and peradventure, so said the agitators, behind this new regulation lay an insidious design against religion. At the same time the land settlement which followed the annexation was bearing hardly upon the Rajput yeomanry, and thus the Hindu sepoy was assailed both by spiritual and temporal

[1] Kaye, i. p. 470.

misgivings. Means also were at hand to provoke the Mohammedan soldiery. Emissaries from the mock court of Delhi or from the deposed King of Oudh were busy among them, stirring up the easily excited fanaticism of a proselytising religion, insisting that by the dethronement of the King the British betrayed their design of sweeping away all Moslem rule from India, and that by their present war with Persia [1] they were proclaiming themselves the enemies of the Prophet. Then to the ignorant, whether Mohammedan or Hindu, it was represented that the introduction of the railway and the electric telegraph was an unholy thing. Thus the promptings of religion and patriotism, nay even the railleries and cajoleries of the bazaars, were used to urge the sepoys to become the saviours of their country, while the numerical inferiority of the British forces left them masters of the situation. Furthermore, were they not the conquerors of the warlike Sikhs; was it not true that the Crimean War had exhausted the British Army; and had not the holy Shah Mahomet Ullah foretold that the century of British rule which began with the victory of Plassey should end in 1857? [2] But in spite of all these strong

---

[1] A successful expedition to Persia composed of troops from Bombay and England had just terminated.

[2] The chupatti, in a manner which has never been explained, was made use of to spread mutinous ideas. A cake brought by a native to a village was there eaten by him in the presence of the villagers, whose duty then was to bake another cake and supply a runner to take it to another village, somewhat after the manner of the "Fiery Cross." Lotus leaves, the symbol of war, were passed among the sepoys, and the expression "*Sub lal hoga*" ("Everything will become red") went round as a sort of password.

influences the sepoys were not ready instruments in the hands of the promoters of sedition, and had it not been owing to a fortuitous circumstance the Great Mutiny might never have broken out.[1]

Towards the close of 1856 the new Enfield rifle cartridge was introduced into India. The bullet was wrapped in greased paper and had to be bitten before loading. Now the flesh of the pig is an abomination to the Mohammedan, and the cow is a sacred animal to the Hindus. It was whispered that the grease used in the cartridges was obtained from these animals, and that a treacherous blow was aimed at the great religions of India. The rumour flew abroad and fanned into flame the embers of sedition that had been secretly smouldering for many a year.[2]

In spite of warnings definite enough, there was no anxiety in Upper India at the beginning of May, 1857. Certainly none was betrayed by the military authorities. The bulk of the white troops were in the Punjaub. Three regiments were in the Hills, and between Meerut and Calcutta were only three British battalions. Most of the arsenals were held by native troops and no preparations for mobilisation existed anywhere. The Commander-in-Chief in Bengal, General the Hon. G. Anson, was at Simla, and the annual migration to the Hills was in full progress when on May 10 the sepoy garrison of Meerut broke out in mutiny.

[1] See "Sepoy Revolt," p.123, etc.
[2] A report was also spread that bone-dust was mixed with the Commissariat flour and that bullock's blood was used to refine sugar.

On May 12 the re-establishment of the Moghul Empire was proclaimed at Delhi, and by the end of the first week in June India was ablaze from the Nurbudda River to the Himalaya Mountains.

When the news of the seizure of Delhi reached Lord Canning, the Governor-General, he at once ordered troops from Burmah to Calcutta, and called on the Governors of Madras and Bombay for reinforcements. He communicated with the Governors of Ceylon, Mauritius, and the Cape of Good Hope, and measures were taken to divert to Calcutta an expeditionary force then on its way to China. Urgent demands were also sent to England.

In the next chapter it is proposed to trace the outbreak of the Mutiny as it occurred in the various districts, and narrate the events which led to the development of two distinct campaigns which had only an indirect bearing on each other. It will then be seen why the military operations were concentrated at first round Delhi and Lucknow, which attracted to themselves all the elements of revolt that lay within their spheres of influence.

CHAPTER II

THE OUTBREAK OF THE MUTINY

I. Bengal

In January, 1857, the troops in Bengal were distributed as follows :

At Barrackpore

| 2nd Ben.N.I. | 43rd Ben.N.I. |
| 34th Ben.N.I. | 70th Ben.N.I. |

At Calcutta and Dum Dum

53rd Foot.	No. 20 F.B. 3/5 Ben.A.
Governor-General's	Detachment 9th Batt.
Body-guard	Ben.A.

At Berhampore	*At Chittagong*
11th Ben.Irr.C.	34th Ben.N.I.(3 Companies).
19th Ben.N.I.	Aracan Battalion.
Detachment 9th Batt. Ben.A.	Detachment 9th Batt. Ben.A.

| *At Midnapore.* | *At Jalpagorie* | *At Raneegunge* |
| Shekwattee Battalion. | 73rd Ben.N.I. | Detachment 63rd Ben. N.I. |

Mutinies at Berhampore and Barrackpore.—When it was noticed, towards the end of the month, that the sepoys under instruction at Dum Dum [1] regarded the new cartridge with suspicion, every effort was made to allay their fears and the issue was immediately stopped. Nevertheless this suspicion spread

[1] The " Rifle Instruction Depôt " was at this place.

rapidly and the native infantry at Berhampore and Barrackpore grew greatly excited. When warned, on February 25, to prepare for exercise next day with blank cartridge of old pattern, the 19th Ben. N.I. (Lieutenant-Colonel W. St. L. Mitchell) showed signs of insubordination, and when night came they broke open the arm-racks and loaded their muskets. They were speedily confronted by the 11th Irregular Cavalry and by two 9-pounders—post guns [1]— loaded with case. When called upon to lay down their arms, they refused to do so until the cavalry and artillery were withdrawn. Then Mitchell, sending back the cavalry to their lines and the guns to the magazine, ordered a general parade to take place at sunrise, and the 19th fell in without a symptom of insubordination. On these events being known, Mitchell was directed to march his regiment to Barrackpore, where it was proposed to disarm them, and the 84th Foot were recalled from Rangoon to Calcutta, where they arrived on March 20. On March 29 a single sepoy of the 34th Ben.N.I., by name Mongol Pandy, posted himself in front of the regimental quarter-guard and began firing at European officers, while the guard looked quietly on. He was eventually overcome, and the jemadar of the guard, Issuri Pandy, made a prisoner.[2] On

---

[1] Post guns were kept at various stations for use on emergency, unattached to any particular company. On this occasion they were manned by golandauz from the 9th Battalion Ben.A. Colonel Mitchell was in command of the station.

[2] Mongol Pandy tried to shoot himself, but he failed and was eventually hanged. Jemadar Issuri Pandy was tried by Court Martial and hanged on April 22. From the names of these men the term " Pandy " came into general use to denote a mutinous sepoy.

March 30 the 19th Ben.N.I., who had reached the neighbourhood of Barrackpore, were disarmed, as were also the 34th on May 2.

The First Reinforcements.—After the outbreak at Meerut a detachment of the 84th Foot were sent to Cawnpore, and on May 24 the 1st European Madras Fusiliers arrived from Madras and set out for Benares. The next arrivals (at Calcutta) were the 35th Foot from Burmah, the 64th Foot and 78th Highlanders, who had accompanied the expedition to Persia,[1] and a detachment of the 37th Foot and 3/8 R.A. (2nd Captain F. C. Maude) from Ceylon. These troops all disembarked during June and July. On June 14 the 2nd, 43rd, and 70th Ben.N.I. were disarmed, but the 73rd Ben.N.I. at Julpagorie and the native artillery companies at Dum Dum and Berhampore remained quietly at their duty.

II. MEERUT

Meerut was the head-quarters of the Meerut Division, commanded by Major-General W. H. Hewitt. The Meerut Brigade, under Brigadier-Commandant A. Wilson,[2] was composed of the following troops:

6th D.Gs. (Carabineers).	1st Batt. 60th K.R.R.	2/1 Ben.H.A. (Captain and Brevet-Major H. Tombs).
3rd Ben.L.C.	11th Ben.N.I.	No. 14 F.B. 3/3 Ben.A. (Captain and Brevet-Major E. W. S. Scott).
	20th Ben.N.I.	European Depôt Ben.A.
		Golandauz Depôt Ben.A.

---

[1] The troops employed in Persia in 1856 were mostly from the Bombay Presidency, whither they were returning when the Mutiny broke out.

[2] Lieutenant-Colonel and Brevet-Colonel A. Wilson was Commandant of the Bengal Artillery as well as Brigadier.

No cartridges of the new pattern were issued at Meerut after the troubles in Bengal, and when some unattested native artillery recruits refused to receive even old-pattern ammunition they were at once paid up and discharged. This was at the end of April. About the same date old-pattern cartridges were served out to the 3rd Bengal Light Cavalry for the usual musketry course, and eighty-five troopers refused to accept them. After a court of inquiry was held the Commander-in-Chief ordered them to be tried by Court Martial. The Court, composed of native officers, condemned them to imprisonment for ten years with hard labour; they were manacled on parade and placed in the town jail on Saturday, May 9. Next day the native garrison [1] rose in mutiny, shot many of their British officers, released the eighty-five prisoners, and fraternised with the rabble of Meerut.

Inactivity of the Garrison.—After some delay the Carabineers, the Rifles, and the field artillery were moved to the native lines, the artillery recruits and the dismounted portion of the cavalry being left to protect the European barracks. Darkness had now fallen, and there was still no sign of the rebels, though sounds of tumult were heard from the town. General Hewitt, after consulting Brigadier Wilson, came finally to the conclusion that the sepoys

---

[1] The 11th Ben.N.I. seem to have joined the Mutiny with reluctance. Five native officers and twenty men remained with the Meerut Garrison and many went to their homes instead of to Delhi. Captain H. C. Craigie's squadron of the 3rd Light Cavalry showed marked loyalty on the night of May 10, and some fifty sowars remained with the garrison.

were threatening the European cantonment, and he marched the brigade back for its protection. Strong pickets were then thrown out and the troops bivouacked on the parade ground.

Escape of the Mutineers.—But not till the following morning did he learn that the mutineers had gone to Delhi and that the telegraph wires were cut.[1] Messengers were instantly sent off to Delhi and Umballa, but the Delhi messenger never reached his destination. No effort of any kind, however, was made to follow up the rebels : the General seemed to consider himself in a state of siege, and all round him pillage and violence raged unchecked.[2] On the 15th 500 Sappers and Miners, Ben.E., arrived from Rurkhi, and the main body broke out at once in mutiny, shot their commanding officer, and made for Delhi.[3] A troop of Carabineers and two guns of 2/1 Ben.H.A. overtook them and killed some fifty or more, but 280 escaped with their arms. The remaining two companies were held together by the adjutant, Lieutenant F. R. Maunsell, and though for a time deprived of their arms, they performed excellent service and were fully trusted afterwards.[4] On May 24 a small village close to

[1] Before the wires were cut a private telegram was sent by a European employée at the post office to her aunt in Agra, warning her not to carry out her purpose of coming to Delhi. This was the only intimation of the Mutiny that reached Agra for some days.

[2] So entirely had all semblance of authority disappeared that it was believed the English in Meerut had been slain to a man (Kaye, ii. p. 173).

[3] This mutiny may have been due to panic (Kaye, ii. p. 178).

[4] They were the only trained men the Bengal Engineers possessed at the Siege of Delhi. Several of them were present at the blowing-in of the Kashmir Gate (cp. iii. p. 121).

Meerut—a nest of robbers and murderers—was effectually dealt with; but nothing more was done by the garrison to crush the revolt until May 27, when, in accordance with an order from the Commander-in-Chief, Brigadier Wilson set out with a field force to effect a junction with troops from Umballa. Early in June Major-General Hewitt was relieved of his command and Major-General N. Penny succeeded him.

III. Delhi

The Moghuls.—The empire of Delhi had faded into a tradition, but " the tradition was still an honoured one and had sunk deeply into the memories of the people." [1] The titular King was a worn-out man of eighty years, ruled in all things by a capable and unscrupulous woman, who for reasons of her own abhorred the British.[2] Emissaries from the palace kept the Mohammedan population in continual excitement, and in this fateful month of May all were looking for some event, they knew not what, that should restore the lost glories of the Moghuls.[3]

Arrival of the Meerut Mutineers.—The garrison of Delhi, commanded by Brigadier H. M. Graves,[4] was in cantonments upon the Ridge about two miles north of the city and was exclusively native. It consisted of the 38th, 54th, and 74th Ben.N.I. and No. 5 F.B. 3/7 Ben.A. (Captain H. P. de Teissier). On May 11

---

[1] Kaye, ii. p. 2.

[2] Queen Zeenut Mahal. Her wish that her son should succeed her husband to the exclusion of other and senior heirs had been frustrated by the British Government.

[3] Kaye, ii. p. 35, etc.

[4] Lieutenant-Colonel and Brevet-Colonel H. M. Graves, 13th Ben.N.I.

a Court Martial parade was held at sunrise, and though there were some murmurs of discontent and sympathy when the sentence of death on Jemadar Issuri Pandy was read out, the British officers appear to have had no suspicion of impending trouble when the parade was dismissed. Yet at this moment the leading troopers of the Light Cavalry had already passed the bridge of boats, and, making their way past Selimgurh,[1] were acclaiming the King of Delhi outside the palace walls. At the request of the King, Captain Douglas [2] commanding the Palace Guard appeared on a balcony and ordered the intruders to depart, and the troopers, whose numbers were rapidly swelling, turned southwards and made their way into the city by the Raj Ghat Gate. Then, having been joined by the scum of the neighbourhood, they set to work to pillage and burn houses and murder Europeans wherever they were found. Meanwhile Douglas went to satisfy himself that the Calcutta Gate of the city was properly secured by the guard of the 38th Ben.N.I., and on his way met the Commissioner, Mr. S. Fraser, who, learning at an early hour of the approach of the mutineers, drove in from his residence at Ludlow Castle. Presently the rioters, surging up from the south of the city, approached the spot and at once were welcomed by the sepoys of the guard, whom the two officers were powerless to recall to their duty. After many narrow escapes Fraser and Douglas reached

---

[1] A fort on the river-front of Delhi (cp. iii. p. 75).
[2] Lieutenant and Brevet-Captain C. R. G. Douglas, 32nd Ben.N.I. The guard was composed of native troops maintained by the King.

the palace, but only to be murdered there in company with other Europeans by some of the miscreants who inhabited that den of infamy. The gates were then opened and the courtyards thronged by the troopers and rabble, whose numbers were presently swelled by the infantry who had hurried along the Meerut road in constant (but unnecessary) dread of pursuit and vengeance.

Loss of the City.—It was between 9 and 10 o'clock a.m. before any news from the city reached cantonments, and even then the state of affairs was only partially realised. Two companies of the 54th Ben.N.I. under Colonel J. P. Ripley were sent off at once to secure the bridge of boats, to be followed by two more under Major F. S. Paterson as soon as a division of 3/7 Ben.A. could be got ready. Troops marching from cantonments to the bridge of boats would enter Delhi by the Kashmir Gate, where the Main Guard was situated contiguous to the city walls; they would then traverse an extensive courtyard from which issued a road leading through the European quarter to the Calcutta Gate. As Ripley entered this courtyard, the rabble, headed by the cavalry troopers, were pouring down the road from the city, and he immediately ordered his men to fire. At the same time the field officer of the day, Captain W. F. N. Wallace, turned out the Main Guard, which consisted of fifty men of the 38th Ben.N.I. But the 54th were not loaded,[1] and

---

[1] Why the 54th were not loaded is a mystery. It is stated that after their officers had been shot by the troopers they were bayoneted when on the ground by men of the regiment.

the guard, with insulting words, refused to act. A pause ensued during which some troopers galloped forward and shot down the officers of the 54th, in front of their own men. Meanwhile Paterson with his two companies and two guns under Lieutenant W. Wilson met Wallace outside the Kashmir Gate and was told by him what had occurred; but when he reached the Main Guard the rabble and the mutinous sepoys had disappeared—only the dead and wounded officers lay about the ground. Wilson then posted his guns so as to command the road into the city, and sentries were placed at various points, while Wallace hurried to cantonments for reinforcements, getting there at 11 a.m.

The Blowing-up of the Magazine.—Being now aware that a serious crisis had occurred, Brigadier Graves sounded the General Assembly, and sent Major H. E. S. Abbott with 200 men of the 74th Ben.N.I. and two guns under Lieutenant R. Aislabie to join Paterson. For some time the officers at the Main Guard awaited events, disturbed only by the sounds of tumult from the city and the arrival of fugitives from the European Quarter. But suddenly at 3 p.m. the rapid firing of guns was heard, followed by an explosion which shook the Guard House, and presently two officers, bruised and blackened with gunpowder, brought the news that the small-arm cartridge magazine in the Arsenal was blown up. One of these officers was Lieutenant G. D. Willoughby, Ben.A., Commissary of Ordnance, who from the river-front of the Arsenal had seen the mutineers passing the bridge of boats. Realising at once that the Arsenal was in danger, he determined to hold it to the last, and if needful to blow up the principal magazine. Under his command were eight Europeans, namely, Lieutenants G. Forrest and W. Raynor (Veteran Establishment), Conductors Buckley, Shaw, Scully, and Crow, and Sergeants Edwards and Stewart, Ben.A. There was also a native establishment, but they received with reluctance the arms that were forced into their hands. A guard of the 38th Ben.N.I. was at the Main Gate, which, as well as the other gates, was strongly barricaded. Nine 6-pounder guns and one 24-pounder howitzer, double loaded with case shot, were placed in commanding positions within the Arsenal enclosure, a party of natives being told off to keep up the supply of ammunition, and a train of gunpowder was laid to the magazine. About noon the first summons to surrender the Arsenal to the King of Delhi was communicated to Willoughby by the soubadar of the guard. No reply was returned then or later when the summons was repeated. At length towards 3 p.m. a large number of sepoys arrived with scaling-ladders which they reared against the wall, and in a twinkling all the natives inside the Arsenal made their escape by scrambling over the wall by the aid of some convenient sheds, and descending on the other side by the ladders. Then the attacking sepoys mounted in their turn and opened a musketry fire on the defenders, but they were kept at bay by a brisk fire of case shot, in spite of the fact that the treacherous natives had managed to hide the priming pockets before they fled.[1] Each gun had been fired four times, and two of the defenders had been wounded, when it was found that the ammunition at hand was expended. On this the enemy pressed into the enclosure and Willoughby gave the word to fire the train. Scully, who had volunteered for this desperate job, rushed from his gun and touched the powder with his portfire. A terrific explosion followed, and five of the gallant nine were never seen again, while hundreds of rebels were blown to pieces or buried in the débris. Willoughby, Forrest, Raynor, and Buckley miraculously escaped. Half-stunned and almost unrecognisable, they hurried through a sally port on the river face. Willoughby and Forrest made their way to the Main Guard, and Raynor and Buckley, taking the Meerut road, eventually reached that station.[2]

The Flight from Cantonments.—When the explosion occurred there was much excitement on the Ridge, especially among the 38th Ben.N.I., who assumed a threatening attitude towards the gunners manning the two remaining guns of Captain de Teissier's Battery at the Flagstaff Tower, where a crowd of fugitives had collected. About this time the Brigadier sent two messages to the Main Guard;

[1] No tubes of any kind were in use in India. Loose powder carried in a leather pocket was used to prime with. It was fired by a port-fire, ignited by a slow match which the field artillery carried alight when action was expected, attached to one of the axle-tree boxes.

[2] Forrest and Raynor were promoted to the rank of Captain and Buckley was given his commission in the Veteran Establishment. Willoughby and his three companions were nominated for the Victoria Cross, but he himself was murdered a few days afterwards on the road to Meerut.

the first recalled Aislabie's guns, the second Abbott and the 74th. But when Aislabie approached the Ridge he was met by a detachment of the 38th who drove him and his European sergeant off, and prevailed on the native gunners to return to Delhi. Captain de Teissier, seeing the guns wheeling round, rode in their direction, but the 38th opened fire on him and shot his horse. In view of the state of affairs at the Main Guard, Abbott did not at once obey his orders, and he was still there when the detachment of the 38th and the guns entered the Kashmir Gate.[1] Fully convinced by this time that he was urgently required on the Ridge, he ordered the guns to follow, and he marched off with some hundred men towards the cantonments; but the gate was immediately shut behind him and the guns did not follow. The detachments of the 38th and 54th then opened fire on the Europeans in the guard house, and the native gunners sent two or three round shot through its walls. Several officers were killed, but others (including Lieutenant Wilson, Ben.A.) escaped through the building, and dropping from an embrasure into the ditch gained the open country.[2] Meanwhile Abbott, hearing the firing, wished to return; but his men would not permit it, and forming round him hurried him back to

[1] The idea that help would come from Meerut was prevalent both on the Ridge and at the Main Guard. Paterson told Abbott he could not hold the Main Guard with the 54th if the 74th left. The Deputy-Collector who was present added his protests. When Abbott did leave he appears to have been almost hustled back to cantonments by his men.

[2] Women who had taken refuge in the Main Guard were let down by belts and handkerchiefs into the ditch (Kaye, ii. p. 94).

THE FLAG-STAFF TOWER

cantonments. But by this time the Brigadier had realised that flight was the only course left, and he prepared to take the road to Kurnal, along which many refugees from Delhi had already set out.[1] A few sepoys and the two guns almost without gunners made a show of following; but the sepoys soon straggled away, and the artillery drivers, after going a few miles, refused to go farther, and later on joined their comrades in Delhi. There was a powder magazine in the cantonments which unfortunately was not blown up.[2] It was plundered by marauding villagers, who carried off some of the powder, but the bulk of it remained for the use of the mutineers in Delhi. On May 12 the re-establishment of the Moghul Empire was proclaimed, and through the length and breadth of Upper India the news went forth that the British had been expelled from the Imperial City.

IV. UMBALLA

Concentration of British Troops.—The news of these events reached Simla on May 13, and the Commander-in-Chief at once ordered a concentration of troops at Meerut and Umballa and the preparation of a siege train at Phillour. Accordingly the Sappers and Miners from Rurkhi set out for Meerut with the result already narrated, and the Sirmur (Gurkha) Battalion from Dehra joined the Meerut

[1] Surgeon S. H. Batson Bengal Medical Establishment, made an attempt during the day to get to Meerut with a letter demanding help. He was disguised as a fakir, but the colour of his eyes betrayed him, nevertheless he managed eventually to escape.

[2] An attempt was made to blow up this magazine, but it was frustrated by the 38th Ben.N.I.

Field Force on June 1. On the 17th the 9th Lancers and 4th Bengal Light Cavalry, the 75th Foot, the 1st and 2nd Bengal European Fusiliers, the 5th and 60th Ben.N.I., 2/3 and 3/3 Ben.H.A. were assembled at Umballa. The *personnel* of the siege train was supplied by 4/6 Ben.A. from Ferozepore, and 1/7 Ben.A. (a native company) took over a 9-pounder equipment at Phillour and prepared to accompany the train, for which a wing of the 9th Bengal Irregular Cavalry and the Nusseeree (Gurkha) Battalion from Jutogh were to act as escort.[1]

Subsequent Movements.—It was General Anson's intention to move at once on Delhi, but he was greatly hampered by the want of transport, dearth of supplies, and the general state of unpreparedness for war of the Army Departments.[2] On the other hand he was materially aided by the loyalty of the Cis-Sutlej chiefs, Patiala, Jind, Nubha, Kurnal, who kept the roads open and protected the communications with the Punjaub.

On the 17th the advance was begun by the despatch of a small column down the Grand Trunk Road. Ere long it was realised that the Bengal sepoys were only an encumbrance, and the bulk of them were successively removed from the force under various plausible pretexts.[3] After great exer-

[1] The Gurkhas refused to leave Jutogh, and from fear of a mutiny on their part a panic occurred at Simla (Kaye, ii. p. 143). Eventually the battalion did good police service.

[2] As is often the case, the soldiers on the spot had to bear the odium due to the politicians elsewhere. Anson was most unjustly blamed for delay (Kaye, ii. pp. 146, etc.).

[3] The 4th Cavalry and 5th Ben.N.I. were disarmed at Umballa; the 60th Ben.N.I. subsequently mutinied at Rohtak.

tions the main body, including the siege train, reached Alipore on June 6; the safety of Umballa being provided for by a wing of the 2nd Europeans, a detachment of Patiala troops, and the loyal native company 1/7 Ben.A.,[1] which handed over its 9-pounder equipment to 3/3 Ben.H.A., in exchange for the 6-pounders of the troop. On May 27 General Anson died of cholera at Kurnal and was succeeded in command of the field force by Major-General Sir H. Barnard, commanding the Sirhind Division.[2] On June 7 Barnard was joined by Wilson with the Meerut Field Force, whose successful march across the Hindun River will be described in Chapter III.

V. THE PUNJAUB

The troops in the Punjaub were thus distributed :

At Lahore and Mian Mir

8th Ben.L.C.	81st Foot.	2/2 Ben.H.A.
	16th Ben.N.I.	3/2 Ben.H.A.
	26th Ben.N.I.	1/4 Ben.A.
	49th Ben.N.I.	2/4 Ben.A.
		3/4 Ben.A.
		4/4 Ben.A.

At Umritsur and Govindghur

59th Ben.N.I.	2/1 Ben.A.
	No. 16 F.B. 5/8 Ben.A.

At Goordaspore

2nd Ben.Irr.C.	Det. 59th Ben.N.I.

---

[1] The company was placed in charge of an ordnance depôt at Umballa.

[2] The command of the field force first fell to Major-General T. Reed, the senior officer in Northern India. He reached Alipore on June 8, too ill to command. He, however, accompanied the advance.

THE OUTBREAK OF THE MUTINY [CHAP. II

At Ferozepore

10th Ben.L.C.　　61st Foot.　　　　No. 19 F.B. 3/6 Ben.A.
　　　　　　　　45th Ben.N.I.　　4/6 Ben.A.
　　　　　　　　57th Ben.N.I.

At Mooltan

1st Ben.Irr.C.　　62nd Ben.N.I.　　4/3 Ben.H.A.
　　　　　　　　69th Ben.N.I.　　4/3 Ben.A.

At Jullundur [1]

6th Ben.L.C.　　1st Battn. 8th Foot　　1/1 Ben.H.A.
　　　　　　　36th Ben.N.I.
　　　　　　　61st Ben.N.I.

At Noospore and Kangra

　　4th Ben.N.I.　　　　　　　　1/7 Ben.A.

At Phillour [1]

　　3rd Ben.N.I.

At Hosheyapore [1]

9th Ben.Irr.C.　　33rd Ben.N.I.　　5/1 Ben.H.A.

At Sialkote

9th Ben.L.C.　　52nd Foot.　　　　3/1 Ben.H.A.
　　　　　　　35th Ben.N.I.　　No. 17 F.B. 3/1 Ben.A.
　　　　　　　46th Ben.N.I.

At Rawal Pindi

16th Ben.Irr.C.　　1st Battn. 24th Foot.　　4/2 Ben.H.A.
　　　　　　　　58th Ben.N.I.
　　　　　　　　Kemaon (Gurkha) Batt.

At Nowshera

10th Ben.Irr.C.　　27th Foot　　Mountain Train Ben.A.
　　　　　　　　55th Ben.N.I.

At Shumshabad (near Rawal Pindi)

　　17th Ben.Irr.C.

---

[1] The troops at these stations belonged to the Sirhind Division.

At Attock
4/8 Ben.A.

At Jhelum
14th Ben.N.I.   39th Ben.N.I.
Local Levies
No. 1 F.B. 4/7 Ben.A.

At Peshawar

5th Ben.L.C.	70th Foot	1/2 Ben.H.A.
7th Ben.Irr.C.	87th Foot	1/3 Ben.H.A.
18th Ben.Irr.C.	21st Ben.N.I.	No. 7 F.B. 1/1 Ben.A.
	24th Ben.N.I.	No. 10 F.B. 1/2 Ben.A.
	27th Ben.N.I.	2/2 Ben.A.
	51st Ben.N.I.	3/2 Ben.A.
	64th Ben.N.I.	4/2 Ben.A.

Punjaub Irregular Frontier Force
Corps of Guides

1st Punjaub C.	1st Punjaub I.	1st Punjaub Battery.
2nd Punjaub C.	2nd Punjaub I.	2nd Punjaub Battery.
3rd Punjaub C.	3rd Punjaub I.	3rd Punjaub Battery.
4th Punjaub C.	4th Punjaub I.	Punjaub Garrison Company.
5th Punjaub C.	5th Punjaub I.	Hazara Mountain Train.
	6th Punjaub I.	
	1st Sikh I.	Ben. Sappers and Miners.
	2nd Sikh I.	
	3rd Sikh I.	
	4th Sikh I.	

In the beginning of May most of the Punjaub Irregular Frontier Force were still on the frontier, but some regiments were soon moved southwards. For example the 2nd Punjaub Cavalry and a wing of the 2nd Punjaub Infantry were sent to Mooltan in June, 1857.

Lahore.—The news of the seizure of Delhi reached Lahore as well as other military stations in the Punjaub on May 12.[1] Sir John Lawrence, the

---

[1] A telegraph message was sent off on May 11 to all stations in the Punjaub. It is said the operator was murdered at his instrument.

Chief Commissioner, happened to be then at Murree, but Brigadier S. Corbett,[1] who commanded at Mian Mir, acted with admirable promptitude, and all the native troops in the cantonment were disarmed next day, the fort in Lahore was occupied by a European guard, and a company of the 81st Foot was despatched to Govindghur, which dominated the sacred Sikh city of Umritsur.[2] The fort and arsenal of Phillour were occupied by British troops sent from Jullundur, and at Ferozepore, where another important arsenal existed, the native troops were disarmed.[3]

Lawrence at Peshawar.—But the Punjaub, our latest conquest, was also our advanced post, and hovering over its northern frontier were the wild tribes who on more than one occasion had dulled the brightness of our prestige. The Sikh might now think of paying off old scores, the Afghan might now see an opportunity; and the question in men's mouths, British and native, was, What news from Peshawar?[4] It was in truth a point of danger, but Lawrence was a man who could look beyond his own province to the broad plains of India, and he realised that Delhi was the focus of the revolt

[1] Colonel S. Corbett, 16th Ben.N.I.
[2] The native company 5/8 Ben.A. was moved into the fort at Govindghur before the arrival of the detachment of the 81st Foot, as it was considered reliable. It remained there quietly until it was discharged after the Mutiny was suppressed.
[3] The 57th quietly laid down their arms, but the 45th broke out in mutiny, attacked the arsenal, looted the cantonments, and were pursued for twelve miles by the 10th Ben.L.C. and two of the guns of 3/6 Ben.A. (Kaye, ii. p. 463).
[4] Kaye, ii. 449.

and that so long as the rebels held that city the existence of British rule was in peril. He was ready " to weaken the Punjaub to strengthen the empire. He would perhaps have sacrificed the Punjaub to save the empire." [1] What, then, had he to rely on ? First, there was the strong British garrison of the Punjaub and the faithful native corps, such as the Guides and the Punjaub Irregular Frontier Force which he himself had helped to raise. Secondly, there was the well-known hostility of the Sikh to the Poorbeah or down-country sepoy.[2] Thirdly, there was the possibility of raising new levies among a war-like people who loved fighting and delighted in plunder. Fourthly, at his right hand stood Colonel Herbert Edwardes and Colonel John Nicholson, the Commissioner and Deputy Commissioner of Peshawar.[3] With such men to think was to act. A movable column—a staunch striking force to be used where necessary—was organised at once ; permission to raise new levies was obtained from the Governor-General, alterations were made in the various garrisons with the object of preventing native co-operation and facilitating European supervision, and on May 16 the Guide Corps, the first of the Punjaub reinforcements, began its historic march to Delhi.

Sepoy Regiments Disarmed.—Still, in the Peshawar

[1] Kaye, ii. pp. 463–4.

[2] After the conquest of the Punjaub the Oudh and Bengali sepoys assumed arrogant airs that were bitterly resented by the Sikhs.

[3] Major and Brevet-Lieutenant-Colonel H. B. Edwardes, 1st Bengal European Fusiliers ; Captain and Brevet-Lieutenant-Colonel J. Nicholson, 27th Ben.N.I. For an appreciation of these officers see Kaye, i. p. 26, and ii. p. 450, etc.

Valley affairs were in a very critical state, and many were watching the balance in which British success and failure were so delicately poised. It was absolutely necessary to draw the teeth of potential mutineers by depriving them of their arms; yet no step could be taken which might precipitate a concerted rising. On the night of the 21st news of an incipient mutiny at Nowshera reached Edwards at Peshawar. He instantly conferred with Brigadier Cotton,[1] and, in accordance with the excellent arrangements made by that officer, four infantry battalions[2] were disarmed at dawn under the wondering eyes of the inhabitants of the turbulent town of Peshawar. The next day a small mixed force[3] under Colonel T. Chute, 70th Foot, struck across the Khyber, and the 55th Ben.N.I. who had mutinied were broken up and dispersed. On June 10 forty mutineers were blown away from guns on the Peshawar parade ground. The disarming parade of May 22 was impressive; the punishment parade of June 10 was decisive. Enlistment in the new levies had been proceeding slowly, but after June 10, both within and without the border, it was brisk beyond expectation.

Mooltan.—Mooltan was a station important from its situation, and as there was no doubt about the mutinous intentions of the 62nd and 69th Ben.N.I.,

---

[1] Lieutenant-Colonel and Brevet-Colonel S. J. Cotton, 10th Foot (Local Major-General, August, '57).

[2] The 21st Ben.N.I. believed, with reason, to be loyal, was not disarmed.

[3] 70th Foot (300 men), Punjaub Irregular Cavalry (250 men), 3 guns 1/2 Ben.H.A. (Lieutenant G. R. Brown), 8 guns Mountain Train Ben.A. (Captain T. Brougham).

Major C. Chamberlain was ordered to disarm them.[1] This he did on June 11 under the guns of the native troop, 4/3 Ben.H.A., which were drawn up in action. In rear of each gun gunners of the European company, 4/3 Ben.A., were stationed with loaded carbines, and a troop of the 1st Punjaub Cavalry were prepared to ride the native gunners down should they show hesitation. After being disarmed the 62nd and 69th remained quietly in their lines, and though the guns were placed in European charge, 4/3 Ben.H.A. continued to form part of the garrison, which was presently reinforced by the 1st Bombay European Fusiliers from Sind.[2]

Mooltan remained quiet till September 1858,[3] when the infantry of the garrison was reduced to the Bombay Fusiliers. It then contained no cavalry, but V Battery, 1/6 R.A. (Major J. Singleton) had arrived from England. The 62nd and 69th still inhabited the native lines, where latterly the men of 4/3 Ben.H.A. were ordered to camp beside them.[4] On September 2, without any ostensible cause, they all rose, and, arming themselves with long poles and sticks, having hatchets, hammers, etc., fastened on the ends, they made a sudden onslaught on the artillery quarter guard with

[1] The brigadier at the station was ill and Chamberlain (who commanded the 1st Punjaub Cavalry) was named for the duty by Sir J. Lawrence.

[2] In Sind there was no sympathy with the mutineers. The 1st Bombay European Fusiliers was sent to the Punjaub with the full concurrence of the Governor of Bombay. Some Bhil troops were sent from Sind at a later date.

[3] It is convenient to deal with the Mooltan rising here; it had nothing to do with the course of the campaign.

[4] It is an open question whether this native troop of horse artillery would have mutinied if it had received different treatment. (See Stubbs, iii. p. 597.)

the evident intention of possessing themselves of the guns. Singleton, who was on the spot, ordered two guns into action with case shot, and scattered one party of the rebels. Another body made a rush to the stables, but they were met and dispersed by the Bombay Fusiliers. Others approached the stables by another route, and succeeded in killing and wounding some unarmed drivers and syces before they were driven off by 2nd Captain A. H. W. Williams, R.A., and a mixed detachment of gunners and infantry. A third party came under shrapnel fire from Singleton and were at once put to flight. Few of the mutineers escaped.

Jullundur.—The British, however, were not invariably successful in carrying out the disarmament of the native troops. At Jullundur Brigadier M. C. Johnstone [1] commanded the station, and though strongly advised by the resolute Commissioner, Major E. J. Lake,[2] to disarm the sepoys, he allowed himself to be persuaded by their commanding officers that they were good and loyal men. Even when the inevitable outbreak occurred on the night of June 7, he displayed lamentable indecision ; and when the cavalry, who were evidently the ringleaders, came within range of the guns of 1/1 Ben.H.A. (Major H. A. Olpherts), Johnstone (who happened to be beside them) forbade them to open fire. Elsewhere, however, a party of these sowars charged two guns of the native troop 5/1 Ben.H.A. (Lieutenant C. W. Maynard), with cries of " Deen ! Deen ! " but Maynard gave them a salvo of case, and emptied eleven saddles. Leaving

---

[1] Lieutenant-Colonel and Brevet-Colonel M. C. Johnstone, 87th Foot.
[2] Captain and Brevet-Major E. J. Lake, Ben.E.

blazing bungalows and a few murdered Europeans behind them, the rebel horse and foot made off for Phillour, induced the 3rd Ben.N.I. to join them, and then turned towards the south. With considerable difficulty Lieutenant M. C. Sankey, Ben.A,[1] obtained the Brigadier's consent to track them, and next day Olpherts's Troop and a detachment of the 8th Foot went in pursuit ; but the rebels had passed through Loodianah, and were safe on their way to Delhi before Olpherts was permitted to cross the Sutlej.

Jhelum.—A serious reverse occurred in July at Jhelum, where the 14th Ben.N.I. were known to be ripe for mutiny. On the 9th of that month a force commanded by Colonel C. H. Ellice, 24th Foot, composed of a detachment of that regiment, the newly raised Mooltani Horse, and a half troop of 1/3 Ben.H.A. (Lieutenant C. H. Cookes), approached Jhelum to disarm them.[2] When Ellice arrived he found that the sepoys had seized the quarter guard, whence they were only ejected with some British loss in men and horses, Ellice himself being dangerously wounded. The rebels then retired to a village where during the afternoon the contest was renewed. On advancing to within a hundred yards of the village, the guns came under a heavy fire. Men and horses began to fall, and one man, Gunner W. Conolly, was hit three times, but stuck to his work till he fainted.[3] The British attack was fairly

[1] Artillery Adjutant at Jullundur.
[2] Ellice is reported to have departed from the plan laid down by Lawrence for his guidance (Kaye, ii. pp. 623, 624).
[3] Conolly was twice wounded during the morning, and in the afternoon was wounded for the third time. He received the Victoria Cross.

beaten off, and the troops were withdrawn, one gun being temporarily abandoned, the wheel horses of both the gun and waggon limber being shot. Sergeant Smith, the No. 1, and five men brought up and hooked in fresh horses, but these also were shot and the gun was left on the field. During the night the rebels fled, and the gun was recovered.[1]

Sialkote.—The greater part of the garrison (including all the Europeans) had been withdrawn from the neighbouring station of Sialkote to join the Movable Column, and a wing of the 9th Bengal Light Cavalry and the 46th Ben.N.I. alone remained under the command of the Brigadier, Colonel F. Brind, Ben.A. Although he had protested against the withdrawal of the white troops, Brind had the most complete belief in the fidelity of his command. Nevertheless when the news of the success of the 14th Ben.N.I. reached Sialkote on July 8, the garrison rose early next morning, murdered the Brigadier, and some other officers, released the jail prisoners, and, followed by all the Hindustanis in the bazaar, set off next day to Delhi. But they never got there.

The Movable Column.—Colonel N. Chamberlain, commanding the Punjaub Irregular Frontier Force, was appointed to command the Movable Column when it was first formed ; but before it was actively employed he was appointed Adjutant-General of the Delhi Field Force, and was succeeded on June 8 by Colonel J. Nicholson.[2] The composition of the

[1] Few of the 14th Ben.N.I. eventually escaped (Kaye, ii. p. 626).
[2] Nicholson was given the rank of Brigadier.

column, which varied from time to time, was then as follows :

9th Ben.L.C. (wing)	52nd Foot 33rd Ben.N.I. 35th Ben.N.I.	3/1 Ben.H.A. (Captain and Brevet-Lieutenant-Colonel M. Dawes). No. 17 F.B. 3/1 Ben.A. (3 guns) (Captain G. Bourchier).

On June 25 Nicholson suddenly disarmed the 33rd and 35th, and, marching to Umritsur, disarmed the 59th Ben.I. also. About the same date the remaining regiments, the 3rd, 4th, and 58th Ben.N.I., were disarmed at their stations and the guns of the native troop 4/2 Ben.H.A. were transferred to European charge. When the news from Sialkote reached Nicholson on July 9, he at once disarmed the wing of the 9th Cavalry, and, being reinforced by some Punjaub Infantry, made a forced march and fell upon the Sialkote mutineers on June 12 at Trimmoo Ghat, where they had crossed the Ravi. The sudden floods in that river protected them till the 16th, on which date they were finally broken up. Nicholson then returned to Umritsur, and on July 24, being reinforced by the Mooltani horse, set out to join the Delhi Field Force.

VI. RAJPUTANA

The Native States.—Rajputana, in addition to the British district of Ajmere, consisted of some twenty tributary states [1] whose rulers maintained

---

[1] These were Ulwar, Banswara, Bhurtpore, Bikanir, Boondi, Dholpore, Dungarpore, Jeypore, Jaisalmir, Jhalawar, Jodhpore, Kerowlee, Kishengurh, Kotah, Lawa, Oodeypore (or Mewar), Pertabgurh, Shahpura, Sirohi, and Tonk.

state troops and contingents.[1] It was administered by the Agent-General, Colonel G. St. P. Lawrence,[2] its regular garrison being exclusively native.[3] When the news of the seizure of Delhi reached Lawrence on May 19 at Mount Abu, he immediately requested reinforcements from Deesa, in the Bombay Presidency, and issued a proclamation to the Chiefs of Rajputana urging them to preserve peace within their borders, and to muster their followers to help the Government. In 1817 the British had delivered Rajputana from the tyranny of the Mahrattas, and, not unmindful of the past, the chiefs almost without exception were steadfast in personal loyalty; but they were unable in many cases to restrain their adherents, many of whom sooner or later joined the ranks of the mutineers. But in these early days there was no suspicion of this danger, and Lawrence's first act was to exchange the sepoy garrison of Ajmere, where his arsenal was situated, for one composed of troops from Jodhpore. Levies from Ulwar, Bhurtpore, and Jeypore also took the field and were of some assistance in preserving order; and in accordance with the request of the Lieutenant-Governor of the North-West

---

[1] Each independent chief had an armed following more or less numerous. In addition, contingents were in some instances maintained, e.g. the Kotah Contingent, the Jodhpore Legion, etc. These bodies were organised in the same way as British sepoys and had British officers.

[2] Brother of Henry and John Lawrence. His head-quarters was at Mount Abu.

[3] Nusseerabad and Neemuch, the only two regular garrisons, were usually composed of Bombay troops. In May 1857 their place was occupied by Bengal troops, as the Bombay troops had gone with the expedition to Persia.

Provinces, the Kotah Contingent and a detachment of Bhurtpore troops were moved up to the Jumna.

Nusseerabad.—The country remained quiet till May 28, when a mutiny occurred at Nusseerabad, where the 1st Bombay Lancers, the 15th and 30th Ben.N.I., and the native field battery No. 6 F.B. 2/7 Ben.A. were quartered. On that day the infantry and artillery broke out, and although the Bombay cavalry[1] took no active part in the revolt, they refused (when ordered to charge the guns) to follow their officers, two of whom were killed and two wounded. Most of the Europeans, however, escaped to Beawar, thirty-five miles distant, and they were followed by a few faithful sepoys[2] and the majority of the Bombay Lancers. After destroying the cantonments the mutineers set out for Delhi, being ineffectively pursued by a thousand Jodhpore troops.

Neemuch.—The garrison of Neemuch consisted of a wing of the 1st Bengal Light Cavalry, the 72nd Ben.N.I., the 7th Infantry, Gwalior Contingent, and 4/1 Ben.H.A., and on June 3 they also rose, murdered some Europeans, burnt the cantonment, and, marching with military order and discipline, went to Deoli. Having sacked this place and carried off the time gun —a 24-pounder—they made for Agra, and on July 5 reached the right bank of the Jumna near that city. Meanwhile the cavalry of the Malwa Contingent,[3] despatched from Mehidpore in Central India to the

---

[1] It was afterwards stated that the Lancers refused to act against the mutineers on condition that their families were unmolested. The regiment afterwards did good service.

[2] See Appendix at end of Chapter II.

[3] See p. 35.

aid of Neemuch, mutinied when fourteen miles from that place, and having murdered their officers, marched back eighty miles in two days to Mehidpore in order to persuade the infantry and artillery of the contingent to join them. Failing in this, they again retraced their steps, and succeeded in joining the Neemuch mutineers between Deoli and Agra.

Reinforcements from Deesa.—On June 12 a wing of the 83rd Foot, the 2nd Bombay Cavalry, the 12th Bom.N.I., and the 2nd Troop Bom.H.A. reached Nusseerabad from Deesa. Neemuch was occupied by troops from Kotah and Jhalawar,[1] and a company of the 83rd was sent to Ajmere. The mutiny of the sepoy garrisons was followed by great disorder in the country, which was infested by various bands of rebels and robbers, whom Lawrence—now appointed Brigadier-General in Rajputana—was not strong enough to repress.

VII. Central India

The Native States.—The Central India Agency comprised a large number of independent states[2] which varied in size and importance, from the wide domains of the Maharajahs Sindiah and Holkar to the petty holdings—sometimes only a few square miles in extent—of some unimportant Rajah or Thakur. It contained the geographical areas of Malwa, Gwalior, and Bundelkhund, and was ad-

[1] On July 18 they were relieved by British troops and returned to their respective states.

[2] The principal states were Gwalior (with its dependencies), Indore (with its dependencies), Bhopal (with its dependencies), Malwa (including Dewas, Jaora, Rutlam, etc.), Dhar, Jhabna, Barwani, Orcha, Panna, Datia, Charkhari, Banda, Rewah, Nagode, and others.

ministered by an Agent-General, political officers being appointed to the more important states. It was divided into a western and eastern region by the Districts of Jhansi and Jubbulpore under the Lieutenant-Governor of the N.W. Provinces, and was bounded on the south by the Saugor and Nurbudda Territories, on the north-west by Rajputana, and on the north-east by the Jumna and the Ganges. The regular garrison was small, the European troops being represented by two companies of Bengal Artillery, but there were numerous bodies of state troops, and the contingents of Malwa, Bhopal, and Gwalior.[1]

Jhansi.—The first outbreak occurred at Jhansi. When the last Rajah died childless in 1854 the state lapsed to the British Crown, and thenceforward the Begum, his widow, became a bitter enemy of the British Government, eagerly joined in the general conspiracy against British rule, and found many fanatical followers in a city which was regarded by Hindus in the same way as Mussulmen regarded Delhi.[2] When the news of the outbreak of the mutiny reached Jhansi a force consisting of the 14th Bengal Irregular Cavalry, the 12th Ben.N.I., and the native field battery No. 18 4/9 Ben.A. (Lieutenant F. H. Turnbull) was equally divided

[1] Holkar's State Troops consisted of 3,200 cavalry, 6,000 infantry, and 24 guns. The Malwa Contingent comprised a regiment of cavalry, one of infantry, and a field battery. The Bhopal Contingent was 255 cavalry, 712 infantry, and 72 artillery. The Gwalior Contingent was 8,200 strong, and was composed of two regiments of cavalry, seven regiments of infantry, four field batteries, and a small siege train. Its main body was at Morar, but it had many detachments, the 7th Infantry being at Neemuch in Rajputana.

[2] Deep offence was given to orthodox Hindus in Jhansi in 1854 when the British Government permitted the slaughter of cattle.

between that city and Nowgong, fifty miles to the south-east. On June 5 the sepoys in Jhansi rose, and were at once joined by the bulk of the inhabitants, the Begum placing herself at the head of the revolt. Murder and riot ran their usual course, but some seventy Europeans took refuge in the Star Fort,[1] and defended themselves successfully for two days; but being induced to leave it by a promise of safe conduct, they were all murdered with the full concurrence of the Begum. On June 7 the troops at Nowgong also mutinied; most of the Europeans escaped from the station, but many perished in the subsequent flight. Among these was Lieutenant S. E. Townsend, Ben.A.

Jalaon and Hamirpore.—During the next few days detachments of the 53rd Ben.N.I. at Jalaon and Hamirpore broke away to join their mutinous comrades[2] at Cawnpore, and their example was immediately followed by three companies of the 1st Ben.N.I.—on detachment from Cawnpore—at Banda. When an attempt was made to disarm them with the aid of the Nawab of Banda, the state troops joined the ranks of the rebels. But soon the Nawab proclaimed that he no longer recognised British authority and the Raos of Kirwi also asserted their independence. The Maharajah of Rewah, however, proved a staunch ally; he offered assistance to the Government, and eventually placed 2,000 men in the field who preserved order. For a time the 50th Ben.N.I. quartered at Nagode remained faithful, but shortly after the outbreak at Banda they also joined the

---

[1] The Star Fort was in cantonments.   [2] See pp. 51, etc.

ranks of the mutineers. During the first half of June British authority was extinguished in Jhansi, Jalaon, Hamirpore, Banda, and Kirwi; farther to the south detachments of the Gwalior Contingent mutinied at Lalitpore and Sipri [1] while the Rajah of Banpore assembled his followers and, after playing a double part, eventually possessed himself of Chanderi and the surrounding neighbourhood.

Gwalior.—At this juncture the attitude of Sindiah was of the first importance. The main body of his contingent was at Morar and was supposed to be so staunch that, when the mutiny broke out at Meerut, the Lieutenant-Governor of the N.W. Provinces had requested that a detachment might be sent to Agra. When the Jhansi news reached Morar on June 11 the main body mutinied. Sindiah's Minister, Dinkar Rao, however, succeeded in saving the Europeans and sending them under escort to Agra; [2] and thenceforward, acting under the advice of this excellent man, the Maharajah proved himself a loyal supporter of the British Government. But he had a difficult part to play. His state troops were in full sympathy with the mutineers of the contingent, and urgent appeals were made to him to declare for "the faith" and lead them against the Feringhis. For the time being he was able to control the turbulent horde and

---

[1] The mutineers marched to Cawnpore and joined the Nana Sahib. They shared in the defeat inflicted on him by Havelock on June 16 (see ch. iv. p. 174).

[2] Sindiah's escort deserted them near the River Chumbal, where 200 ghazis were waiting to murder them. They escaped through the good offices of a Thakur, Buldeo Sing, who protected them and handed them over to the friendly Rajah of Dholpore, who sent them in safety to Agra.

keep them in the vicinity of his capital, but the arrival of mutineers from Mhow and Indore at the end of July greatly increased the difficulty of the situation.

Indore.—The head-quarters of the Central India Agency was at Indore, Holkar's capital, and in cantonments at Mhow, 12 miles distant, were quartered a wing of the 1st Bengal Light Cavalry, the 23rd Ben.N.I., and No. 8 F.B. 2/6 Ben.A. (Captain T. J. W. Hungerford). As June wore away, it was realised that the sepoys were not to be trusted, and the Governor-General's Agent, Lieut.-Colonel Durand,[1] assembled a special force for the protection of the Residency. This consisted of a picked body of Holkar's state troops, the bulk of the contingent of Bhopal, the Begum of which state was staunch in her loyalty, and a corps of Bhils from Malwa. But on July 1 Holkar's troops turned their arms against the Residency they were called upon to defend, and were joined by many of their comrades and by the Bhopal Contingent with the exception of the artillery and a few Sikh sowars. These and the Bhils alone remained faithful, and they fought so well that eventually Durand and the Europeans with him were able to escape from Indore to Hoshungabad.

Hungerford at Mhow.—Up to July 1 the Mhow garrison remained quiet under the command of Colonel J. Platt,[2] who, unlike the Agent, had a firm belief in its loyalty; indeed it was with great difficulty that Captain Hungerford obtained permission to carry out certain measures of precaution. On the

[1] Lieutenant-Colonel H. M. Durand, Ben.E., was acting in the absence of Sir H. Hamilton.
[2] Lieutenant-Colonel and Brevet-Colonel J. Platt, 32nd Ben.N.I.

evening of July 1, however, the sepoys rose and shot both Platt and his adjutant. At the first alarm, Hungerford disarmed the sepoy guard at the fort and then led his battery, 2/6 Ben.A., to the parade ground, which was lighted up by the glare of blazing bungalows. Presently a bullet or two whizzed over the guns, and Hungerford at once unlimbered and opened a fire of case shot upon the sepoy huts. Groans and a rushing noise were then heard, and it was found that the native garrison were in full flight towards Indore. Finally, in company with some of Holkar's troops and a thousand ghazis who joined them on their march, they made their way to Gwalior, where they arrived on July 31. Hungerford next turned his attention to the defences of the fort, where, well supplied with provisions and aided by an efficient body of civilian volunteers, he could maintain himself till relieved. On July 5 Holkar sent a deputation to Hungerford to express his sorrow for what had occurred at Indore and to offer his co-operation in maintaining order.

Saugor.—The Saugor garrison, under Brigadier Sage,[1] consisted of the 3rd Bengal Irregular Cavalry, the 31st Ben.N.I., the 42nd Ben.N.I. (less two companies at Damoh), and No. 4 F.B. 1/3 Ben.A. (Captain W. B. Marshall). The 52nd Ben.N.I. was at Jubbulpore with a detachment at Mandla, while Seoni, Narsingpore, Hoshungabad, and Betul were held by the 28th Mad.N.I. Early in June Sage detached a small force under Major Gaussen[2] and relieved

---
[1] Colonel W. Sage, 22nd Ben.N.I.
[2] Captain and Brevet-Major D. Gaussen, 42nd Ben.N.I.

Lalitpore; but the passes into Bundelkhund were found to be occupied by the Rajah of Banpore, who, with the Rajah of Sharghur, now assumed an actively hostile attitude, and Gaussen was recalled to Saugor. On June 29 the fort at Saugor was occupied by 1/3 Ben.A. and two days later the 3rd Irregular Cavalry and the 42nd Ben.N.I. broke into mutiny. On the other hand the 31st Ben.N.I. remained loyal and even attacked the mutineers, a desultory combat taking place, in which the Brigadier would not permit the European gunners to take part. Next day the rebels made off to the north, after failing to prevail upon the two companies of the 42nd at Damoh to join them. The 31st remained steadfastly loyal, but it was not thought prudent to permit them to enter the fort. All this while the 52nd Ben.N.I. at Jubbulpore showed no sign of insubordination, and did good service in preserving order in the immediate neighbourhood.

As Gaussen fell back the two rajahs followed him up, and, having lost all his 42nd men by desertion, he reached Saugor on July 12 with only a remnant of the loyal 31st; leaving the forts of Rahatghur and Garakota in the hands of the enemy.

Thus from the Nurbudda to the Jumna British authority had for the time disappeared, except in the actual garrisons of Mhow, Saugor, Damoh, and Jubbulpore.

VIII. ROHILKHUND, AGRA, AND NORTHERN DOAB

Loss of Rohilkhund.—There were no white troops in Rohilkhund, and the garrison, consisting of the 8th

Bengal Irregular Cavalry, the 18th, 28th, 29th, and 68th Ben.N.I., and No. 15 F.B. 6/8 Ben.A. (Captain G. Kirby), was distributed between Bareilly, Shahjehanpore, and Moradabad. By the end of May they had all mutinied, murdered some of their officers and many civilians, and had gone to Delhi under the artillery soubadar Bukt Khan, who afterwards commanded the rebel force in that city. A pensioner of Government, one Khan Bahadur Khan, became the King of Delhi's "Viceregent" in Rohilkhund, which remained for more than a year in a state of anarchy.

Mutinies round Agra.—The garrison of Agra, commanded by Brigadier Polwhele,[1] consisted of the 3rd Bengal European Regiment, No. 21 F.B. 2/5 Ben.A. (Captain E. A. C. D'Oyly), and the 44th and 67th Ben.N.I., each of which had a detachment—three companies in all—at Muttra. One wing of the 9th Ben.N.I. was at Alighur, the other was distributed between Bulandshahr, Manipuri, and Etawah. The news of the Meerut outbreak reached Mr. J. R. Colvin, Lieutenant-Governor of the N.W. Provinces at Agra, on May 11, and in accordance with his urgent request Sindiah sent two regiments of cavalry and a battery (Lieutenant A. Pearson, Ben.A.)—contingent troops —to Agra. The Rajah of Bhurtpore sent a small force of Jâts to Muttra, and the Kotah Contingent set out from Deoli to march to the Jumna.

On May 21 the 9th Ben.N.I. mutinied at Alighur, an example immediately followed by its detachments;[2] and the whole regiment made off to Delhi.

[1] Lieutenant-Colonel and Brevet-Colonel T. Polwhele, 36th Ben.N.I.
[2] The station of Manipuri was saved by the gallantry of Lieutenant

Etawah was meanwhile occupied at Colvin's request by an infantry regiment of the Gwalior Contingent. On May 30 the sepoys at Muttra mutinied and were joined by the Jâts, and next day the 44th and 67th Ben.N.I. were disarmed at Agra. Sindiah's cavalry and artillery were then sent out to Alighur to restore order, but before the end of June they had all mutinied, without, however, molesting their officers, who were able to make their way back to Agra.[1] The insurrection at Gwalior (June 11) was at once followed by the mutiny of the contingent infantry at Etawah, and at the same time news reached Agra that the mutinous brigade from Neemuch was marching upon the city. The European regiment and battery were then moved into the fort, where the majority of the residents also took refuge, a body of volunteers were enrolled, and some native levies were raised, only to be speedily disbanded as untrustworthy. On July 2 the Neemuch rebels were reported to be at Futtehpore Sikri, 25 miles distant from Agra, and the Kotah Contingent was ordered to attack them. This contingent had hitherto behaved well and performed good service in preserving order round Muttra after the mutiny there; but when it received the above orders, far from attacking the Neemuch Brigade it promptly joined them, thus

---

C. A. de Kantzow, 9th Ben.N.I., who was placed in command of a body of police for special service (Mall., i. p. 159).

[1] Lieutenant A. Pearson, Ben.A., had a miraculous escape (Mall., i. p. 295). He and Lieutenant J. H. Lamb (a refugee from Gwalior) were attached to 1/5 Ben.A.

bringing the rebel force up to 5,000 men and six guns.[1]

Action at Sassiah, July 5.—On July 5 Polwhele marched out with the 3rd Europeans, D'Oyly's Battery, and a few volunteers, in all five hundred men and six guns, and about 1 p.m. encountered the rebels at Sassiah, 5 miles from Agra on the Futtehpore Sikri road. The village of Sassiah was held by their infantry, with three guns, well protected by accidents of ground, on each flank. Their cavalry was drawn up in rear, and in their front was an open sandy plain. Across this plain Polwhele advanced, his infantry in the centre and three guns on each flank, until to within 800 yards of the village. He then ordered the 3rd Europeans to lie down and the guns to shell the village. But the rebel guns answered the fire with deadly effect. They blew up a limber and wagon, disabled one of our 9-pounders,[2] killed three gunners, dangerously wounded D'Oyly and Lieutenant J. H. Lamb, and caused much loss among the horses. The six well-sheltered rebel guns were more than a match for the six British guns in the open, and a bayonet charge was needed to save the situation; but this was not delivered until 2/5 Ben.A. had expended all its ammunition and the enemy's cavalry was

[1] A loyal gunner, by name Malthura, had time to spike the guns before they could be removed, while Assistant Surgeon C. Mathias, aided by his servants and others, "strewed in the sand their powder, ammunition, and case shot" (Mall., i. p. 268). Thus the only guns the rebels had were those of 4/1 Ben. H.A.

[2] This gun in the subsequent retreat to Agra had to be abandoned, as many of the native drivers had deserted. It was recovered next day.

hovering on our flanks. At last the 3rd Europeans were sent forward, and the enemy was driven from the village, but there was neither cavalry to follow them up nor shells to harass their retreat. The rebels rallied and renewed the attack; but fortunately their gun ammunition also ran short, and they were driven off by our infantry. Polwhele then marched back to Agra unmolested;[1] but the rebels, following him up, murdered some white people who had not taken refuge in the fort, burnt and looted the cantonments, and next day marched towards Delhi, where they were received as conquerors with a salute of guns. After their departure order was restored in the city of Agra, but the district around had practically thrown off British rule.[2]

Futtehghur.—Meanwhile the important fort and arsenal at Futtehghur had fallen into the hands of the rebels. The 10th Ben.N.I. formed the garrison, and the arsenal was in charge of Major A. Robertson, Ben.A., who had three or four European conductors and some native gunners under his command. On June 15 the sepoys and gunners rose on the arrival of the 41st Ben.N.I., which had mutinied at Seetapore. Previously some of the non-combatants [3] had quitted the place, and on the 18th thirty-three adult males

[1] There were many men wounded, but records are wanting. D'Oyly, although his wound was mortal, continued to direct the fire from a limber upon which he had been lifted (Stubbs, 315 n.). Lamb died from his wounds.

[2] Some successful skirmishing, however, took place in August in the Aligurh district, where De Kantzow was at work with his police. Mr. Colvin died in Agra, September 9, 1857.

[3] Of these a few found protection with a friendly rajah. All the others perished on the way, or were murdered on reaching Cawnpore,

(officers and civilians) held the fort against the rebel attack. They maintained an almost hopeless contest for sixteen days and then attempted to escape by boat, accompanied by the remaining women and children. Many lost their lives in this attempt, and most of the others reached Cawnpore only to be butchered on their arrival.[1] The mutineers then passed into Oudh, after proclaiming the Nawab of Ferruckabad, who was joined by some 30,000 adherents, ruler of the district.

Hansi, Hissar, and Sirsa.—The 4th Bengal Irregular Cavalry and the Harriana Light Infantry, quartered at Hansi, Hissar, and Sirsa to the northwest of Delhi, mutinied during June and joined the rebels in the imperial city. Thus the whole of the Upper Doab, from Cawnpore to the Cis-Sutlej States, was in the hands of the rebels except Agra, which, though safe, was isolated.

IX. OUDH AND CAWNPORE

Oudh in 1857.—Though the annexation of Oudh in 1856 had been peaceably carried out, the ruling classes—Mohammedans of Moghul descent—were deeply dissatisfied with the new state of things whereby many of them lost more or less reputable means of livelihood. A focus of sedition was soon formed in the province; whilst the late King,[2] who

[1] Major Robertson died of his wounds in a village in Oudh.
[2] The Nawabs of Oudh broke away from the Delhi Empire after the death of Aurangzib, and most of the urban population of the province were Mohammedans of Moghul descent. The Nawab was given the title of King by the British in 1817. He was deposed, February, 1856, for persistent misgovernment and oppression.

had been deported to Calcutta, became a centre of intrigue in the British capital. The rustic population consisted for the most part of Hindu Rajputs living in clans with a feudal organisation, and possessing strongholds of a formidable description.[1] Being in a state of chronic rebellion against the exactions of the King and his agents, they in general regarded the British annexation with friendly feelings, until the subsequent land administration, carried out by the British officials often without sympathy and sometimes without justice, provoked grave discontent. At the same time the country-side was troubled by lawless bands of discharged soldiers of the King, for whom room could not be found in the new Irregular Force and the Police.[2]

Sir Henry Lawrence.—Thus in March, 1857, when Sir Henry Lawrence[3] assumed the government of the province, he was confronted by a state of affairs the reverse of satisfactory. He, however, at once applied remedial measures. Under his strong hand disorder was quelled, and the great landowners—the Talukdars—and their followers began once more to regard the British Government with friendly feelings, while the Mohammedans found more em-

[1] These strongholds, or forts, were scattered all over Oudh, and were always surrounded by dense bamboo thickets, and situated in almost impenetrable jungle, through which secret means of access was provided. They were often armed with guns.

[2] The Oudh army was disbanded in 1856 and a new levy with British officers—the Oudh Irregular Force—was raised. Mounted and Foot Police were also organised.

[3] Lieutenant-Colonel and Brevet-Colonel Sir H. M. Lawrence, Ben.A. At the outbreak of the Mutiny he was appointed Brigadier in Oudh, and was promoted Major-General about the time of his death in July.

A "FORT" IN OUDH

ployment and less cause for complaint.[1] But though the towns were quiet and revenue poured freely into the district treasuries, the new Chief Commissioner was not blind to the impending danger of a great military revolt.[2] Quickly and quietly measures were taken to meet the coming trouble which, among the great officials of India, Lawrence alone appears to have foreseen.

Garrison of Oudh.—The troops in Oudh were distributed as follows :

At Lucknow

7th Ben.L.C.	32nd Foot	No. 9 F.B. 4/1 Ben.A.
2nd Oudh Irr.C.	13th Ben.N.I.	(2nd Lieutenant E. P. Lewin).
Mounted Police	48th Ben.N.I.	No. 2 F.B. 2/8 Ben.A.
	71st Ben.N.I.	(Captain A. P. Simons).
	4th Oudh Irr.I.	2nd Battery Oudh Irr.A.
	7th Oudh Irr.I.	(Lieutenant D. C. Alexander).
	3rd Reg. Mil. Police	3rd Battery Oudh Irr.A.
		(Lieutenant St. G. Ashe).

At Fyzabad

15th Ben.Irr.C.	22nd Ben.N.I.	No. 13 F.B. 5/7 Ben.A.
(1 Squadron)	6th Oudh Irr.I.	(Captain and Brevet-Major J. Mill).

At Secrora

1st Oudh. Irr.C.	2nd Oudh Irr.I.	1st Battery Oudh Irr.A.
		(2nd Lieutenant J. Bonham).

At Sultanpore

15th Ben.Irr.C.	8th Oudh Irr.I.	1st Reg. Mil. Police.

At Seetapore

41st Ben.N.I.	10th Oudh Irr.I.
9th Oudh Irr.I.	2nd Reg. Mil. Police.

---

[1] " Sepoy Revolt," p. 42, etc.

[2] A certain Moulvie—generally known as the Moulvie of Fyzabad—was preaching a religious war and was promptly imprisoned. During the first mutinies in Oudh he escaped and became one of our most active enemies.

At Durriabad
5th Oudh Irr.I.   1st Reg. Mil. Police.

At Salone
1st Oudh Irr. I.

At Gonda
3rd Oudh Irr.I.

Mutinies in the Province.—At Lucknow on May 3 the 7th Oudh Irregular Infantry refused to use the new cartridge and were at once disarmed; but though at the time there was no further insubordination, the suspicions of Lawrence were in no way allayed. A commanding position enclosing some 500 yards square round the Residency was selected, and its entrenchment was begun. The Machi Bhawan, a large building half a mile to the westward, was cleared out to serve as a refuge for the Europeans in case of desperate emergency. The walls were strengthened by the addition of earthworks, guns were mounted in it to overawe the city, and it was provided with a reliable garrison.[1] In addition, the southern end of the Mariao Cantonments, which contained the sepoy lines, was occupied by a detachment of British troops.[2] The works of defence were still in progress when the news of the seizure of Delhi reached Lucknow. It now became evident

[1] Lawrence at once separated the Sikhs from the other sepoys, and summoned from their homes two bodies of pensioners; one of old trained British sepoys and one of Oudh artillerymen. Both these companies of auxiliaries gave staunch and loyal assistance during the siege ("Lucknow and Oudh," p. 82).

[2] The European Barracks were a mile to the east of the Residency, the Mariao Cantonments some three miles to the north, both outside the city.

that, if the mutiny spread, the position of Cawnpore, 50 miles south-west of Lucknow, would be one of great peril, and Lawrence sent a company of the 32nd Foot, 240 troopers of the 2nd Oudh Irregular Cavalry, and three guns of the 3rd Oudh Battery (Lieutenant Ashe) to reinforce the garrison there; but on the 30th the commandant of that station sent the British infantry back, and the Oudh cavalry went on to Allahabad. The same night the 48th Ben.N.I. broke out in mutiny, and were joined by many of the 71st, a few of the 13th, and about half of the 7th Bengal Light Cavalry. Spreading over the Mariao Cantonments, they burnt bungalows and murdered a few officers, but the dispositions of his troops made by Lawrence prevented the mutineers from entering the city where the 4th and 7th Oudh Irregular Infantry were quartered. These regiments remained quietly at their duty, the police preserved order during the night, and next morning the rebels were driven off northwards and made for Seetapore. During the following days revolt spread with great rapidity; Oudh, except on the Nepaul frontier, was girt by a chain of rebellion; and the garrisons of all the out-stations joined the ranks of the rebels. Europeans were in flight all over the province; many British officers were murdered; others escaped to Lucknow, protected by their own men in some cases.

Bonham's Escape.—The escape of Lieutenant J. Bonham, Ben.A., from Secrora was a notable example which is thus described by Kaye: " One officer, however, still remained when all the rest had abandoned the cantonment. This was Lieutenant John Bonham of the Artillery, who commanded a battery of the Oudh Irregular Force. He was a young soldier of great promise, and, for his years, of commanding influence. He kept his company together after the other branches of the service had revolted. But, although his courage commanded the respect and admiration of his men, it could not secure their fidelity. The Commissioner had gone. All the infantry officers had gone. Everything portended that British authority was utterly extinguished. For a little while, however, the infantry, without their own officers and subdued by the quiet gallantry of the young artilleryman, asked him to take command of them. They made the most solemn promise never to leave him, and swore to protect him with their lives. Upon this Bonham consented to take command, and at once made his arrangements to march the infantry and guns into Lucknow. But these professions of loyalty were but short-lived. The infantry soon became insolent in their demeanour, and there was a marked change in the bearing of the gunners. Still Bonham clung to his guns; and when the infantry advanced upon them, ordered his men to fire. The order was not obeyed, and when Bonham himself went to the battery to do the work, his men pointed their carbines at him. There were some, however, among them who were eager to serve him, and who implored him to leave the place with the European non-commissioned officers. They brought him horses and they brought him money, and urged him to fly for his life. So at last

... he started for Lucknow."[1] He was accompanied by Farrier Sergeant James Bewsey, Assistant Farrier James Millar, Ben.A., and Sergeant-Major Court, 2nd Oudh Infantry, and after some exciting adventures they reached the capital in safety.

With the exception of the 41st Ben.N.I., who marched from Seetapore on Futteyghur, the mutineers remained in the neighbourhood of their stations, and Lawrence was still unmolested in Lucknow; but with the exception of the capital, Oudh was lost.

Cawnpore.—Cawnpore was the head-quarters of the Cawnpore Division commanded by Major-General Sir Hugh Wheeler. At the time of the Meerut outbreak a weak company of Bengal Artillery, 1/6 Ben.A., and some invalids of the 32nd Foot were the only white troops there, and the headquarters of the 7th Batt. Ben.A. (Major G. Larkins), 1/7 and 1/8 Ben.A., the 2nd Bengal Light Cavalry, and the 1st, 53rd, and 56th Ben.N.I. formed the native garrison. Towards the end of May, Wheeler was reinforced by Lieutenant Ashe's half battery from Lucknow, and by weak detachments of the 84th Foot and 1st Madras European Fusiliers from Calcutta.

The Nana Sahib.—And now two fatal steps were taken at Cawnpore. For one of them, Mr. C. G. Hillersdon, the magistrate, must certainly share the responsibility with the Divisional General, who after fifty years' service in India had a belief, amounting to infatuation, in the loyalty of the sepoys. To the north of Cawnpore lay the village of Bithur, where in a large mansion close to the Ganges lived

[1] Kaye, iii. p. 476, etc.; Stubbs, iii. p. 342, etc.

the Mahratta prince, Dundu Punt, commonly called the Nana Sahib, who represented the faded glories of the Peshwa.[1] Government still allowed him an armed retinue of 500 men and three guns, and he was in possession of a considerable private income; but the large pension which his immediate predecessor had enjoyed lapsed on the death of Bajee Rao in 1852. The Nana was the Rao's adopted son, and under a smiling exterior he nourished a bitter hatred of the British, who, he considered, had robbed him of his rights. His hypocritical professions of friendship did not deceive everybody,[2] but he gained Hillersdon's confidence, and in compliance with his request a detachment of 200 men and two guns marched in from Bithur on May 26. They were under the command of one Tantia Topi—formerly the playmate and now the trusted officer of the Nana—and they were sent to reinforce the native guard over the Treasury.

The Entrenchment.—This act was bad enough, but the next was disastrous. Forced by the pressure of events to make provision for the safety of the Europeans, Wheeler had to select a position where, in the event of a sudden and temporary *émeute* (nothing more serious being expected), they

---

[1] The Mahrattas were finally defeated in 1818, and their leader, the Peshwa Bajee Rao, threw himself on the mercy of the British. He was removed to Bithur, where he lived, in the enjoyment of an ample government pension which lapsed on his death as he had no direct heir.

[2] The suspicion of Mr. M. R. Gubbins, Financial Commissioner of Oudh, backed by the opinion of Sir Henry Lawrence, was communicated to General Wheeler—perhaps too late ("Cawnpore," by G. O. Trevelyan, p. 66).

could maintain themselves until the arrival of reinforcements. Such a position was presented by the Arsenal, which stood on the north of the town near the river—a large walled enclosure, containing the magazine and other buildings, which could have been made impregnable against native attack, and whence guns would have commanded the Treasury, the Jail, and the Public Offices. But this suitable position was not chosen by General Wheeler because a sepoy guard was posted over it, and this guard he refused to remove. Now at this moment there stood in the open maidan to the south of the town two large unfinished buildings of thin brick, one with a thatched roof, originally intended for a hospital and some new barracks in course of construction. The two large buildings were now surrounded by a rectangular trench, dug with difficulty in the sun-baked soil, and a parapet about 5 ft. high, which was not bullet-proof at the crest. Here and there intervals in the parapet were left for guns which would have to be worked practically in the open, while on every point of the compass an assailant could find cover both for infantry and artillery. This wretched redoubt armed with a few light guns, stored with inadequate supplies of ammunition and food, and with a well which was totally unprotected, was the spot where the British were destined to say their last word in the defence of Cawnpore.

Mutiny of the Native Troops.—On May 21 the women and children entered the fatal entrenchment. On June 5 the bulk of the sepoy garrison broke out,[1] and after an orgie of plunder and murder, in which the Nana's followers willingly joined, the mutineers set out for Delhi. But now the Nana saw his opportunity. Not only was he rich with the spoil of the plundered treasury, but the abandonment of the Arsenal placed at his disposal priceless stores of heavy guns and unlimited ammunition. He fully realised that a titular Peshwa would get no welcome from the Moghuls; but the spark of Mahratta nationality was not extinct and might burst into flame were that Peshwa to display his standard at the head of a conquering army. The nucleus of such an army was at hand. By the bait of unlimited plunder and the offer of a gold bangle to each sepoy, the Nana persuaded the rebels to return to Cawnpore, and then, for ever casting the mask aside, he sent word to General Wheeler that he intended to attack him. Meanwhile the white troops, numbering some 400 men and a few faithful sepoys, manned the futile defences; and it was no sudden *émeute*, no temporary rising they had to contend with; they were confronted by as terrible a crisis as Britons had ever been called upon to face.

[1] The cavalry were the first to rise, but their old soubadar major bravely defended the colours and the treasure which were in the quarter guard, where he was found next day lying wounded. He was afterwards killed in the entrenchment. An hour later the 1st Ben.N.I. broke out and were followed next morning by the 56th. The 53rd remained quietly in their lines until Ashe (by Wheeler's command) opened fire on them, the native officers and such men as wished to declare their loyalty having previously been called in. All the officers and about 100 men joined the Europeans, the rest of the regiments followed the rebels. No attempt was made to molest the British officers of any of the native regiments (Mowbray Thomson, p. 38, etc.). All the golandauz apparently joined the mutineers.

X. ALLAHABAD, BENARES, AND THE LOWER DOAB

Allahabad.—Allahabad, situated on the Grand Trunk Road half-way between Calcutta and Umballa, with its large arsenal and its fort commanding the junction of the Jumna and the Ganges, was a place of the highest strategical importance. Its garrison was almost exclusively native, a few invalids of the Bengal Artillery being the only European troops. They were quartered in the fort together with the Ferozepore Regiment (Sikhs) and a company of the 6th Ben.N.I. The rest of the 6th and a squadron of the 2nd Oudh Irregular Cavalry were in the cantonments $2\frac{1}{2}$ miles distant. When the Meerut-Delhi news reached Allahabad, little anxiety was felt, as the Sikhs were considered to be reliable; and the 6th Ben.N.I., regarded as the model regiment of the Bengal Army, was loud in its professions of loyalty and insistent in its demands to be led against the mutineers. As a mere measure of precaution the European women and children were moved into the fort and a corps of civilian volunteers was organised.

Benares.—But at Benares, 70 miles to the eastward, where Brigadier Ponsonby[1] was in command, the situation was regarded by the British authorities with deep apprehension. Benares, the sacred city of the Hindus, held 200,000 inhabitants, and the native garrison was to the European in the proportion of ten to one. Until the end of May, indeed,

[1] Major and Brevet-Colonel G. C. Ponsonby, 11th Bengal Light Cavalry.

the only European troops were the half of No. 12 F.B. 2/3 Ben.A. (Captain W. Olpherts); but at this time Colonel Neill[1] arrived from Calcutta with a detachment of the 1st Madras European Fusiliers, and a company of the 10th Foot was sent from Dinapore. But even with these additions the white troops only reached a total of 250, while the native garrison, consisting of the 13th Bengal Irregular Cavalry, the 25th[2] and 37th Ben.N.I., and the Loodianah Regiment (Sikhs) numbered over 2,000 men.

The Disarming Parade.—When the news of a mutiny at Azimghur reached Benares on June 4 it was determined to lose no time in disarming the sepoys, and a general parade was ordered for 5 p.m. in front of their lines. In view of the fanatical crowds close at hand in the city, of the large number of native and the small number of European troops, the proposed operation demanded the greatest judgment. On other occasions as critical as this regiments about to be disarmed plainly saw that means were present to enforce the command to lay down their arms. They could see white troops with loaded rifles threatening their flanks, or cavalry with weapons in hand ready to ride them down, or they were suddenly confronted by a line of guns and gunners with their matches lighted. But on this occasion no such steps were taken. While the rest of the garrison were still on their way from their private parades, the 25th and 37th were ordered to put their arms in the

[1] Major and Brevet-Lieutenant-Colonel J. G. S. Neill.
[2] A wing of the 25th Ben.N.I. was in Rangoon.

racks.¹ There were no signs of resistance, and the men were in the act of carrying out their orders when they became aware of the approach of the European troops, headed by the Brigadier.² In a moment they broke out in mutiny, and began to fire on their own officers and the approaching British. The fire was at once returned by our infantry, and Olpherts with the greatest promptitude sent a salvo of case shot into the rebels, who fled to their huts and continued to fire from their shelter. The huts were somehow set on fire, and a few round shot from the guns speedily turned out the mutineers, who fled across the plain pursued by our infantry. This affair was still in progress when the Loodianah Regiment and the 13th Irregular Cavalry made their appearance. The Sikhs at once got greatly excited, and with shouts and yells began firing in all directions. Olpherts' guns were just then limbered up and absolutely without support. The Sikhs were some 80 to 100 yards distant, their bullets began to fall among the gunners, and at any moment a sudden onslaught might be made on the battery. Without hesitation Olpherts unlimbered, swung his guns round, and opened a fire of case. Three times the Sikhs tried to rush the guns, and three times were driven back. They then fled, accompanied by the greater part of the cavalry; and the handful of white troops remained masters of the situation, but at a cost which might have been saved.

¹ The racks, known as "bells of arms," could be locked up and secured; but might be, and sometimes were, broken open.

² It is said that the gunners of Olpherts' Battery were seen to be carrying case shot and cartridges in their hands.

Mutiny at Allahabad.—With the object of preventing an irruption of mutineers from Benares, the bridge of boats at Allahabad was now guarded by a detachment of the 6th Ben.N.I. with two guns (Lieutenant T. N. Harward), and somewhat in their rear Lieutenant A. H. Alexander was posted with a picket of the Oudh cavalry. On the evening of June 6 Harward was ordered to withdraw his guns to the fort, but was prevented doing so by the sepoys, who insolently told him they would have to be taken to cantonments. Harward ran at once to the picket, and, Alexander having mounted him on a spare horse, the two officers led the sowars towards the bridge. But their efforts were unavailing: the sowars at the last moment refused to charge the sepoys. Harward, escaping by a miracle, made his way to the fort; but Alexander was murdered, and the sepoys dragged the guns into cantonments, where the 6th N.I. were already in open mutiny. This model regiment had been paraded a few hours before to hear the thanks of the Governor-General for its loyal professions, but shortly before the arrival of the guns it broke out into sudden mutiny and displayed the greatest ferocity in murdering its officers and any Europeans who came in its way. A few officers, however, escaped to the fort.

Defence of the Fort.—Here matters were in a critical condition. The Ferozepore Regiment had by this time heard of the fate of their countrymen at Benares, and the issue was in their hands. While yet it was doubtful, Lieutenant W. C. Russell, Ben.A., the Commissary of Ordnance, laid a train to the principal magazine and prepared to blow it up if need arose. But happily the Sikhs were commanded by an excellent officer, Lieutenant J. Brasyer,[1] under whose influence they remained staunch, and the company of the 6th Ben.N.I. was disarmed and expelled from the fort.

On this, Allahabad, where a self-appointed representative of the King of Delhi was dressed in a little brief authority, gave itself up to riot and disaster, and "two great waves of rebellion, the one from Allahabad, the other from Cawnpore, swept away British authority in the lower Doab."[2]

Relief of Allahabad.—Meanwhile Colonel Neill had succeeded Ponsonby in the command at Benares, and with the help of capable Civilians, Mr. H. C. Tucker and Mr. F. Gubbins, and of some of the leading inhabitants, ensured the tranquillity of the city. His next step was to recover Allahabad. Pushing on detachments in advance, he himself arrived at the fort on June 11, and next day opened fire with its guns on a neighbouring suburb which harboured the scum of the population. The menace of a general bombardment aided by the effect of a few spirited sorties sufficed to tranquillise the town; the King's Agent vanished and the mutineers fled to Oudh; but anarchy continued to reign in the lower Doab. More reinforcements from Calcutta might now be expected, and Neill began to prepare for an advance on Cawnpore. On June 30 the first

[1] Brasyer received his commission from the ranks of the Bengal Artillery.
[2] Kaye.

body of troops was set in motion, and the same day Major-General H. Havelock arrived from Calcutta to take command.

XI. BENARES DISTRICT, BEHAR, AND CHOTA NAGPORE

In this region the garrison was distributed as follows :

At Gorruckpore
12th Ben.Irr.C. (1 Squadron)

At Azimghur
17th Ben.N.I.

At Jaunpore
Det. Loodianah Regiment (Sikhs)

At Ghazipore
65th Ben.N.I.

At Dinapore[1]
10th Foot, 7th Ben.N.I., 8th Ben.N.I., 40th Ben.N.I.,
No. 11 F.B. 4/5 Ben.A.

At Baosi and Deoghur
32nd Ben.N.I.

At Rohni
5th Ben. Irr.C.

At Segowlie
12th Ben.Irr.C.

At Hazaribagh
Detachments of 8th and 40th Ben.N.I.

Mutinies in the District.—On June 3, as a party of the 12th Bengal Irregular Cavalry who were escorting treasure from Gorruckpore to Benares passed

[1] With a fatuity it is difficult to realise, no steps were taken to meet the danger to our lines of communication caused by the preponderance of native troops at this station which was close to the fanatical city of Patna; and, as the days went by, the sepoys, though suspected, were not disarmed.

through Azimghur, the treasure was looted by the 17th Ben.N.I., who mutinied, shot some Europeans, opened the jails and handed Azimghur over to the mob. The 17th then marched off to Fyzabad, whither they were followed a few days later by the Loodianah Regiment from Jaunpore. Meanwhile the cavalry loyally escorted their own and some other officers safely to Benares, whence they were sent back to Gorruckpore.

XII. Madras

When the news of the outbreak at Meerut reached Madras the European Fusiliers were at once despatched to Calcutta, whither they were followed in August by infantry and artillery under Brigadier Carthew,[1] while a column under Colonel Fischer[2] marched northwards from Cuttack. A force of three arms also, known as the Nagpore Movable Column, assembled at Kampti. The Madras Army[3] proved loyal, and the Presidency was only troubled by sporadic disorder which was dealt with by small field forces.

XIII. Bombay

Energetic Action of the Governor.—In Bombay also the Governor, Lord Elphinstone, instantly recognised the gravity of the crisis. He ordered the 64th and 78th Foot, who were returning from Persia and were yet at sea, to proceed direct to Calcutta ; he hired and sent to Mauritius and the Cape

[1] Lieutenant-Colonel M. Carthew, 1st Madras Fusiliers.
[2] Lieutenant-Colonel T. M. Fischer, 18th Mad.N.I.
[3] One cavalry regiment caused some trouble regarding pay.

of Good Hope suitable vessels for the transport of troops; he at once complied with George Lawrence's call for help from Deesa; he authorised the Resident to send troops from Sind to Mooltan, and he assembled at Poona a force of the three arms which marched for Mhow on June 8.

Risings in the Presidency.—The Bombay Army as a rule was staunch, but the Presidency was not free from danger. Conspiracy rife in the capital was held in check, but open mutiny occurred in the Mahratta country south of Poona. The 27th Bom.N.I. and some local levies were at Kohlapore,[1] the 29th at Belgaon, and the 28th at Dharwar. The three regiments and the petty chiefs of the district had plotted a concerted rising for which Kohlapore was to give the signal. On July 31 the 27th Ben.N.I. broke out, but the local levies remained loyal; the telegraph gave timely warning to Belgaon, and Dharwar—a possibility the rebels had forgotten—and the mutiny was already suppressed when Colonel Jacob,[2] despatched from Poona to command the district, arrived at Kohlapore. He brought with him two guns of the 4th Troop Bom.H.A., and he disarmed the 27th without further trouble.

In addition to the large field forces that operated in Rajputana and Central India the Bombay Government was occasionally obliged to send out troops to

[1] Kohlapore is some 130 and Belgaon 190 miles south of Poona. Dharwar is 42 miles east-south-east of Belgaon, where the depôt of the 64th Foot and 3/1 Bom.A. were in garrison.

[2] Colonel G. Le G. Jacob, 27th Bom.N.I.

[3] The 28th and 29th were preserved from actual mutiny by the timely action of the authorities.

check disorder. Thus, in the spring of 1858, 2/4 and 4/4 Bom.A. (Captain G. P. Sealy) accompanied a small column to the Satpura Hills to deal with a tribe of insurgent Bheels who on April 11 were signally defeated at Amba Pawnee, where Lieutenant C. E. Basevi, Bom.A., was seriously wounded.

In the following month the chief of Nargund [1] murdered the magistrate and attempted to rouse the Mahrattas to rebellion. But his fort—a very strong one—was taken and he was hanged after being defeated by Colonel Malcolm [2] and a small mixed force which included two guns of 8/14 R.A. (Captain L. G. Paget).

From time to time other expeditions of a minor character were undertaken.

XIV. HYDERABAD

Until Central India was pacified (June, 1858), Hyderabad was a potential danger. Fortunately the Nizam, well advised by an astute minister, was loyal to his treaties, and the Resident [3] was a capable man; but the Deccan was the resort of adventurers and desperadoes and the city of Hyderabad was a focus of disorder. Soon after the Meerut outbreak some of the Nizam's troops attempted to provoke a rising; but not only was this mutiny instantly stamped out, but in a little while it was found possible to send the bulk of the Hyderabad Contingent to reinforce the Bombay troops marching upon Mhow.

---

[1] Nargund is in the South Mahratta country.
[2] Colonel G. Malcolm, South Mahratta Horse.
[3] Major C. Davidson, 66th Ben.N.I.

Shorapore and Chicumba.[1]—In February, 1858, however, the Rajah of Shorapore, a feudatory of the Nizam, attempted to disturb the tranquillity of the Deccan, but was defeated and taken prisoner by Colonel G. Malcolm, who in this expedition was accompanied by 3/1 Bom.A. (2nd Lieutenant H. T. Vachell).

Towards the end of that year, also, a larger force of Arabs and Rohillas collected in Berar, and Brigadier Hill[2] with the Hyderabad Contingent (again assembled in the Deccan after the Central India Campaign), a detachment of the 12th Lancers, and B Troop Mad. H.A. (Captain A. T. Cadell) defeated and scattered them at Chicumba on January 5, 1859. In this action 2nd Captain J. B. Swete, Mad.H.A., was severely wounded.

XV. The Task of the British

The occupation of Delhi was the outward and visible sign of our supremacy in Upper India, and its loss was a heavy blow to British prestige. The Punjaub was thereby cut off from direct communication with Calcutta, and a wedge of revolt was driven into the heart of the British possessions. But the leading part now assumed by the Moghuls was viewed by their fellow-plotters with alarm; for ready as they were to overthrow the British power,

---

[1] These minor affairs, as well as the events at Kohlapore and Amba Pawnee, had nothing to do with the main campaign and can be conveniently dealt with in this place.

[2] Colonel W. Hill commanded the Hyderabad Contingent, which consisted of 5 regiments of infantry, 5 of cavalry, and 5 companies of artillery.

they had no wish to restore the ancient tyranny of Delhi. Jealousy and suspicion thus paralysed the counsels of the rebels, who happily had neither competent leaders nor a common design; and after the seizure of Delhi mutiny hung fire for three whole weeks. How this breathing-space was utilised by the British has already been shown; and the task that now lay before them was the capture of Delhi and the relief of Cawnpore and Lucknow. In the first instance their base was the Punjaub, and their sources of supply and lines of communication were safe so long—but only so long—as their ultimate success was believed in by the inhabitants of the states through which these lines ran. In the second instance their base indeed was the sea; but the long lines of communication were an ever-present anxiety, and no man could say what danger might flash from the dark clouds beyond the Jumna. Many months were to pass before anything could be done to dispel those clouds and revive British authority in the region from which it had passed away.

On June 17 General Sir P. Grant, Commander-in-Chief in Madras, arrived in Calcutta to succeed General Anson.

APPENDIX TO CHAPTER II
NATIVE FIDELITY

THE reader will doubtless have noticed the cases of fidelity of native troops recorded in the text. One who had spent many years in India determined to collect and place on record all such instances, but he died before his intention was carried out. The following notes on the Mutiny at Nusseerabad were drawn up by the wife of Captain C. A. Timbrell (who was in command of 2/7 Ben.A.) for the proposed work, but were never used. Her daughter—the child referred to—has very kindly placed them at the author's disposal.

In May, 1857, Captain Charles W. Timbrell was in command of a field battery of the Bengal Artillery at Nusseerabad. There were also at that station two regiments of Bengal Native Infantry and one of the Bombay Cavalry. There was not an English soldier in the station. After the news of the outbreak at Meerut on May 11 and at Delhi the following day much anxiety was of course felt, but all remained quiet at Nusseerabad till May 28. The Brigadier and all the officers were aware that if the artillery proved false, as they had done at Meerut and Delhi, there would be no chance of saving life or property at this station; but as the men declared, when Captain Timbrell spoke to them, that " they would remain faithful to their officers and the Government as the artillery had always done hitherto," there appeared some hope that they would not fraternise with the sepoys, and it was thought best not to appear to suspect them of treachery.

On May 28, at 4 p.m., the alarm was given that the men were all in a state of mutiny. When Captain Timbrell reached the artillery lines he was horrified to see one of the infantry regiments keeping guard over the guns and his own men standing quietly by looking on. They all said they did not wish to murder their officers, but refused any longer to obey orders.

The order was then given for the Bombay Cavalry to charge the guns; the officers advanced, but the men stood still. The sepoys fired the guns, and two of the Bombay officers were killed, and Captain Timbrell's horse was shot under him. At this critical moment six of his men rushed forward and rescued him from his perilous position, and never left him afterwards; they were one family, a father and five sons. It was impossible to get another horse, as the station was now in an uproar and all property was seized and destroyed or set on fire by the mutineers and villagers. Every bungalow was burnt.

The Brigadier, finding it impossible to restore order or offer any resistance, gave orders for all the officers to leave the station with himself and the Bombay Cavalry, who had remained neutral. The officers' names were called over, and three were missing, and it was supposed that all three had been shot; in fact, it was reported three had been seen to fall.

The artillery lines were at some distance from the others, and when confusion increased it was impossible to communicate with each other personally; so Captain Timbrell was counted one of the three killed, and was thus left alone with his men—eight more had joined him. Captain Timbrell had sent his two subalterns to acquaint the Brigadier at the outset, and to ask for the cavalry to be sent to his relief, but neither of them was able to get back to him.

As soon as it got dark the mutineers left for Delhi, and the station was deserted. Then these few faithful men disguised Captain Timbrell and conducted him over the hills to Ajmere, where they arrived on the morning of the 29th, and he was heartily welcomed and congratulated on his wonderful escape; they were quite prepared to hear of the destruction of Nusseerabad, as the flames had been seen for miles round.

The men remained with Captain Timbrell in the fort at Ajmere, by permission of General Lawrence, till November, when all the Bengal officers were ordered back to their own Presidency, and Agra was the station to which they all marched. Some of the rajahs provided an escort of some of their own troops, and the fourteen men also accompanied the party; in fact, they were now called " Captain Timbrell's body-guard," one, by name Ali Mahomed, was so devoted he took on himself the duties of a bearer to Captain Timbrell's little girl.

On reaching Agra the Government decided that these men should be taken into the police, and otherwise rewarded for their fidelity. Captain Timbrell was very ill all the winter in the fort at Agra, and in April was ordered to the hills, when the same men obtained leave to escort him and his wife and child to Mussoorie, and he saw them again on his way down to Calcutta in December, 1858, when he was sent to England on sick leave.

Eleven ladies and seven children escaped from Nusseerabad when the alarm was given under the escort of the junior officers, and rejoined their husbands about a fortnight later.

CHAPTER III

THE DELHI CAMPAIGN

I. The Occupation of the Ridge

Brigadier Wilson on the Hindun, May 30 and 31.— On May 26, in accordance with his orders, Brigadier A. Wilson[1] set out from Meerut with the following force :

6th D.Gs. (Carabineers) (2 squadrons)	60th K.R.R. (wing) Ben. Sappers and Miners (2 companies)	2/1 Ben.H.A. (4 guns) (Captain and Brevet-Major H. Tombs) No. 14 F.B. 3/3 Ben.A. (Captain and Brevet-Major E. W. S. Scott) 100 Artillery Recruits with 2 18-pounder guns (Lieutenant A. Light)
	O.C.A., Major M. Mackenzie	

On May 30 he reached the River Hindun, where he found the bridge intact at Ghazi-ud-din-Nuggur, and encamping his main body on the left bank, he pushed his advanced pickets across the river. A strong body of mutineers, issuing from Delhi, drove these pickets back, while at the same time they bombarded the British camp with heavy guns. At the first alarm two companies of the

[1] When Wilson left Meerut, detachments of the 6th Dragoon Guards and 60th Rifles and two guns of 2/1 Ben.H.A. (Lieutenant T. P. Smith) were left in garrison. There were also 400 artillery recruits in the depôt.

60th, supported by a troop of the Carabineers and 3/3 Ben.A., made sure of the bridges, the key of the position, while 2/1 Ben.H.A., supported by a squadron of the Carabineers, galloped to the right along the bank of the river to open a flanking fire on the enemy; at the same time Lieutenant Light brought the two 18-pounder guns into action and the enemy's artillery was speedily silenced. The 60th Rifles then advanced and drove back the rebels, who, abandoning five heavy guns [1] and a large supply of ammunition, retreated towards Delhi pursued by the Carabineers. There is no record of their loss. The British casualties were 11 killed and 15 wounded.

Wilson retained his original position, and next day the mutineers returned to the attack, occupying a ridge about a mile from our advanced pickets, upon which their guns opened fire. Our guns at once replied, and an artillery duel prevailed for two hours, after which the enemy's fire appeared to slacken; at the same time the 60th Rifles cleared the rebels out of a village which would have interfered with our advance, and Wilson, ordering his whole line forward, carried the ridge with his infantry. The rebels retreated to Delhi, taking their guns with them. Their loss is unknown; our casualties, including Lieutenant H. G. Perkins, Ben.H.A.,[2] were 12 killed and 12 wounded. Next day (June 1)

[1] Two 24-pounder guns, one 18-pounder gun, one 10-inch and one 8-inch howitzer.
[2] Perkins, who belonged to Tombs' Troop, had shortly before been sent to find a ford across the river. He was killed in the act of mounting his horse as the troop was limbering up to advance to cross the river by it.

Wilson was reinforced by the Sirmur (Gurkha) Battalion, 490 strong, under Major C. Reid, and marching northwards, crossed the Jumna at Baghpat and joined Sir Henry Barnard, at Alipore, on June 7.

Barnard's Advance on Delhi.—The force now assembled under Barnard was composed as follows:

6th D.Gs. (Carabineers) (2 squadrons).	60th K.R.R. (wing).	2/1 Ben.H.A. (4 guns) (Captain and Brevet-
9th Lancers.	75th Foot (7 companies).	Major H. Tombs). 2/3 Ben.H.A.
	1st Ben.E.F. 2nd Ben.E.F. (6 companies).	(Captain and Brevet- Major E. K. Money). 3/3 Ben.H.A.
	Sirmur Batt. Jind Contingent.[1]	(Captain and Brevet- Major F. Turner).
	Ben. Sappers and Miners	3/3 Ben.A. No. 14 F.B. (Captain and Brevet- Major E. W. S. Scott). Artillery Recruits with 2 18-pounder guns (Lieutenant A. Light).

O.C.A. Brigadier A. Wilson.

Matériel	*Siege Train* [2]	*Personnel*
8 18-pounder guns. 4 8-inch howitzers. 4 8-inch mortars. 6 5½-inch mortars.		4/6 Ben.A. (Captain and Brevet- Major E. Kaye).

Total enclusive of Siege Train and Jind Contingent:
600 Cavalry   2,400 Infantry   22 Guns

Action of Badli-ki-Serai, June 8.—At 2 a.m. on June 8 Barnard marched from Alipore to attack the rebels who were in position at Badli-Serai. They were encamped about a serai on the Grand

---

[1] Contingents of varying strength and varying fighting value, supplied by loyal rajahs and chiefs, joined the British forces from this time onward.

[2] The siege train from Phillour reached camp on June 6, and it was joined, two marches from Umballa, by the Head-quarters of the 2nd Bengal European Fusiliers.

Trunk Road, and on a small hillock 150 yards in front of it five heavy pieces were placed in a sandbag battery flanked by well-posted field guns, their main position extending from the serai to the village of Badli, 550 yards to the westward. The road excepted, the ground was very swampy, and the right flank of the serai was protected by an extensive jheel.

Brigadier Grant,[1] commanding the cavalry, was ordered to take three squadrons of the 9th Lancers and 2/1 and 3/3 Ben.H.A. by a circuitous route and fall upon the enemy's left flank. The infantry was formed in two brigades, the 75th Foot and 1st Bengal European Fusiliers, under Brigadier Showers,[2] and the 60th Rifles, the 2nd Bengal European Fusiliers, and the Sirmur Battalion under Brigadier Graves.[3] Attached to these two brigades were 2/3 Ben.H.A., four guns of 3/3 Ben.A. and an improvised heavy field battery (two 18-pounder guns and two 8-inch howitzers) under Major Kaye. The baggage and park were left at Alipore protected by two guns of Scott's Battery, a company of the 2nd Europeans, and the Jind Contingent. It was Barnard's intention to turn the left of the rebels with Grant's Cavalry, while Showers supported by the artillery attacked their front and Graves inclined to his left to cut off their retreat.

As dawn broke, the enemy's camp-fires were

[1] Lieutenant-Colonel and Brevet-Colonel J. H. Grant, 9th Lancers. Always known after the Mutiny as Sir Hope Grant.
[2] Lieutenant-Colonel St. G. D. Showers, commanding the 2nd Bengal European Fusiliers.
[3] Brigadier H. M. Graves, who commanded the troops at Delhi on the outbreak of the Mutiny.

visible; but when Showers deployed his brigade into line, Grant's column was not yet in sight, and the rebels opened a heavy cannonade on our advancing troops. Barnard's guns at once came into action, Kaye in the centre, Money and Scott on his right and left, but they were unable to silence the enemy's well-sheltered artillery. Men began to fall; Kaye's bullock-drivers ran away and one of his waggons was blown up. Barnard then ordered a bayonet charge, and the 75th Foot, supported by the 1st Europeans, stormed the sandbag battery. Then Graves' Brigade, which had to march round a jheel, threatened the rebel right and Tombs' Troop[1] came into action on their left flank. The enemy then fled, leaving their guns and their camp in our possession; but the most important work of the day was yet to be done.

Seizure of the Ridge.—Pressing on in pursuit, Barnard reached a point four miles from Delhi where a road, branching to the left from the Grand Trunk Road, led direct to the Delhi Cantonments; following this road himself with Graves' Brigade, a squadron of the 9th Lancers, and 2/3 Ben.H.A., he sent Brigadier Wilson with Showers' Brigade and the rest of the cavalry and artillery along the Grand Trunk Road to the Subzi Mundi, while the Sirmur Battalion kept up connection between the two columns. On reaching the Najafghur Canal Barnard

[1] It will be remembered that Major Money had 9-pounder guns which he took over from 1/7 at Umballa (page 21). Grant's column encountered heavy ground, and the 9-pounder guns could not get forward. Tombs, however, with his 6-pounders was able to proceed and came up in time to take part in the battle (Knollys, p. 64).

found the bridge only partially destroyed, but when the guns crossed they came under the accurate

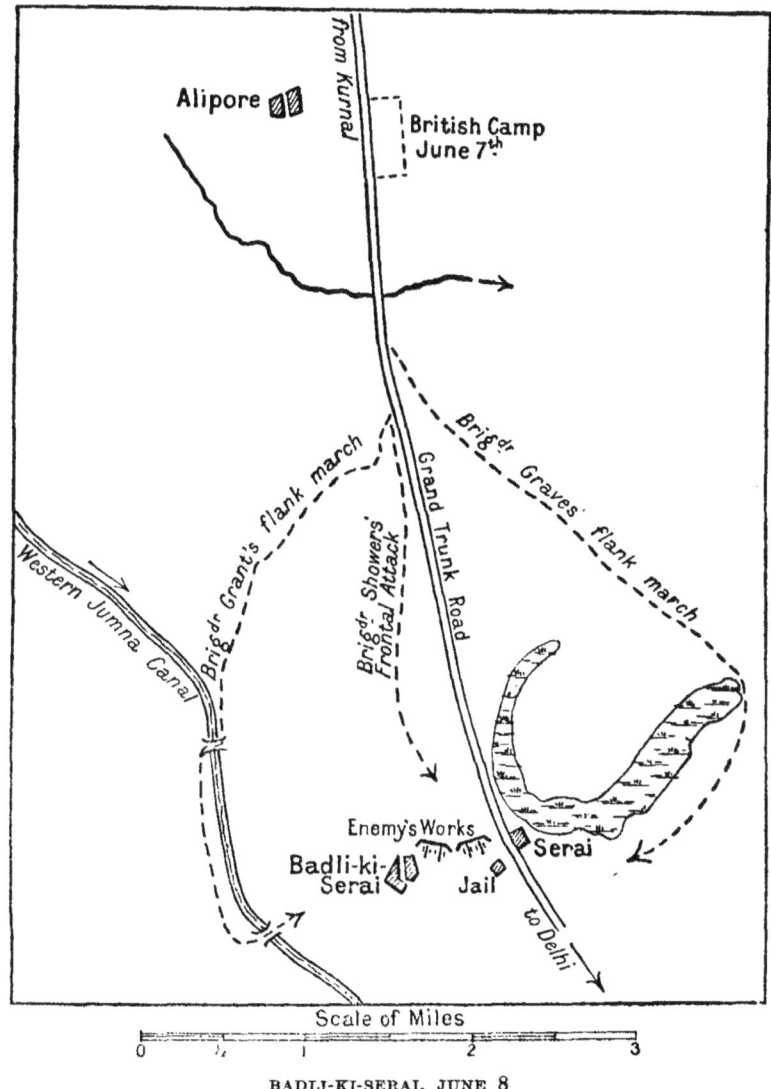

BADLI-KI-SERAI, JUNE 8

fire of three of the enemy's guns posted at the Flagstaff Tower. Money, however, galloping forward quickly silenced them and they were forthwith taken by the 60th Rifles and the 2nd Europeans. Showers' Brigade now swept along the Ridge to Hindu Rao's House, which Wilson had already reached, capturing on his way an 18-pounder gun. During these operations the guns of Delhi kept up a desultory fire which blew up one of Scott's limbers and killed a few men.

As he had cleared the Ridge of the enemy and his troops were worn out by heat and fatigue, Barnard determined to encamp on its reverse slope, where his rear would be covered by the Najafghur Canal; but before his tents were erected a fierce cannonade was opened, about 2 p.m., from the city walls, and a body of mutineers issuing from Delhi threatened the British right. Our weary troops at once re-occupied the crest of the Ridge and drove back the attackers, who for the rest of the day left them in peace. Our men had been under arms for sixteen hours in the summer heat of India, but once more the flag of Britain flew within sight of the Imperial City.[1]

The loss of the enemy on June 8 is unknown, but 26 guns, including 18- and 24-pounders, were captured during the day. Our casualties amounted to 4 officers (including Colonel Chester,[2] the Adjutant-General) and 49 men killed, and 14 officers and 118 men wounded.

II. STRENGTH OF THE REBELS IN DELHI AND THE BRITISH ON THE RIDGE

The Defences of Delhi.—The eastern face of Delhi was (at this season of the year) washed by the

[1] Kaye, ii. p. 196.
[2] Major and Brevet-Colonel C. Chester, 23rd Ben.N.I.

waters of the Jumna, which, swollen by the melting of the snows, completely filled its channel and afforded ample protection to this side of the city; the other sides being guarded by bastioned fronts, forming a rough arc of which the river was a chord. The bastions themselves were small, holding some twelve guns; the curtains as a rule only carried a musketry parapet, but here and there towers for a single gun were provided for additional flank defence. The walls were of massive masonry with a command of some 16 feet above the plane of site, and a berm (from 16 to 30 feet wide) ran round them. In front was a ditch from 16 to 20 feet deep and about 25 feet wide with a masonry scarp and an unrevetted counterscarp protected by a very short glacis. The ordnance on the walls comprised 24-pounder and 18-pounder iron guns, 10-inch howitzers, 10-inch and 8-inch mortars and pieces of smaller calibre, and there was an unlimited supply of ammunition in the Arsenal.[1] Projecting ravelin-wise from the river-front lay Selimghur, an ancient fort armed like the bastions, and commanding the bridge of boats which gave the rebels free access to the left bank of the Jumna. With the exception of a zone stretching from the Subzi Mundi to Metcalfe's House, a mere fraction of the environs, the mutineers could enter or quit the city at will.

Strength of the Rebels and their Reinforcements.—The composition of the Delhi garrison when the

---

[1] The gallant exploit of Willoughby and his companions did not destroy the Arsenal; they blew up the small-arm cartridge magazine, but very extensive supplies of war stores remained intact.

Ridge was first occupied, and the reinforcements which reached the city during the siege, were as follows :

Mutineers in Delhi, June 8, 1857

3rd Ben.L.C. (*a*).	38th, 54th, and 74th Ben.N.I.	No. 5 F.B. 3/7
4th Ben.Irr.C. (*b*).	11th and 20th Ben.N.I. (*a*).	Ben.A.
	9th Ben.N.I. (*c*).	Ben. Sappers and
	44th and 67th Ben.N.I. (*e*).	Miners (*d*).
	45th Ben.N.I. (*f*).	
	5th Ben.N.I. (*g*).	
	Harriana Battalion (*b*).	

Nusseerabad Brigade, June 16 and 17

1st Bom.L.C. (detachment).	15th and 30th Ben.N.I.	No. 6 F.B. 2/7 Ben.A.

From Jullundur and Phillour, June 21 and 22

6th Ben.L.C.	3rd, 36th, and 61st Ben.N.I.

From Rohilkhund, July 1 and 2

8th Ben.Irr.C.	18th, 28th, 29th, and 68th Ben.N.I.	No. 15 F.B. 6/8 Ben.A. 2 6-pounder post guns from Shahjehanpore.

From Jhansi and Nowgong during July

14th Ben.Irr.C.	12th Ben.N.I.	No. 18 F.B. 4/9 Ben.A.

Neemuch Brigade, end of July

1st Ben.L.C. (wing).	72nd Ben.N.I. 7th Inf. Gwalior Contingent.	4/1 Ben.H.A.

Other arrivals on a small scale were constant, and by the middle of August the insurgents were 30,000 strong.

(*a*) From Meerut; (*b*) from Hansi, Hissar, and Sirsa; (*c*) from Alighur; (*d*) from Rurkhi; (*e*) from Muttra; (*f*) from Ferozepore; (*g*) deserters from Umballa.

Though at first there was only one company of artillery in the rebel ranks, a sufficient number of trained gunners [1] was supplied by the magazine and

---

[1] It was noticed at once that the Delhi guns were very well served. Writing to Lord Canning, on June 13, Sir H. Barnard, referring to the

gun lascars. Furthermore the ranks of the defenders of Delhi were strengthened by many native soldiers on furlough, by customs chuprassis who had deserted their posts, by police and jail guards, and by a miscellaneous rabble who, though far from formidable in the field, could do much mischief when firing from behind walls.[1]

The British Position.—The last outlying spur of the Aravelli Mountains runs past the western side of Delhi and disappears in the Jumna about two miles above the city. From the walled enclosure known as the Eedghur, three-quarters of a mile west of the Garstin Bastion, the ground falls towards Kishangunge and then rises again to Hindu Rao's House, leaving a gap in the spur through which pass the West Jumna Canal and the Grand Trunk Road. From the culminating point by Hindu Rao's House the spur is represented by a rocky ridge with a command of from 40 to 60 feet above the river. This portion of the spur was occupied by the British, their front being nearly parallel to their lines of communication to the north through Kurnal and Umballa. Their left could be secured by pickets thrown out towards the river, and in this region their operations were favoured by a series of ravines running down the eastern slopes of the Ridge; but their right was dangerously exposed—for the suburbs of Kishangunge, Telawarra, and Subzi Mundi, con-

rebel artillerymen, remarks, " Their practice is excellent—beats ours *five* to *one*" (Kaye, ii. p. 529). It must be remembered that our heavy guns were at this time manned by recruits and partially trained men.

[1] Lieutenant H. W. Norman (*State Papers*, i. p. 439),

taining many strong and defensible buildings, offered covered positions easily accessible from the city whence sudden irruptions could be made. Their front was secured by pickets along the crest of the spur, who were exposed to heavy fire from the city walls; but the camp was pitched on the reverse slope, where the abandoned cantonments shielded it from view, and its rear was protected by the wide and deep Najafghur Canal, which also afforded an excellent supply of water.

Strength of the British and their Reinforcements.—The British force was very small for the work before it, and the provision for artillery and engineer services was manifestly inadequate. The rebel guns were superior in numbers and in weight of metal, and it was quite impossible to make regular approaches on account of the scarcity of sappers and the impossibility of withdrawing infantry from the defences to furnish working parties. For many days all the besiegers could do was to maintain their position. The following statement shows the strength of the British and the reinforcements which reached them during the progress of the siege:

Present on Ridge, June 8
(Detail as given on p. 70.)

British Reinforcements, June 9

| Guides Cavalry | Guides Infantry |
| (3 troops). | (6 companies). |

Major H. Olpherts's Detachment, June 22

| 2nd Punjaub C. (wing). | 75th Foot (1 company). 2nd Ben.E.F. (2 companies). 4th Sikh I. (detachment). | 1/1 Ben.H.A. (4 guns) (Captain and Brevet-Major H. Olpherts). 5/1 Ben.H.A. (2 guns) (Lieutenant G. A. Renny). |

Reinforcements from Punjaub, June 22 to July 3

5th Punjaub Cav. (1 squadron).	Head-quarters, 8th Foot. ,, 1st Punjaub I.	1/1 Ben.H.A. (2 guns). 61st Foot. 5/1 Ben.H.A. (2 guns). Detachments from 4th Batt. Ben.A. Sikh Companies Ben.A.[1]

Reinforcements from Punjaub, July 25

1st Punjaub Cav. (1 Squadron)

Hodson's Horse [1] (200 sabres).   Kumaon (Gurkha) Batt.

Punjaub Movable Column, August 14

Mooltani Horse.[1]	52nd L.I. 61st Foot. 2nd Punjaub I.	No. 17 F.B. 3/1 Ben.A. (Captain G. Bourchier).

Reinforcements from Punjaub, etc., from September 7

Siege Train

8th Foot.	6 24-pounder guns ; 8 18-pounders guns.
60th K.R.R.[2]	6 8-inch howitzers ; 4 10-inch mortars.
4th Punjaub I.	4 8-inch mortars ; 6 5½ inch mortars.
Beluchi Batt.[3]	Detachment of 4th Batt. Ben.A.

Effective Force, July 3 . . . 6,600.
August 14 . . 8,000.
September 14 . 8,748.

[1] Newly raised units.   [2] From Meerut.
[3] The Beluchi Battalion was ordered from Hyderabad, Sind, to Mooltan in June, and it went thence to Delhi.

Note.—After the capture of Delhi there was a redistribution of troops and more of the Punjaub regiments, including new levies, reached the city. At the end of September, 3/1 Ben.H.A. (Lieutenant C. Cooksworthy) had also arrived there.

III. THE DEFENCE OF THE RIDGE

Occupation of the Ridge.—On June 8 infantry pickets, with two field guns attached to each, were established at Hindu Rao's House on the right, at the Flagstaff Tower on the left, and at the Observatory and the Mosque situated between them.[1] Three 18-pounder guns were posted on a convenient hillock called the Mound,[2] on the right flank of the camp, and a cavalry and artillery picket were established close by. During the night the engineers made an emplacement in front of Hindu Rao's House for an 18-pounder gun and 8-inch howitzer—the Centre Battery; two other emplacements—the Right and Left Batteries—were made at some distance on both flanks; and on the 9th a few siege guns and mortars were brought to these positions.[3] These emplacements were called the Ridge Batteries. The picket at Hindu Rao's House, known as the

[1] Hindu Rao was a Mahratta noble. Since his death, a few years previous to the Mutiny, his house had been left empty. It was a roomy and convenient edifice with good approaches from the cantonment and the city, and afforded good shelter and accommodation for a considerable body of troops.

The Observatory was of irregular structure, ill-lighted and ill-ventilated, but still a serviceable building, as it afforded good support to the advanced position on our right.

The Mosque, though in ruins, was well suited for an outpost. It had good masonry walls, and afforded both shelter and accommodation.

The Flagstaff Tower, a double-storied circular building, afforded a fine view of the country lying between the Ridge and the walls of the city, and was sufficiently strong to afford good shelter to troops (Kaye, ii. p. 517, etc.).

[2] Also known as the General's Mound.

[3] Writing to Lord Canning, Barnard says (June 13), "We have six heavy guns in position." On June 16 he added, "My commanding artillery officer can only man six guns" (Kaye, ii. pp. 529 and 539 n).

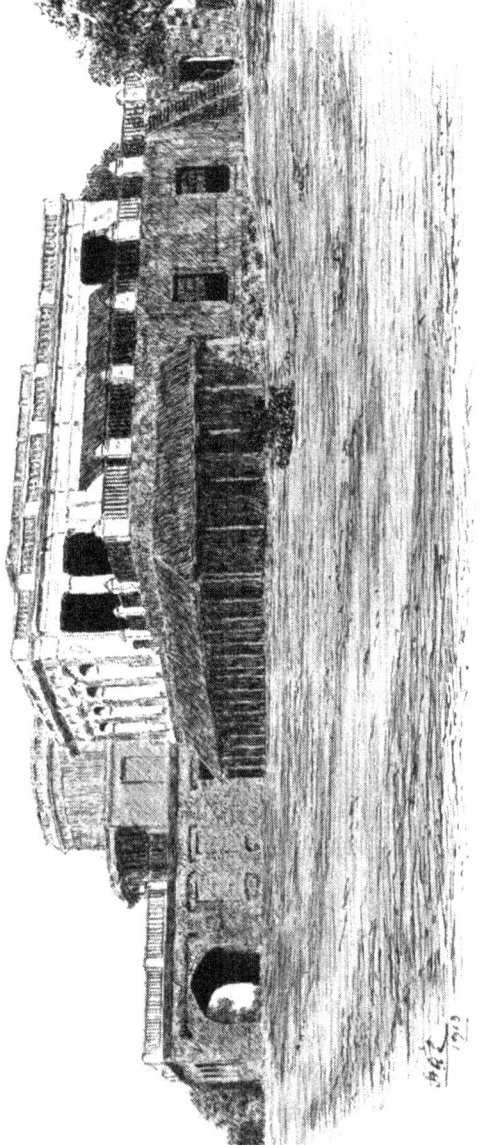
HINDU RAO'S HOUSE

Main Picket, was occupied throughout the siege by Major Reid with the Sirmur Battalion, reinforced by the 60th Rifles and occasionally by other troops. Two guns of 3/3 Ben.A. (Lieutenant M. Elliot) were permanently attached to Reid.

Rebel Attacks, June 9, 10, *and* 11.—On June 9 the rebels renewed their attack on our right flank and threatened the Main Picket. The Guide Corps had just arrived in camp, having marched from the Peshawar frontier, a distance of 580 miles, in twenty-two days; yet they at once turned out in support of Reid, and the enemy were driven back with loss. Nothing daunted, the rebels made similar attacks on the 10th and 11th, only to meet with similar defeat.

Rebel Attacks, June 12 *and* 15.—On June 12 the rebels made a determined attack before daybreak on the Flagstaff Tower. Two guns of 3/3 Ben.H.A. (Lieutenant H. P. Bishop) were posted there and a picket of the 75th Foot (Captain E. W. J. Knox) was being relieved by one furnished by the 2nd European Fusiliers when the enemy suddenly advanced from the ravines. In the ensuing conflict Knox was killed and Bishop wounded, but a reinforcement of the 2nd Fusiliers doubling up from camp charged with the bayonet [1] and aided by the fire of the guns drove the rebels back towards the Kashmir Gate. To guard against further attacks of this nature a picket of 300 infantry was established at Metcalfe's House.[2]

---

[1] The Fusiliers were commanded by Captain A. Boyd, and one of the officers, Lieutenant T. Cadell, gained the Victoria Cross.

[2] The picket was eventually divided into three portions: one of 150

On this day also two attacks made on the British right were beaten off without difficulty, and on June 15 the rebels attacked the pickets at Metcalfe's House without success.

Constant Artillery Fire.—Meanwhile the enemy kept up a constant bombardment from the city walls to which our weaker artillery replied, their fire being increased when that of the rebels was especially troublesome. Our 18-pounders, firing at from 1,200 to 1,500 yards, produced little effect; though sometimes, with the aid of the howitzers, they silenced for a time the guns in the Mori Bastion. During the night our mortars bombarded the city; and when the trunnions of some of the 8-inch howitzers gave way, these pieces were sunk in the earth and used as mortars. We had not a large supply of ammunition for the heavier guns; and for the captured 24-pounders we had practically none except the enemy's 24-pounder shot, which were recovered when possible and returned to them.

Kishangunge, June 17.—On June 17 the rebels opened a cannonade more than usually severe, to divert our attention from Kishangunge, where batteries were being made to enfilade the Ridge. Barnard, however, determined to destroy their works, and ordered two columns to co-operate for this purpose. One column was commanded by Major Tombs, Ben.H.A., and was composed of two

men on a mound to the west of the compound, with a detached party in a house commanding the road leading to the Kashmir Gate; a second of 50 men midway between this mound and the river; and a third of 150 men in the stables close to the river (Lieutenant H. W. Norman, *State Papers*, ii. p. 441).

companies 60th Rifles, four companies 1st Europeans, four guns of 2/1 Ben.H.A. (Lieutenant W. Wilson), a few Guides Cavalry, and some Sappers. The other was under Major Reid with the Sirmur Battalion and four companies of the 60th Rifles. The two columns met at the cross roads a quarter of a mile north of the West Jumna Canal. Tombs followed that leading direct to the Eedghur and Reid took the left-hand or eastern road. Skirmishing through enclosed ground, Tombs found the rebel left occupying the high ground about the Eedghur. A hot musketry encounter had endured for a time when the enemy began to waver under the fire of the guns, and Tombs at once rushed the position. The gate of the Eedghur was blown open, thirty-five of its defenders were killed and a 9-pounder was captured. Meanwhile on the rebel right Reid was engaged in a close encounter, and after battering down the gates of two serais he forced his way into the centre of Kishangunge, where he destroyed some uncompleted works and killed some sixty mutineers.[1] The British loss was small.[2]

Ochterlony Gardens, June 19.—On the 17th and 18th the Nusseerabad Brigade reached Delhi. The arrival of reinforcements was always the signal for renewed activity on the part of the rebels, and on June 19 they threatened every point of the British positions. While our attention was engaged all along

[1] Though these works were destroyed, the rebels continually brought guns to Kishangunge and Telawarra up to the day of the assault.

[2] Tombs had two horses shot under him (making five since the beginning of the campaign) and was wounded slightly.

our front, a sowar galloped into camp with the intelligence that a large body of rebels, accompanied by guns, had made their way unperceived round our right flank, and were advancing against our rear. Brigadier J. H. Grant at once assembled all the light troops available [1] and set out at the head of a squadron of the 9th Lancers, two guns of 2/3 Ben.H.A. (Major Money), two guns of 3/3 Ben.H.A. (Lieutenant Bishop), and two guns of 3/3 Ben.A. (Major Scott). The Brigadier found the enemy in position half a mile in rear of the Ochterlony Gardens, and came under a brisk fire from their artillery. Advancing his six guns right up to them and holding his cavalry ready to charge, Grant succeeded in checking the rebels so long as the light lasted, but when darkness came on they began to work round his right flank and Scott's guns were in jeopardy until extricated by a timely charge of the 9th Lancers.[2] But reinforcements were rapidly arriving. Turner with the remaining guns of 3/3 Ben.H.A. and Tombs with 2/1 Ben.H.A., supported by squadrons of the Carabineers, 9th Lancers, and Guides, came into action some distance on Grant's left, and detachments of the 60th Rifles, 75th Foot, 1st Europeans, Sirmur Battalion, and Guides pushed forward. But before the arrival of the infantry, darkness favouring the rebels on this flank also, Tombs's guns were in imminent danger, and were only saved by a gallant dash to the front of a hand-

---

[1] The artillery were so fully employed on picket duties that complete troops or batteries were seldom available for any sudden emergency.

[2] Grant was unhorsed and his life was saved by Privates T. Handcock and J. Purcell, 9th Lancers, and his orderly Raper Khan. Handcock got the Victoria Cross, Purcell was killed at a later period of the siege.

ful of troopers of the Guides led by Captain H. Daly. The arrival of the infantry relieved the situation, and the rebels, too numerous and too well covered to be attacked, retired unmolested into Delhi. Their loss is unknown. The British casualties were 3 officers and 19 men killed and 7 officers and 70 men wounded. The loss in horses was heavy, namely, 25 killed and 35 wounded.

The Racecourse Picket.—In order to protect the rear of the camp from attacks of this kind, an emplacement for two 18-pounder guns was made on the Racecourse, and a cavalry and infantry picket was established beside it.

Subzi Mundi, June 21.—On June 21, the mutineers from Jullundur and Phillour having entered the city, definite information reached Barnard that an attack might be expected on June 23, the centenary of Plassey. Major H. A. Olpherts, Ben.H.A., with a reinforcement of 850 men and 6 guns was marching from Umballa in charge of treasure; and on June 23 orders reached him at Rhai to leave the treasure under guard at Alipore, make a night march on Delhi, and fall upon the rear of the rebels, should they be attacking the British Camp. Olpherts reached the vicinity of Delhi without incident; but as he was marching into camp a furious cannonade was opened from the city walls, and guns brought by the rebels into Kishangunge enfiladed the Ridge, while their infantry swarming in the Subzi Mundi threatened the Mound Picket. The Main Picket was hardly pressed also; but the 60th Rifles, the Sirmur Battalion, and the Guides stood their ground, and Elliot's two guns did good service.¹ The heavy ordnance in the Right and Centre Batteries (Captain R. C. H. B. Fagan and 2nd Lieutenants R. T. Hare and W. R. Craster) came into action, but were unable to silence the well-covered guns in Kishangunge; and detachments of the 1st and 2nd Europeans failed to drive the mutineers from the Subzi Mundi. Brigadier Showers then assembled all available troops, including the infantry, who had just completed their twenty-two miles' march under Major Olpherts, and the rebels, suffering severe loss, were driven back into Delhi. The British casualties were 1 officer and 38 men killed and 4 officers and 117 men wounded.

New British Pickets.—The right of our position was now further strengthened by two infantry pickets, posted in strong buildings in the Subzi Mundi on each side of the Grand Trunk Road, some four hundred yards from the Right Battery. A connecting trench, including an emplacement for three 9-pounder guns called the Crow's Nest, was begun connecting the pickets with the Battery, and the Samee House ² was also occupied as an advanced post.

[1] Lieutenant Elliot's gun and howitzer of 3/3 Ben.A. were worked in the open at the Main Picket, and their service was one of great danger. On this day they were under fire from the Subzi Mundi buildings on one side and under shot and shrapnel fire from the city walls on the other. Early in the day the gun was disabled, right wheel smashed and trail loop-eye shot off. The howitzer had a shot through the axletree bed, but was serviceable. Lieutenant Elliot placed his men under cover of the rocks as well as he could, and with two others worked the piece for several hours. Sergeant Newson served the ammunition, Gunner D. McCarthy loaded, and Elliot laid and fired (Stubbs, iii. p. 266 n.).

[2] The Samee House was an old temple, about 900 yards from the Mori Bastion, and was known to the Army as the "Sammy House."

Rebel Attacks, June 22–30.—During the remainder of June hardly a day passed without an attempt on some part of our line, and on the 27th a simultaneous attack on both flanks caused us a loss of 13 men killed and 1 officer and 48 men wounded; but in every instance the rebels were successfully repulsed.

British and Rebels Reinforced.—On July 1 the Rohilkhund mutineers marched into Delhi across the bridge of boats, with bands playing English tunes, and the regimental colours flying; and a day or two later those from Jhansi and Nowgong entered the city by the southern gates. Between June 26 and July 3 British reinforcements brought up the effective force at Delhi to 6,600 men, a large number of sick and wounded having been sent to Meerut and Umballa.

Affair of July 4.—On July 3 after a demonstration against the British right some 6,000 rebels marched on Alipore, and forcing a picket of the 5th Punjaub Cavalry to fall back on Rhai, they plundered the town. The firing being heard in camp, Major T. Coke, 1st Punjaub Infantry, started at 2 a.m. on July 4 to follow up the rebels. With him were 300 cavalry (9th Lancers, Carabineers, Guides, and Kohat Horse); 800 infantry (61st Foot, 1st Punjaub Rifles); 2/3 Ben.H.A. (four guns); 3/3 Ben.A. and two guns of the native troop 5/1 Ben.H.A. There was a fear that the enemy were operating against our communications with the Punjaub, but at sunrise it was discovered that they were returning to Delhi with their loot. In attempting to intercept them Coke was so much hampered by the swampy nature of the ground that although he got within range of them with his guns, the rebels were able to withdraw their artillery and regain Delhi with little loss, although obliged to abandon their plunder. On his way back to camp Coke was attacked by a fresh body of mutineers who issued from Delhi; but he drove them back, inflicting a loss of 80 killed, his own casualties being 3 men killed and 23 wounded.

Brigadier Wilson succeeds to the Command.—On July 5 Sir Henry Barnard died of cholera. Major-General T. Reed, who, incapacitated by sickness from undertaking any active duty, had nevertheless accompanied the army in its march from Alipore, now assumed nominal command, but on July 17 relinquished it to Brigadier A. Wilson, in whose favour three senior officers were passed over. He was promoted to the rank of Major-General.[1]

Cavalry Raid, July 9.—On July 9 the cavalry picket at the Mound was furnished by a weak troop of the Carabineers (Lieutenant J. Stillman) and two guns of 2/1 Ben.H.A. (Lieutenant J. Hills) and a detachment of the 9th Bengal Irregular Cavalry, who were on the right front on vedette duty, watching the Grand Trunk Road. About 10 a.m. a sowar came in to say that a body of rebel cavalry were on the west of this road, and Hills sent him on to camp

---

[1] When Wilson assumed command he gave great relief to the troops by the systematic manner in which he arranged the various duties. He also, as far as possible, put a stop to the practice of following up the rebels close to the city walls when they were driven off after an attack. By so doing considerable loss had been incurred by fire from the ramparts ("Forty-one Years in India," i. p. 198).

to report to Major Tombs. He then ordered his guns to advance to their alarm post, a small breastwork not far from the Mound, which Stillman ascended to reconnoitre the movements of the enemy, who, he imagined, were about to make a demonstration on the rear of the camp. Neither he nor Hills had

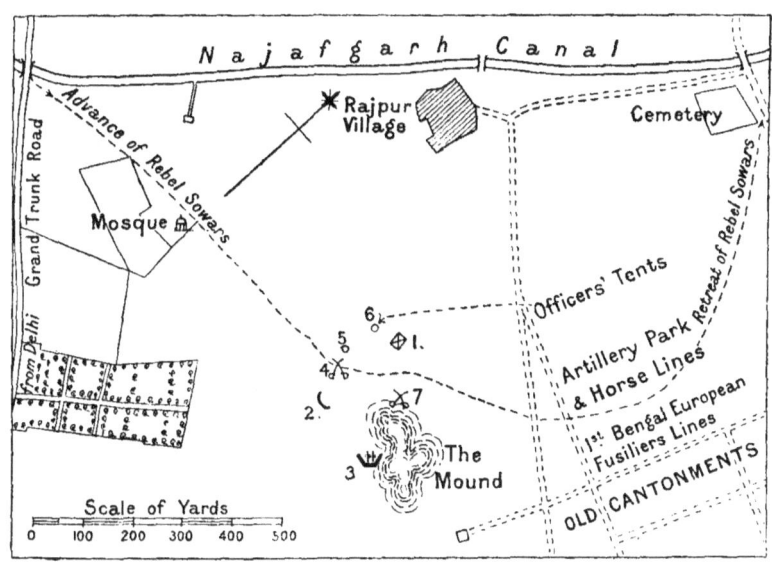

CAVALRY RAID, JULY 9, 1857

1. Position of picket.
2. Breastwork (alarm post).
3. 18-pounder guns on mound.
4. Hills's charge.
5. Hills's combat.
6. Point from which Tombs fired.
7. The final encounter.

the least suspicion of their real design. The guns, escorted by the Carabineers, were advancing at a walk when some sowars, at first taken for men of the 9th, came round the corner of a small mosque within the line of our pickets. Suddenly the leader waved his sword, and his party of about 120 charged the guns. On Hills giving the order " Action right," the guns were unlimbered, and the gunners rushed for the ammunition, but before they could load the sowars were among them, and they were forced to seek refuge under the guns or to run up the Mound for safety. The Carabineers, 32 young soldiers without an officer, broke and galloped back to camp, mixed up with the gun teams and limbers, and followed by the triumphant rebels, who rode straight for the lines of the native troop 5/1 Ben.H.A., who they hoped would join them. Meanwhile Hills had charged the advancing rebels single-handed. It did not occur to him that he was putting himself in the line of fire, for his mind was dominated by the one idea of giving his men time to load, and never did an artillery officer do a more heroic deed. He cut down the first man he encountered, struck a second, and was then ridden down horse and all. Untouched by the hoofs that thundered over him, he rose to his feet, and had just regained his lost sword when three men, two of whom were mounted, came at him. Hampered as he was with a heavy cloak soaked in rain, he wounded the first man with his pistol, and catching the lance of the second man with his left hand, he wounded him with his sword. The third man (on foot) then came up, and, wrenching Hills's sword from his hand, threw him on the ground.

Meanwhile Tombs having received and sent on the report that rebel cavalry were about, walked down from camp to visit Hills's picket. As he approached his destination he became suddenly aware of a confused mass of horsemen galloping past on his left, while some thirty paces distant he saw his subaltern lying at the mercy of a sowar. In a moment Tombs's pistol was levelled on his left arm, and the rebel fell dead, shot through the breast. Then the two officers ascended the Mound and were watching the movements of the enemy, when Hills caught sight of his own pistol in the hands of a rebel approaching on foot. He and Tombs immediately went to intercept him; but the man came on with great courage, cut at Hills, who parried the blow, and then slashed through the wadded cap cover worn by Tombs. Turning on Hills again, he gave him a severe wound on the head, and was then run through the body by Tombs who had been saved by his head-dress.[1]

The guns of 5/1 Ben.H.A. were parked near the right of the camp, and as the raiders approached they called upon the native artillerymen to join them; but happily the men of the 5th Troop were true to their salt, and even shouted to those of Olpherts's Troop, which was unlimbered close by, to fire on the rebels, though they themselves were in the line of fire. Lieutenant Renny and Captain Fagan, Ben.A., who was writing in his tent when the incursion took place, assembled in haste a few gunners and fell upon the invaders, who were also attacked by a party

[1] Tombs was one of the finest soldiers the army has ever produced. He was not only brave in himself, he was the cause of the bravery that was in other men. Time after time he distinguished himself by conspicuous and always useful courage. On this occasion he was given the Victoria Cross.

In 1857 the present General Sir James Hills-Johnes was twenty-four years old, and of a very slight build. It is said he tried to use his fists when he lost his sword. He received the just reward of the Victoria Cross.

of the 1st Europeans. After 15 of the sowars had been killed the remainder made their escape across the canal over the bridge by the cemetery.

Simultaneously with this attack, a cavalry demonstration in the neighbourhood of the Racecourse Picket was checked by the advance of a detachment of the 2nd Europeans and the steady behaviour of a troop of the 9th Lancers.[1]

Subzi Mundi, July 9.—While this affair was in progress a heavy cannonade was maintained against us all along our front and a persistent musketry fire was opened from the enclosures on the banks of the West Jumna Canal. The Adjutant-General, Brigadier N. Chamberlain,[2] taking 900 infantry (8th Foot, 60th Rifles, 61st Foot, and 4th Punjaub Infantry) and 3/3 Ben.A. (Major Scott), then advanced through the Subzi Mundi, and Major Reid supported him on the left with such infantry as could be spared from the Main Picket. The insurgents were quickly cleared out of the gardens and other enclosures, but here and there obstinate resistance was offered by them in brick serais which were only carried with the help of the steady fire of 3/3 Ben.A.[3] There were 500 rebels killed on the spot, our casualties being 1

[1] Lieutenant Light was in command of the 18-pounders at the Racecourse Picket, and wrote as follows to Colonel Grant: " The bold front and steadiness of the troop of the 9th Lancers under Lieutenant Martin saved my guns. Nothing could have been steadier, for the rebels came close up and circled round us, and Martin kept wheeling and facing them, at one time not 50 yards from them " (Knollys, p. 81).

[2] Appointed on the death of Colonel Chester.

[3] Lieutenant H. W. Norman states that " our success was greatly aided by the admirable and steady practice of Major Scott's battery under a heavy fire, 11 men being put *hors de combat* out of its small complement " (*State Papers*, i. p. 454).

officer and 51 men killed and 8 officers and 163 men wounded.

Subzi Mundi, July 14.—July 14 was a heavy day in the batteries round Hindu Rao's House, where the gunners were at work from 8 a.m. till 7 p.m. The mutineers again came out in great force; again opened a hot musketry fire against our right, while the guns on the city walls kept up an incessant cannonade; and again a British force advanced through the Subzi Mundi, supported as before on the left by Major Reid. Chamberlain accompanied, but Brigadier Showers commanded the main column, which consisted of two half troops of 2/3 and 3/3 Ben.H.A. (Major Turner and Major Money), the 1st Europeans and the 1st Punjaub Infantry, and a small detachment of native cavalry. About three in the afternoon the rebels, withdrawing their field guns, began to fall back on Delhi and were pursued to within 600 yards of the walls, casualties being caused in our ranks by grape shot from the bastions. The enemy's loss was over a thousand;[1] the British casualties were 17 men killed and 16 officers, including Chamberlain, and 177 men wounded.

Subzi Mundi, July 18.—On July 18 an almost similar attack was made; the Ridge batteries were bombarded and the Subzi Mundi attacked; and the rebels were driven over the canal by Lieutenant-Colonel J. Jones, 60th K.R.R., with a small mixed force, including 4 horse artillery guns.[2] During

[1] Round the Samee House alone 80 dead bodies of mutineers were counted, and for hours carts were seen taking corpses to the city.

[2] The author is unable to identify these guns.

the day we lost 12 men killed and 3 officers and 66 men wounded. The rebel loss is not recorded, and does not appear to have been serious.

Improvements in the Defences of the Ridge.—By this time the incessant exertions of the engineers, under Captain A. Taylor, Ben.E.,[1] had levelled the old serais and walls and cleared the enclosures for some distance round the posts held by the Subzi Mundi pickets, while the breastwork connecting them with the Right Battery was now complete. The bridges over the West Jumna Canal had all been destroyed for several miles on the west of the Subzi Mundi, with the exception of two, namely, that leading to Kishangunge, and another which was watched by a cavalry picket from Azadpore, two miles from camp. As there was a possibility of the rebels flooding the main ditch round the city, and as cavalry could pass to our rear over the Phulchadar Aqueduct, three and a half miles from the Cabul Gate, this aqueduct was blown up with the result that the canal bed became dry from that point to the city.[2] The bridge over the Najafghur Canal at Bussye (eight miles from camp) was also broken down. The Ridge defences daily became more formidable and captured guns were used to supplement those already in position; the Samee House also was greatly strengthened. The work in the batteries was incessant, and the *personnel* composed of " the

---

[1] Captain, afterward General Sir Alexander Taylor, G.C.B., arrived at Delhi in July and infused new life into the engineer operations.

[2] As Delhi had plenty of wells the water supply of the city was but little affected.

shadows of five companies,"[1] recruits from Meerut, and newly raised Sikh gunners, was quite inadequate for the task. A few days after their loyal behaviour on July 9, ten of the youngest gunners of 5/1 Ben.H.A. deserted, and it was therefore deemed advisable to deprive the troop for a time of its guns;[2] the gunners, however, were detailed for duty in the mortar batteries.

Metcalfe's House, July 23.—On July 23 a force accompanied by field guns, issuing from the Kashmir Gate, attacked the pickets at Metcalfe's House as well as that stationed at the Mosque. The rebels had good cover and kept changing the position of their guns, which it was found impossible to silence. It was therefore determined to occupy their attention with a brisk fire from the Ridge and at the same time to attack them in flank. Brigadier Showers accordingly with some 800 infantry (8th and 61st Foot, 1st Europeans, and 1st Punjaub Rifles) and two half troops of horse artillery (2/3 and 3/3 Ben.H.A.) worked round the right flank of the attackers, supported by 280 men of the 75th Reg. from the pickets at Metcalfe's House. Although Showers got unperceived within a few hundred yards of the rebels, they succeeded in withdrawing their guns, and some sharp infantry fighting followed amongst the garden enclosures before they finally retreated to Delhi. Their loss is not recorded. Our casualties were 1 officer and 12 men killed and 5 officers and 34 men wounded.

[1] This is the expression used by Stubbs, iii. p. 292.
[2] The guns were returned to the troop after the fall of Delhi; it did good service later on (see chap. ix.).

Rebel Attacks, August 2.—The next few days were unmarked except by the usual heavy gun fire on both sides and the usual skirmishes at the advanced pickets. At the end of the month the Neemuch Brigade entered Delhi, and on August 1 a large force of mutineers moved out of the city to repair the bridge at Bussye. A sudden flood, however, having destroyed their works, they returned towards Delhi and occupied Paharipore and Kishangunge in company with fresh troops from the city. At nightfall a determined attack was made on our position, especially at the Crow's Nest and Samee House, where the insurgents pressed close up to the defences, only to be repulsed by musketry and case shot. All that night and far into the next day the contest continued, but our losses were small because the men were now well covered. The rebels, on the contrary, lost heavily, and during the afternoon of the 2nd gave up the attack. Our casualties were 1 officer and 9 men killed and 36 men wounded.[1] That morning the Kumaon Battalion with a large store of ammunition and treasure marched into camp from Rhai.

Rebel Pickets driven in, August 12.—During the early days of August the rebels made persistent attacks against Metcalfe's House, and posted a picket in front of the Kashmir Gate. The British pickets were well able to defend themselves from attack, but the reliefs coming from the Ridge were danger-

[1] Lieutenant-Colonel and Brevet-Colonel H. Garbett, Ben.A., was wounded on August 8 and died at Simla in the following January from the effects. He, with his adjutant, Lieut. J. A. Angelo, had arrived at Delhi on July 5, and succeeded Wilson in command of the artillery.

DEFENCES OF THE RIDGE WHEN COMPLETED

Note.—The Racecourse Picket is outside the limits of this map. (See Map II. p. 144.)

The guns in the batteries are put in conventionally and do not represent the actual numbers.

ously exposed, and on August 12 Brigadier Showers delivered a sudden attack on the rebel position with three columns. On the right Major Coke with 450 men (75th Foot, 2nd Europeans, and 1st Punjaub Infantry), on the left Major G. O. Jacob with 350 men (1st Europeans), and some Guides cavalry advanced on the Kudsia Bagh, Showers himself being in the centre with a squadron of the 9th Lancers and two guns of 1/1 Ben.H.A. In reserve were 300 men (8th Foot, Kumaon Batt., 4 Sikh Infantry), and four guns of 1/1 Ben.H.A. Neither centre nor reserve was called upon to act, for Coke and Jacob completely surprised the rebel picket, capturing 4 guns of 4/1 Ben.H.A. (the native troop that mutinied at Neemuch) and sweeping the rebels from the Kudsia Bagh. Our loss amounted to 19 men killed and 8 officers, including the Brigadier, and 85 men wounded. Over 250 of the rebels were left dead on the field. From this date the pickets at Metcalfe's House were free from close attack, though a battery established by the rebels on the left bank of the Jumna caused them some annoyance.

The Movable Column.—On August 14 Brigadier J. Nicholson marched into camp with the Punjaub Movable Column, which brought up the number of effectives to 8,000 rank and file of all arms. There were 1,535 sick and 304 wounded in hospital. A siege train with large supplies of ammunition was now on its way from Ferozepore, and on the 24th was still a few marches distant. It had but a slender escort, and a large force of rebels with eighteen guns set out on that day towards Bahadurghur to intercept it. Next morning Nicholson marched in pursuit at the head of the following troops :

9th Lancers	61st Foot	1/1 Ben.H.A.
(1 squadron).	(420 men).	(Captain F. F. Remmington).
Guides Cavalry	1st Ben.E.F.	2/1 Ben.H.A. (4 guns)
(1 squadron).	(380 men).	(Lieutenant W. Wilson).
2nd Punjaub C.	1st Punjaub I.	2/3 Ben.H.A.
(1 squadron).	(400 men).	(Captain C. H. Blunt).
Mooltani Horse	2nd Punjaub I.	
(1 squadron).	(400 men).	
	Ben. Sappers	O.C.A., Captain and Brevet-
	and Miners	Major H. Tombs.
		S.O., Lieutenant M. C. Sankey.

Battle of Najafghur.—After a difficult march, Nicholson approaching Najafghur found the rebels about 4 p.m. in a position facing north and extending from a village near the bridge over the Najafghur Canal on their right to the town of Najafghur on their left, a distance of one and a half miles. A nullah swollen by the rains and crossed at a deep ford on the road by which the British were advancing, ran along their front, and two villages were occupied in rear of their right and centre. A brief reconnaissance disclosed the facts that a strong serai in the centre of their front line, and some rising ground adjoining it, formed the key of their position. Four heavy guns were posted there, and 9 field guns were in action between the serai and the bridge. Leaving a small mixed force to act as rear guard, Nicholson crossed the nullah and formed line to the front with the 61st Foot, the 1st Europeans, and the 2nd Punjaubis, 4 guns being on the right and 10 on the left, and the Lancers and Guides in support. His artillery having fired a few rounds, Nicholson led the infantry to within thirty yards of the rising ground near the serai, and then, after a single volley, carried the key of the position with the bayonet. At the same time the 1st Punjaubis, under Lieutenant W. H. Lumsden, detached to the right, cleared the enemy out of Najafghur, and Nicholson, changing front to the left, swept back the main body of the enemy, who retreated across the canal bridge. Their guns, posted on the right bank, opened a heavy fire upon the bridge, but our artillery was rapidly pushed forward and soon reduced them to silence. Lumsden, in conforming to the general movement, found Nagli —the village in rear of the rebel centre—occupied by a desperate band of sepoys, who, seeing their retreat cut off, held their ground with great determination. Lumsden and 11 men were killed and 26 were wounded, when the 61st Foot, sent back to reinforce the Punjaubis, reached Nagli; and the British regiment lost 1 officer and 5 men before the rebels, few in number, but fighting with the courage of despair, were all slain. The rebel horse, about a thousand strong, more than once during the action made a show of charging, but were on each occasion checked by the fire of our guns. The British cavalry were held in reserve. Nicholson bivouacked at the bridge, which was blown up after all our troops had passed over, and marching at dawn on the 25th he returned to Delhi the same evening. His casualties were 2 officers and 23 men killed and 3 officers and 67 men wounded. The enemy's loss is not recorded, but 13 guns and a large quantity of ammunition were captured.

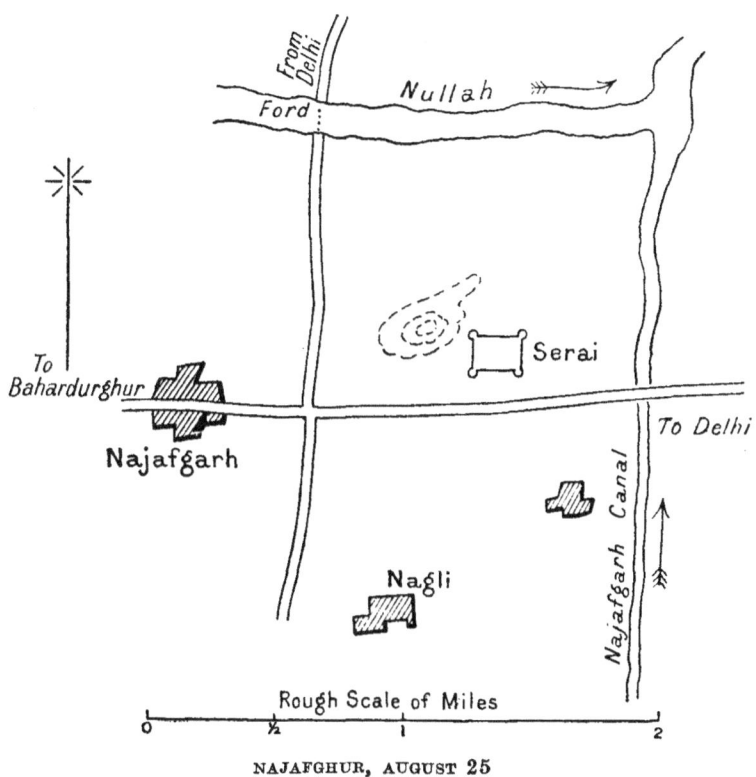

NAJAFGHUR, AUGUST 25

Rebel Attack, August 25.—While Nicholson was marching back to Delhi the rebels, possibly under the impression that few of our troops were left in camp, attacked both flanks of the Ridge, and after suffering severely from our artillery fire were driven back with a British loss of 8 men killed and 13 wounded.

The Last British Reinforcement.—The remainder of August passed without incident, and by the beginning of September all possible reinforcements, including the 4th Punjaub Infantry, the last regiment John Lawrence could spare, had arrived, as well as the siege train from Ferozepore. On September 6 the total effective force was 8,754, of which 443 cavalry, 2,294 infantry, and 580 artillery were white troops. The sick and wounded in hospital numbered 2,977. The time had now come to assume the offensive.

IV. THE CAPTURE OF DELHI

The Question of the Assault of Delhi.—Though excellent recruits were daily joining the British Army in the Punjaub, and though the chiefs of the Cis-Sutlej states were loyally protecting our lines of communications, Sir John Lawrence and civilians of lesser rank were fully, almost painfully, alive to the truism that the bonds of alliance in war can only be forged by mutual interest; and they felt that so long as the prestige of the possession of Delhi rested with the rebel cause, there was a very present danger of the snapping of those bonds.[1] But they ignored the military difficulties of the situation and were blind to the military dangers that might arise after even a successful assault on the city. It is futile now to inquire what might have occurred if a John Nicholson had been at the head of the army that

[1] Lord Roberts, in "Forty-One Years in India," quotes the Punjaub Administration Report, 1857 and 1858, as follows: "If Delhi were not taken, and that speedily, there would be a struggle not only for European dominion but even for European existence within the Punjaub itself." He goes on to point out that in the Punjaub a spirit of restlessness was on the increase; even the most loyally disposed were speculating on the chances of our being able to hold our own and doubting the advisability of adhering to our cause (i. p. 215). In addition there was great unwillingness to enlist in the new levies shown by the inhabitants of the tract of country between the Sutlej and the Ravi, and there was a Mussulman conspiracy and some abortive risings at Lahore.

conquered at Badli-ki-Serai; but of this we may be certain, there would have been no vacillation. The civilians would have been bluntly told that the time was not yet ripe and there would have been no "letting ' I dare not ' wait upon ' I would.' " Barnard, however, gallant soldier though he was, was apt to be turned by every wind of doctrine. Some of the young hot-heads in the Engineers persuaded him when he arrived on the Ridge that Delhi would fall to a *coup de main*. " The gates," said they, " have not been bricked up, and some of the bridges are intact; rebel reinforcements are on the way, but have not yet reached the city, and now is the time." [1] To this Barnard assented. The assault was fixed for June 12, and all was ready, but Brigadier Graves, some say by accident, some say by design, failed to appear with 300 men at the appointed place; the attempt was postponed and Barnard accepted the Brigadier's excuses.[2]

A new project was now prepared, and a Council of War met in General Reed's tent on June 15, where it appeared that military opinion was in favour of waiting for a reinforcement of a thousand men, and that civilian opinion deprecated delay. The Council adjourned, and Brigadier Wilson, commanding the artillery, drew up a paper in which he pointed out that, in consequence of the great extent of Delhi, he dreaded the success of an assault as

---

[1] It will be noted that this attempt, if carried out as was arranged, would have been made just before the arrival of the Nusseerabad Brigade, the first large rebel reinforcement.
[2] Kaye, ii. p. 528.

much as its failure; that our small available force of 2,000 British bayonets would be swallowed up in so large a city, while serious losses must be expected from the ordnance, numerous and unsubdued, mounted on the walls; that as long as we held the Ridge we kept the whole of the insurgents in and around Delhi; that, after a successful assault by us on the city, these same insurgents would spread all over the country, and that we were too weak to furnish movable columns and follow them up. In these views General Reed concurred, and pointed out that a failure at Delhi would be followed by the loss of the Punjaub. He summed up by saying that " the military reasons for waiting for a sufficient force to ensure success far outweighed any political inconvenience that might arise, inconvenience which would all be remedied by certain success in the end."[1] The decision rested with Barnard, and on June 16 he decided to postpone the assault, to which decision he for a time adhered.

On June 27 Brigadier N. Chamberlain, a welcome addition to the staff, arrived in camp and took up the duties of Adjutant-General. The British had been further reinforced; the question of an assault was again mooted, and it was determined to deliver a *coup de main* on July 2. Lieutenant-Colonel R. Baird-Smith, Ben.E., just appointed Chief Engineer of the Delhi Field Force, hurried post-haste from Rurkhi to be present at it, and he arrived on July 3, only to find the idea abandoned, as a great rebel attack was expected on that day.

[1] Kaye, ii. p. 537.

On assuming command of his department, Baird-Smith realised at once that close attack by siege batteries was impossible, not only because there were few guns heavier than 18-pounders available, but because there was a great scarcity of gun ammunition. He, however, thought the city might be taken after blowing in the gates, and he has left the opinion on record that a *coup de main* would have been successful if delivered in the early days of July.[1] But the heavy fighting which then ensued thinned our ranks, and by the 17th both Baird-Smith and Chamberlain realised that a *coup de main* was impossible.

Brigadier Wilson's Views.—During the short tenure of Reed's command (July 9–17) there was no talk of an assault, and it only remained to hold on till the arrival of reinforcements; " but the question had arisen and had been freely discussed at head-quarters whether until we could appear before Delhi in greater strength, it would not, both on military and political grounds, be a wiser course to relax our hold and employ our eager troops in other parts of the country."[2] This question, which did not originate with Brigadier Wilson, was in the air when he assumed command, and Baird-Smith, learning that it was likely to be laid before him, resolved to anticipate the formal reference. An opportunity arose on the very day Wilson took up his appointment, and in an interview between the two officers the Chief Engineer strongly insisted on the necessity of holding fast to Delhi at all costs. Before they parted he got instructions to submit proposals for

[1] Kaye, ii. pp. 572, 573.   [2] *Ibid.*, p. 589.

a second siege train, and to hurry on the supply of gabions, fascines, and other engineer stores. Two of his letters display Wilson's mind at this period. On July 18 he wrote to John Lawrence to the following effect: "I will hold on here to the end, but I must be reinforced, and that quickly. Send me one British and two Sikh regiments, or I shall be obliged to retire on Karnal with disastrous results."[1] To this Lawrence replied six days later by the despatch of the Movable Column. The other letter was to the Lieut.-Governor of the North-West Provinces, who, having heard a rumour that the evacuation of the Ridge was contemplated, made an earnest protest against such a course. It ran thus: "It is my firm determination to hold my present position and to resist every attack to the last. . . . Reinforcements are coming up under Nicholson. If we can hold on till they arrive we shall be secure."[2]

On August 10 the second siege train set out from Ferozepore, and on the 13th the Movable Column marched into camp and Lawrence had thereby spoken almost his last word in the matter of reinforcements. But Wilson, though his position on the Ridge was now secure, felt grave doubts as to his power to capture Delhi, and he was weighed down

[1] The original letter, written in French, is given (Kaye, ii. p. 589 n).
[2] Kaye, ii. pp. 593, 594. That the position of the British on the Ridge was not secure until the arrival of Nicholson's column is confirmed by that officer himself in a letter written on August 12 to Herbert Edwardes, where he says: "I think Wilson has hitherto had considerable cause for anxiety. Had the enemy had the enterprise to detach a strong force to his rear, he could not have sent more than 500 or 600 men against them" ("Life of Sir A. Taylor," by Miss Taylor, ii. p. 275).

by the thought of the appalling results should he attack and fail: the siege train, when it arrived, might admit of offensive operations against the city, but he had little hope of a successful assault until he received support from the south.[1] He proposed writing in this sense to Lord Canning, who was urging him to hesitate no longer, and on August 20 he communicated this intention to Baird-Smith; but that officer strongly deprecated such a course, and submitted to the General, there and then, a plan of attack which the Engineers had prepared.

Baird-Smith and Alexander Taylor.—Almost from the day of his arrival on the Ridge, Baird-Smith was confined to camp by the combined effects of a distressing illness and a troublesome wound; and though his advice and support were always at the service of his general, the active duties of the engineers were perforce superintended by the second in command, Captain Alexander Taylor. This officer, naturally strong and athletic, was in robust health, and while he enjoyed a professional reputation many an older soldier might have envied, he also possessed the whole-hearted confidence of his chief. Ever since July 17, the day of his first interview with Wilson, Baird-Smith had been considering the close attack on Delhi, while his active subordinate worked out the details on the ground. On August 20 the project was complete.

The Proposed Plan of Attack.—That portion of the northern front of Delhi lying between the Kashmir

---

[1] The letter to Baird-Smith, setting forth Wilson's views, is given *in extenso* (Kaye, iii. pp. 551, 552).

and Water Bastions was selected for the main attack. For action in this quarter had two obvious advantages : the left flank of an assailant would be covered by the Jumna ; and if he succeeded in breaching the walls, he would find himself in a comparatively open part of Delhi where troops could be rallied for a further advance. On the other hand, the triangular space marked out by Ludlow Castle and by the Mori and Water Bastions—the extreme points of the northern front of Delhi—was the ground on which breaching batteries must be placed, and this ground was in possession of the enemy. Furthermore, it was intersected by ravines and covered with trees and undergrowth, and no map of it existed.[1] True, the long curtain between the Mori and Kashmir Bastions was practically without heavy ordnance ; but if the rebels foresaw where the attack was going to be delivered, and if that attack were carried out by the stereotyped method of parallels and approaches, they would be able to supplement the fire of the bastions by mounting heavy guns along the curtain and crush any breaching batteries the British might attempt to make. A similar result might follow if the British, supposing they could have spared the men to do it, took possession of the required ground by open force. Taylor, therefore, set about his work with the utmost secrecy. Noticing

[1] General Sir F. Maunsell, K.C.B. (formerly Lieutenant F. Maunsell, Ben.E.), distinctly states, in *R.E. Journal*, July, 1911, that the plans of Delhi kept at Rurkhi were in his charge in 1857 ; that they showed no details of the ground upon which the siege batteries were built; and that, such as they were, these plans were not brought on from Rurkhi.

one day from the Ridge that the rebel picket at Ludlow Castle was in the habit of quitting its post without being properly relieved, he determined to watch for an opportunity and see for himself the lie of the unknown land. He found that on this their vital flank, the rebels were unexpectedly lax in vigilance, and thenceforward, sometimes by day, sometimes by night, and often at imminent risk, he made a patient but complete reconnaissance. He was helped by other engineers, especially by Lieutenant Thomason,[1] who had been quartered at Delhi; but he kept his own counsel, and only discussed his plans with Baird-Smith. At length the project was complete in every detail; the sites of the batteries were determined, trees were marked, and all arrangements made to guide troops and guns quietly and swiftly to their destination.

In preparing his plans for submission to his chief Taylor realised that the essence of success lay in seizing the required ground by surprise, while the attention of the rebels was diverted as long as possible from the real point of attack. The first steps, therefore, would be to quietly occupy Ludlow Castle and the Kudsia Bagh at a moment when the rebels were not on the watch, and to begin—and if possible complete in a single night—a siege battery which would protect subsequent operations, at the same time deceiving the enemy as to our real intentions. A site for this work (No. 1 Siege Battery) was selected 700 yards north of the Mori Bastion, and it was such that, should the rebels attempt to mount

[1] Lieutenant C. S. Thomason, Ben.E., afterwards General R.E.

fresh ordnance, they must do so under fire. On the left front of the Samee House, in rear of this site, and 950 yards from the Mori Bastion, an emplacement (the Samee House Battery) for light guns had already been made in order to distract the attention of the rebels whilst No. 1 Battery was in course of construction, and to encourage them in the belief that the west front of the city was threatened. These two works would constitute the Right (or Auxiliary) Attack.

When the bombardment opened the Samee House Battery would protect the right flank of our siege works and also fire on the Mori Bastion. No. 1 Battery, with its heavy metal, would eventually crush this bastion, and at the same time would protect the right flank during the construction of the proposed breaching batteries and afterwards supplement their fire. These batteries (No. 2 and No. 3 Siege Batteries) were to be placed in the grounds of Ludlow Castle and in rear of the Kudsia Bagh to breach the Kashmir and Water Bastions. A mortar battery (No. 4 Siege Battery) was to be placed in rear of No. 3. These three batteries would constitute the Left (or Main) Attack. A wide and deep ravine, passing between No. 1 and No. 2 Batteries and running into the Jumna just south of the Kudsia Bagh, would form a sort of rough parallel, giving ample accommodation for infantry guards, hospital dhoolis, and supplies of all sorts.

The two breaches were to be assaulted as soon as they were reported practicable, and at the same time the Kashmir Gate was to be blown open to admit of the passage of field guns. An attack was also to be made by field troops on the Lahore Gate.

The Plan Accepted by Wilson.—When Baird-Smith laid this project before Wilson he urged its acceptance by every argument in his power, and was supported by Chamberlain, the wounded Adjutant-General. But in the days that followed August 20, though his proposed letter to Lord Canning was not sent, the mind of Wilson, whose health was beginning to fail, was racked by the sense of his responsibilities, which was not relieved by the victory of Najafghur nor by the arrival of the siege train which marched safely into camp on September 5. Then came a pressing letter [1] from John Lawrence urging the dangers of delay, and the proposed project seemed the only way out of the difficulties of the British. Certainly Wilson had no alternative to propose; but he was of opinion that the chance of success was gravely prejudiced by the heavy fire the working parties would be exposed to, and he doubted the practicability of seizing the coveted ground by surprise and without loss. However, on the first of these points he yielded to the judgment of the Chief Engineer, and he gave way on the second when Nicholson, after a nocturnal visit to Ludlow Castle, corroborated Taylor's contention that the operation was feasible. Wilson still felt that the success of an assault on Delhi " lay on the hazard of a die "; but he was willing to accept that hazard, and on the night of September 7 the

---

[1] The letter is given *in extenso* in " Forty-one Years in India," i. p. 213.

proposed occupation of Ludlow Castle and the Kudsia Bagh was successfully carried out and ground was broken for No. 1 Siege Battery.[1]

Lieutenant F. Maunsell, Ben.E., was appointed Director of the Right Attack, and Lieutenant W. Greathed, Ben.E., of the Left.

The Samee House and No. 1 Siege Battery.—On the evening of September 6 four 9-pounder guns and two 24-pounder howitzers were placed in the Samee House Battery. They were supported by four 18-pounder guns in the Right Ridge Battery and by two 9-pounders in the Crow's Nest. These guns were in charge of Captain F. F. Remmington, who was appointed to the duty although he was commanding 1/1 Ben.H.A., there being no other artillery officer available.[2]

The same night the engineers filled 16,000 sandbags and concealed them near the site of No. 1 Siege Battery, which was traced by Maunsell on the afternoon of the 7th; and when darkness fell the working

---

[1] This paragraph is based on the pencil memorandum made by Wilson on the plan of attack (Kaye, iii. p. 554n), on details given by Miss Taylor in "Life of Sir A. Taylor," and on Sir F. Maunsell's remarks in the *Nineteenth Century and After*, October, 1911.

[2] With Remmington were Lieutenant H. J. Evans, M. M. Fitzgerald, M. Elliott, and R. Aislabie. Remmington was in command of 1/1 Ben.H.A., Major H. Olpherts being on the sick list. The artillery attached to the siege train were numerically unequal to the task of manning the siege ordnance, and both officers and men of the horse artillery were detailed to assist in this duty up to the morning of the assault. Efficient detachments were also supplied by the Carabineers and 9th Lancers, and several officers of the line, who had learnt the work in the Ridge Batteries, proved very useful in the advanced trenches. The two field batteries (3/1 and 3/3 Ben.A.) were little interfered with; one battery supported the pickets, the other was held in reserve in camp.

parties were on the ground, the Sirmur Battalion which supplied the covering party was in position, and camels were coming in with stores. Later on bullock carts brought down ammunition, and finally guns drawn by twenty pairs of bullocks made their appearance. Early in the night two rounds of shrapnel were fired from the Mori Bastion and several men were hit; but it was evident that the enemy had no conception of what was going on, as a single ineffective round fired at a later hour was the only further offensive action taken by them. When dawn broke the battery was finished and the ammunition stored, but only one gun was mounted, and the platforms for the others had still to be laid. The rebels at once displayed the greatest activity. A heavy fire was opened from the Mori Bastion; a trench was made outside the city walls, whence musketry supported by fire from field guns assailed the working parties, and cavalry issuing from the Lahore Gate threatened our flanks. The Samee House Battery now came into play with excellent effect, and a howitzer worked in the open behind the parapet of No. 1 Siege Battery reinforced the fire of its single gun. With dogged persistence the battery was completed after some hours' work, and fired its first salvo at the Mori Bastion.[1] Its right portion had the profile of

[1] "If the Engineer Corps had not been so perfectly organised, if the little group of trained Sappers had been less efficient, if the many slightly trained human agents whom Lieutenant Maunsell had no choice but to employ had not been forced to rehearse the rôle they were to play that night so often and so carefully, no courage, no determination, and no gallantry would have made it possible to erect

an elevated battery, and contained five 18-pounder guns and one 8-inch howitzer; the left portion was "half sunken," and contained four 24-pounder howitzers; a trench about 100 yards long connected the two portions. Major J. Brind commanded on the right, Major F. Turner on the left, the personnel being supplied by 2/3 and 3/3 Ben. H.A.[1]

Major Brind's special duty was to silence the Mori Bastion, which about noon presented a very dilapidated appearance; but the rebel guns, though often temporarily silenced, were worked with great persistence, and up till the morning of the 14th (when the assault took place) Brind had no rest. Nevertheless, with but a meagre relief of officers and men, he continued a steady and very efficient fire. Turner's object was to ruin the defences generally and to keep down as far as possible the fire of the Kashmir Bastion. His howitzers remained in action until the 10th, when the gabions took fire and the left portion of No. 1 Siege Battery was destroyed; but fortunately this happened only a few hours before the time fixed for removing the howitzers to Siege Battery No. 2. Major Turner, through sickness, was obliged on this day to resign his command to Major E. Kaye.

No. 2 Siege Battery.—No. 2 Siege Battery was traced on the night of the 7th, under the personal

that battery (No. 1) in so short a time" ("Life of Sir A. Taylor," by Miss Taylor, i. p. 298).

[1] With Major Brind were Lieutenants A. Bunny, H. P. Bishop, and E. Salwey. With Major Turner were Lieutenants W. J. Gray, E. H. Hildebrand, T. H. Salt, A. Gillespie, H. Chichester, and Riding-Master S. Budd.

superintendence of Captain Taylor, by Lieutenants Greathed, Lang, and Thackeray, Ben.E., and consisted of two portions connected by trenches; the left being close to the garden of Ludlow Castle, and the right about 120 yards distant on ground behind the cemetery. It was completed during the next three days, and was armed during the night of the 10th. Major J. H. Campbell [1] commanded the left portion, armed with nine 24-pounder guns, which were designed to effect a breach in the curtain between the Water and Kashmir Bastions at its junction with the right flank of the latter. On the right were seven 8-inch howitzers, two 18-pounder guns, and the four 24-pounder howitzers removed from No. 1 Battery. They were intended to destroy the masonry parapet of the Kashmir Bastion, to dismount its guns, and strip off the light musketry parapet for 200 yards on each side of the breach. Both portions opened fire on the morning of September 11, and by nine o'clock—after three and a half hours' work—all the guns in the Kashmir Bastion were silenced, and crumbling walls bore witness to the efficacy of the fire, which was maintained till the night of September 13. On the evening of the 11th Campbell received a severe grape-shot wound, and was succeeded by Captain E. B. Johnson from the artillery staff.

No. 3 Siege Battery.—The original site for No. 3 Battery, behind the Kudsia Bagh, proving on

[1] With Major Campbell were Lieutenants T. A. Dirom, F. S. Roberts, and W. R. Craster. Presumably the officers who served in the left portion of No. 1 Siege Battery accompanied Major Kaye to No. 2. The author is unable to state what company served in this battery.

further examination to be unsuitable, Taylor proposed to place the battery inside the Custom House enclosure, the wall of which would serve in some sort to protect the working parties; for the building which the rebels had neither occupied nor destroyed was only 160 yards distant from the Water Bastion. The necessary permission to carry out this bold project being obtained, the dangerous work began on the night of the 10th. The working parties, under Lieutenants Medley and Tandy, Ben.E., were only unarmed native pioneers, yet the record of sieges contains no better example of passive courage. Thirty-nine of them were killed or wounded during the night, but the work continued.[1] About midday on the 12th the battery was armed and ready to fire, and the perilous duty of unmasking the embrasures in broad daylight was carried out by Lieutenant Greathed and a party of sappers. The battery held six 18-pounder guns, which were to breach the left face of the Water Bastion, and these were supported by twelve $5\frac{1}{2}$-inch mortars in an emplacement a little in rear. The guns were commanded by Major E. W. S. Scott, to whom this post of honour was assigned, and were manned by 4/6 Ben.A. and the newly raised Sikh Company of artillery. The mortars were under Captain C. H. Blunt.[2]

[1] "The establishment of Major Scott's Battery (No. 3) within 160 yards of the walls, to arm which heavy guns had to be dragged from the rear under a constant fire of musketry, was an operation that can rarely have been equalled in war" (Lieut. H. W. Norman's "Narrative of the Siege of Delhi," State Papers, i. p. 481).

[2] The officers with Major Scott were Captain R. C. H. B. Fagan, Lieutenants A. H. Heath, Sir W. Stirling-Hamilton, R. T. Hare, T. Ryan, and J. Evans (9th Lancers). With Blunt were Lieutenant

Scott opened fire about 1 p.m., and, although he could only employ four guns at a time (as the embrasures required constant repair), by midday on the 13th the wall was pierced, and by sunset a practicable breach twenty yards wide was opened. On the 13th two guns of 3/1 Ben.A. under Lieutenant G. Cracklow helped to demolish the musketry parapet on either side of the breach, and Blunt kept up a continuous bombardment on the *terre-plein* of the bastion and its vicinity. Four of the enemy's heavy guns were rendered unserviceable, and the carriages of all his light guns were destroyed.

No. 4 Siege Battery.—No. 4 Siege Battery was made in rear of the Kudsia Bagh. It was commanded by Major H. Tombs, and manned by the native gunners of 5/1 Ben.H.A.[1] It contained four 10-inch and six 8-inch mortars, and maintained a continuous bombardment of the northern part of the city and its defences from the morning of September 11 until the assault.

Action taken by the Rebels.—The guns in the bastions were soon overpowered, as already narrated, but the Selimghur Fort was untouched by our projectiles, and kept up a continuous fire on the vicinity of the Custom House and Kudsia Bagh. By breaking a hole in the curtain between the Kashmir and Mori Bastions the rebels brought a fresh gun to bear, which, in addition to a brisk fire of rockets, was the only effort they made in the city to supplement

H. M. Cadell (prostrated by fever) and Cornet F. Trench, 7th Bengal Light Cavalry.

[1] Lieuts. G. A. Renny and H. J. Evans were with Major Tombs ; on September 14, the day of the assault, Renny was in command.

the guns of the original defence.[1] But outside the walls they took effective steps. It was not until the place was captured that the great strength of the Kishangunge and Telawarra suburbs was fully realised. There in sheltered positions the rebels installed a 10-inch mortar, 24-pounder, 18-pounder, and other guns, which bombarded the Ridge and brought an enfilade fire to bear on No. 1 and No. 2 Siege Batteries—a fire which all Remmington's efforts were powerless to subdue. On the night of the 10th they made a musketry trench ninety yards in front of the curtain between the Water and Kashmir Bastions, and next morning opened a hot musketry fire which they maintained till the assault; Captain R. C. H. B. Fagan, Ben.A., who was killed in No. 3 Siege Battery, was one of its first victims.

Musketry fire also was kept up incessantly from the walls. A few sorties were attempted, which were always repulsed without difficulty; but the rebel skirmishers, creeping out from Telawarra and finding cover in the broken ground, fired on No. 1 Siege Battery and on the positions held by Remmington, who had to employ case-shot to keep them at bay.

The Assault.—On the night of the 13th the breaches were examined by the engineers and reported practicable,[2] and the place being now ripe for assault, the following plans were adopted. At dawn on the 14th two infantry columns were to assault simultaneously the two breaches, while a third was to

---

[1] An attempt to make a *terre-plein* behind the long curtain was not persisted in.

[2] The breach in the Kashmir Bastion was reconnoitred in broad daylight by Lieutenant A. M. Lang, Ben.E.

enter the city after blowing in the Kashmir Gate. A fourth column, supported by the Cavalry Brigade, was at the same time to attack Kishangunge and enter Delhi by the Lahore Gate. A fifth column as a reserve was to await events near Ludlow Castle. Engineer officers were detailed to guide each column of attack, and it was intended that an officer and sixty men of the artillery should accompany each of the first three columns. Field guns were to be included in the fourth.

At dawn on September 14, Nos. 1, 2, and 3 Columns were drawn up in a concealed position in the Kudsia Bagh, where the 60th Rifles were waiting to cover their advance. But delay arose from various causes, and the sun was up before the assault was delivered. The composition of the columns was as follows :

No. 1 Column under	No. 2 Column under	No. 3 Column under
Brigadier J. Nicholson.	Brigadier W. Jones.[1]	Brigadier G. Campbell.[2]
75th Foot (300 men).		
1st Ben.E.F.	8th Foot (250 men).	52nd L.I. (240 men).
(250 men).	2nd Ben.E.F.	1st Punjaub I.
2nd Punjaub I.	(250 men).	(500 men).
(500 men).	4 Sikh I. (50 men).	Kumaon Batt.
Details from 4th Batt.		(260 men).
Ben.A. (Lieutenant		
R. Aislabie).		Engineer Party.
		Lieutenant D. C. Home, Ben.E.
		Lieutenant P. Salkeld, Ben.E.

Our siege batteries were still keeping up a fire so effective that no hostile piece on the rampart dared to come into action ; but when Nicholson gave the signal, our guns suddenly ceased firing, and the

---

[1] Lieutenant-Colonel and Brevet-Colonel W. Jones, 61st Foot.
[2] Lieutenant-Colonel and Brevet-Colonel G. Campbell, 52nd Foot.

60th Rifles dashed forward. Then the heads of the attacking columns emerged from their concealment and at once came under a heavy musketry fire.[1]

No. 1 Column.—The 1st Europeans, led by Nicholson in person, escaladed the left face of the Kashmir Bastion, while the 75th Foot and the 2nd Punjaubis stormed the breach, and then, sweeping along the ramparts to their right, fought their way past the Mori Bastion to the Kabul Gate. A portion of the column occupied some of the houses at this point; the remainder pushed along 200 yards nearer the Burn Bastion, and found themselves facing a narrow lane commanded by musketry fire and two guns. Repeated charges failed to force a passage, and men were falling fast when Nicholson, rushing to the front to inspire the hesitating troops, was mortally wounded. A retreat was then made back to the Kabul Gate, and No. 1 Column came under the orders of Brigadier Jones.

No. 2 Column.—Simultaneously with the advance of No. 1 Column, No. 2 stormed the breach at the Water Bastion, and, having taken possession of the neighbouring defences, turned to its right and, following the track of No. 1, pressed on in its support. But touch was lost between the columns. No. 2 then fell back to the Kabul Gate, No. 1 came up from the city on its left, and Jones assumed command of both columns.

No. 3 Column.—On the advance of Nos. 1 and

---

[1] There were cases of sick men getting out of hospital to take part in the assault. Colour-Sergeant Kennedy of the 2nd Europeans was seen mounting the breech his carbine slung and one arm and his body still in bandages.

THE LANE WHERE NICHOLSON WAS WOUNDED

2 Columns, the engineer party attached to No. 3 Column ran for the Kashmir Gate, where the bridge, the footway being removed, consisted of transverse beams only. Lieutenant Home, accompanied by Sergeants J. Smith and Carmichael, and Havildar Madhoo carrying powder bags, to which fuses were attached, made their way across and succeeded in nailing them against the right leaf of the double gate.[1] The wicket was open and they were under a heavy fire. Carmichael was shot dead and the Havildar was wounded, but Home and Smith slipped down into the ditch unhurt. Then Lieutenant Salkeld at the head of the firing party rushed forward, slow match in hand, and was at once shot through the leg and arm; but before he fell he passed the slow match to Corporal F. Burgess, who fell mortally wounded just as he succeeded in lighting the fuse. Havildar Tiluk and Sepoy Ram Het were also shot dead. At this moment the gate was blown in, and Home, in the ditch, ordered Bugler Hawthorn, 52nd Light Infantry, to sound the regimental call.[2] The leading company of that regiment scrambled across the remains of the bridge, charged through the gateway, and occupied the main guard, followed by the rest of Campbell's command. No. 3 Column then fought its way forward to the Begum Bagh, the main body occupying the Dariba, and a detachment the Kotwali in the Chandni Chauk. Campbell now was in sight of the Jumma Musjid, but its arches were bricked up and its gate barricaded. It could defy

[1] The left leaf was bricked up.
All the British survivors of the party received the Victoria Cross.

the efforts of unsupported infantry, and Campbell had no powder bags nor engineers. No artillery detachment accompanied his column, and field guns could not enter by the Kashmir gate until the bridge had been repaired. He therefore awaited events in the position he had gained; but when three hours had passed without the arrival of support, and he learnt that No. 1 and No. 2 Columns had not been able to advance beyond the Kabul Gate, he fell back on St. James's Church.

No. 4 Column.—The 4th Column under the command of Major C. Reid, Sirmur Battalion, was composed of detachments of the 60th Rifles, the Sirmur Battalion, the Guides Infantry, and parties from other regiments, European and Native, who could be spared from the Main Picket, 860 in all. The contingent lately supplied by the Maharajah of Kashmir (2,000 men and 4 guns) was also placed under Reid's orders, and 1/1 Ben.H.A. was detailed to support him.

In order to carry out the duty confided to him Reid formed the following plan.[1] Ordering Captain H. A. Dwyer with 400 men and four guns of the Kashmir Contingent to occupy the Eedghur before dawn, he himself determined to march his British troops direct on Kishangunge, holding the main body of the Kashmir Contingent, under Captain R. C. Lawrence,[2] in reserve. Having reached Kishangunge

[1] See Malleson, ii. App. A., where Major Reid's plan of attack is given in his own words.
[2] Captain R. C. Lawrence (a brother of Henry, John, and George Lawrence) was in political charge of the Kashmir Contingent, sometimes called the Jummu Auxiliary Force.

and captured the heavy guns he knew to be there, he hoped to turn them on the rebels, while at the same time the Kashmiri guns at the Eedghur would come into action. By this cross fire of artillery he would drive the enemy out of Kishangunge and Paharipore, and, leaving a reserve to hold these suburbs, he would march down the dry bed of the canal to the Kabul Gate, which might be expected to be in British possession, and entering the city would wheel to his right and occupy the Lahore Gate. But General Wilson forbad any movements of troops on our right until the Kashmir Gate had been blown in, and Dwyer, late in starting, instead of reaching the Eedghur unopposed, in accordance with Reid's expectations, was attacked when halfway there by 2,000 sepoys, who drove the Kashmiris back and captured their guns.

Defeat of No. 4 Column.—The sun was well up when Reid with his main column approached the bridge across the West Jumna Canal, leading to Kishangunge, and he at once came under a heavy fire. His field guns had failed to appear; but the infantry went forward in columns of fours, and hugging the walls which lined the road were fairly sheltered, a handful of the 60th Rifles covering the advance. On the enemy's side of the bridge the road was flanked to our left by a loopholed serai, to our right by a loopholed wall, and a breastwork was made across it. Our troops captured the breastwork with a rush, but were unable to hold it; for they were assailed by overpowering numbers of the enemy in front and on both flanks, in addition to which the retreating Kashmiris pouring in from the right flank increased the confusion. No. 4 Column then retreated across the bridge, and Reid himself was seriously wounded. Captain D. D. Muter, 60th Rifles, who commanded our advanced troops, and Captain Lawrence, who commanded the reserve, both considered themselves the senior officer, and for a time contradictory orders were issued and the situation became very grave. But Lieutenant H. J. Evans opened a brisk shrapnel fire over the heads of our beaten troops from the 9-pounder guns in the Crow's Nest, and the rebels, advancing on the Subzi Mundi, were speedily checked; while Lieutenant C. Hunter, with two guns of 2/3 Ben.H.A., came into action farther to the right and drove back some infantry and cavalry making towards our rear. No. 4 Column was re-formed on the Subzi Mundi picket, and the right of the Ridge was safe. But the rebels remained in undisputed possession of the Kishangunge and Telawarra Suburbs, and might have delivered a telling blow against the right flank of our main attack on Delhi—a danger obviated by the presence of the Cavalry Brigade.

The Cavalry Brigade.—The Brigade under the command of Brigadier J. H. Grant was composed as follows :

9th Lancers.	1/1 Ben.H.A. (3 guns)
Guides Cavalry.	(Lieutenant A. H. Lindsay).
1st Punjaub C.	2/1 Ben.H.A. (4 guns)
2nd Punjaub C.	(Lieutenant W. Wilson).
5th Punjaub C.	
Hodson's Horse.	O.C.A., Captain and Brevet-Major H. Tombs.

The brigade was formed up at dawn near the Samee

House Battery and remained there until the Mori Bastion was in our hands, when it advanced close to the city walls, and, feeling towards Telawarra for No. 4 column, at once came under a heavy fire. The successful rebels who had driven that column back had found excellent cover in some ruined mud hovels within 250 yards of the Kabul Gate, and assailed Grant with guns and musketry. Tombs ordered forward Lieutenant Lindsay's half troop to fire on the huts, but this officer, unsupported by infantry, could barely hold his own, and when grape fire opened upon him from a heavy gun at the Lahore Gate, he was forced to retire and to use case shot repeatedly to keep the rebels from his guns. He was now joined by Lieutenant Wilson's four guns, but the enemy were in such force that until the arrival of infantry the guns were in great jeopardy. During the two hours that this contest endured the cavalry remained drawn up ready for action exposed under most trying conditions to a galling fire.[1] At length some infantry arrived on the scene, first a handful of the Guides from the Main Picket and then the Beluchi Battalion from the Reserve, and the rebel fire at once began to slacken; Tombs then withdrew his artillery in good order after Lieutenant

[1] " The conduct of all my force," says Grant, " both European and Native was admirable—the unflinching coolness and steadiness of the 9th Lancers being especially conspicuous. Nothing daunted by their numerous casualties these gallant soldiers held their trying position with a patient endurance ; and on my praising them for their good behaviour they declared their readiness to stand the fire as long as I chose ... Tombs' Troop lost 27 men out of 48 and 19 horses, the 200 men of the 9th Lancers had 42 men and 61 horses killed or wounded ; and the Guides Cavalry, which was in support, 15 men and 19 horses " (Knollys, pp. 125, 126).

G. B. Traill (1/1 Ben.H.A.) had spiked two of the enemy's guns, an 18-pounder and a 12-pounder. Bourchier's Battery (3/3 Ben.A.) joined Grant, who, being able to protect himself, took up a position near Ludlow Castle.

The Reserve. — The Reserve under Brigadier J. Longfield [1] was composed as follows:

60th K.R.R. (200 men).	4th Punjaub I. (550 men.)
61st Foot (250 men).	Jind Contingent (200 men).
Beluchi Batt. (200 men).	

The 60th Rifles allotted to this column formed no part of it, for in the early hours of the morning they were detached to cover the attack of Nos. 1, 2, and 3 Columns, and the Beluchi Battalion, which had been placed as a guard over the siege batteries, was sent to support the Cavalry Brigade as already narrated. The remainder of the column—followed later by the Beluchis—entered Delhi, and were posted in the northern quarter of the town; two of the guns of 3/3 Ben.A. being detached on picket at the Kabul Gate. Wilson and his staff witnessed the assaults from Ludlow Castle, and when the north-east corner of the city was captured and the bridge at the Kashmir Gate repaired, they rode into Delhi, headquarters being established in Skinner's House.

The Work of September 14.—Thus on September 14 a footing was gained along the northern part of Delhi, from the Kabul Gate to the Water Bastion; the enemy holding the city south of a line drawn through the Burn Bastion and the Arsenal. Our loss that day in Europeans was 170 killed, 572 wounded;

---

[1] Lieutenant-Colonel and Brevet-Colonel J. Longfield, 8th Foot.

SEPT. 14, 1857] DEARTH OF ARTILLERYMEN 127

in native troops 103 killed and 310 wounded. Total casualties : 1,155.[1] The rebel loss is not recorded; it must have been very great, for no quarter was given.

As all the siege guns were in action until the last moment, the relief of the detachments of field artillery who were serving them was very complicated; as a matter of fact some were not relieved at all, and some were relieved too late. Thus it came about that while in any case Captain Remmington must have retained his important command as there was no one to replace him, the gunners of his troop were so badly required at the Samee House that half were not relieved at all; and the other half were relieved so late that a half troop which joined the hard-pressed Reid arrived with drivers only.[2] The artillery party for Nos. 1 and 2 Columns should have been supplied by 2/3 and 3/3 Ben.H.A.; but the gunners were not relieved in time for this purpose. When dismissed from the trenches they returned to camp, and eventually the two troops, 2/3 and 3/3 Ben.H.A., joined Lieutenant-Colonel Dennis,[3] 52nd Foot, who was in charge of the camp, whence Lieutenant Hunter proceeded to the assistance of No. 4 Column as already related. None of the columns of attack on this day

[1] This number is based on Colonel Chamberlain's estimate ("State Papers," i. p. 375). It is said the returns of the artillery were incomplete (Stubbs, iii. p. 290), and that the losses of the 8th Foot were understated.

[2] It was intended that the whole of 1/1 Ben.H.A. should have been with Reid, but he was joined by a half troop which arrived with drivers only. None of the gunners were relieved in time. This half troop, after Reid's defeat, apparently joined the Cavalry Brigade.

[3] Major and Brevet-Lieutenant-Colonel J. L. Dennis, 52nd Foot.

were actually supported by guns, and the only column accompanied by a detail of gunners was No. 1.

Capture of Delhi.—Though a footing was gained in the city, there was yet much to be done. Next day, September 15, there were skirmishes along the line held by us, but no progress was made; mortars, however, were brought into the city to bombard Selimghur, upon which some guns in the Water Bastion were turned by Lieutenant T. H. Salt, Ben.A. On the 16th the Arsenal was captured by the 61st Foot, 4th Punjaub Infantry, and Beluchi Battalion; but the rebels set fire to some thatched houses adjoining its walls, and at the same time occupied a covered position to fire on any one attempting to extinguish the flames. It seemed impossible to get at the rebels to eject them, when Lieutenant G. A. Renny, Ben.H.A., climbed up the Arsenal wall, and being handed $5\frac{1}{2}$ mortar shells with lighted fuses by Lieutenant R. T. Hare, Ben.A., and Lieutenant Vicars, 61st Foot, he threw them down amongst the rebels, who fled speedily. Simultaneously Lieutenant Thackeray, Ben.E., climbed up the wall and threw buckets of water handed up to him upon the flames.[1] The Arsenal then became the left pivot of our line. Next day, the 16th, Kishangunge, containing 5 heavy guns, was evacuated by the rebels. During the 17th and 18th a persistent fire of mortars was directed on Selimghur and the Palace, and an advance on the Burn Bastion was made by sapping through the houses under the

---

[1] Lieutenant G. A. Renny and Lieutenant E. T. Thackeray both received the V.C. for their exploits.

THE KASHMIR BASTION, AFTER THE ASSAULT

direction of Captain A. Taylor, Ben.E. On the 19th Brigadier Greathed,[1] in command of a mixed force of European infantry and two guns of 3/1 Ben.A. (Lieutenant H. E. Harrington), made a fruitless attempt to capture the Lahore Gate. He advanced by a lane which avoided the Burn Bastion and led into the Chandni Chowk; but the rebels were prepared, and suddenly brought a 6-pounder gun into action. One of Harrington's guns was then sent to the front, and the infantry were ordered to charge under cover of the smoke as soon as it was fired. But the men hesitated—they thought the narrow lane would turn out to be a trap—and the gun for some minutes was left unsupported. Then Greathed rapidly threw a detachment into some houses flanking the lane, and under their fire the baffled force was withdrawn in good order. In the meantime Taylor made his way into two or three houses which commanded the Burn Bastion, and that work was occupied without opposition. Next day Lieutenant F. S. Roberts, Ben.A., and Lieutenant A. M. Lang, Ben.E., working under Taylor's direction, by a happy accident discovered that the Lahore Gate could be easily captured by surprise. This was immediately carried into effect, and shortly afterwards the Garstin Bastion was occupied; whilst the cavalry, moving round by the Eedghur, discovered a large deserted camp on the south of Delhi. They then entered the city by the Delhi Gate and took possession of the Jumma Musjid, and about the same time the gates of the Palace were blown in; but by this time the bulk of

[1] Major and Brevet-Lieutenant-Colonel E. H. Greathed, 8th Foot.

the mutineers had fled southwards along the banks of the Jumna. On the 21st the King was captured, and next day his two sons and a grandson were seized and shot by Captain W. S. R. Hodson of Hodson's Horse. Thus the Moghul Dynasty was swept away and the British flag waved over Delhi.

A Notable Siege.—The siege had lasted over a hundred days at the hottest period of the year, and for a time the besiegers were themselves besieged. In order to maintain their position on the Ridge they had to engage in thirty different combats, always against a greater force and often opposed to troops ten times their strength, who had all the advantage of ground and superior artillery. The city itself had an enceinte of seven miles, and the fortifications had recently been modernised by our own engineers; the rebels were well provided with heavy guns and ammunition, and, in addition to the dense fanatical population, Delhi contained at the termination of the siege at least 30,000 trained soldiers. The British effective force never exceeded 9,000,[1] and when the city fell there were nearly 3,000 sick and wounded who needed protection. The heroic episodes of the Magazine and the Kashmir Gate will ever be associated with the name of Delhi; but gallantry was not confined to any branch of the service, and each arm won for itself a niche in the Temple of Fame. To build a siege battery in the few dark hours of a sultry night, to seize ground actually in possession of the enemy and to construct breaching batteries thereon within musket shot of

[1] See p. 79.

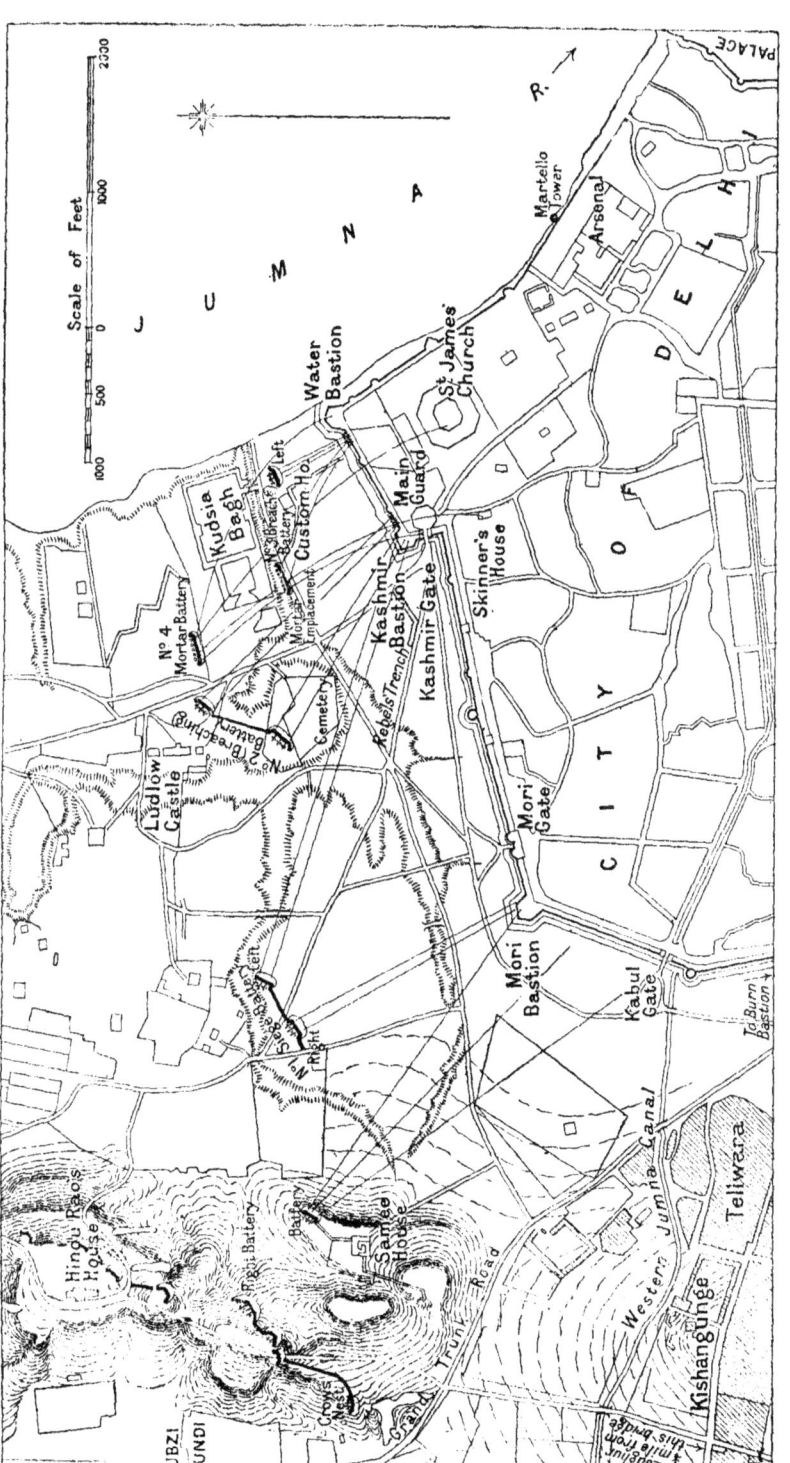

THE SIEGE BATTERIES.

The guns in the batteries are put in conventionally and do not represent the actual numbers. Where no guns are shown no guns were in action.

hostile walls, were triumphs for the engineers which have never been surpassed. The reader will remember the splendid devotion of the Bengal Artillery throughout the siege and the difficulties under which they maintained their fire during the actual bombardment; he will bear in mind how the infantry in open day stormed a place defended by vastly superior numbers, and after a six days' struggle possessed themselves of an immense city; how the cavalry at a critical period of the assault remained stoically under fire to distract the attention of the enemy from the real point of attack.

But the siege of Delhi was not remarkable solely for the triumph of the few over the many. The probable length of the attack is the only thought that troubles men engaged in an ordinary siege; but throughout the siege of Delhi men's minds were filled with a foreboding that, on the slightest reverse, the Punjaub would rise and pour down upon them from the north, the hordes of Central India would burst upon them from the south, our troops in Oudh and in the Doab would be scattered to the winds; and that Upper India, if not all India, would be lost to the British Crown. Yet from first to last the besiegers fought like men without a care, encouraged by the unwavering determination of the civil officers, and fired by the example of their own leaders.

V. The Pursuit of the Rebels

Greathed's Column.—Immediately after the capture of Delhi, Major-General Wilson went on sick leave and was succeeded by Major-General N. Penny.

It was now of the first importance to follow up the escaped rebels, but the efforts of our troops were directed more to the restoration of civil authority and to the punishment of offenders than to the destruction of the rebel army, the greater part of which passed away unseen to reappear later on in Rohilkhund, Oudh, and Central India. It is true that on September 21 a force under Brigadier Greathed, 8th Foot, was despatched southwards to open up communications with Cawnpore and Lucknow and to follow the mutineers who had fled from Delhi; but it saw little of the beaten rebels, and eventually was for a time diverted from its purpose by a wholly inadequate cause.

It was composed as follows:

9th Lancers.	8th Foot	1/1 Ben.H.A. (5 guns).
Detachments of:	75th Foot.	(Captain F. F. Remmington).
1st Punjaub C.	2nd Punjaub I.	
2nd Punjaub C.	4th Punjaub I.	2/3 Ben.H.A. (5 guns)
5th Punjaub C.		(Captain C. H. Blunt).
Hodson's Horse.	Ben. Sappers and Miners.	No. 17 F.B. 3/1 Ben.A. (Captain G. Bourchier).
		O.C.A., Capt. and Brev.-Major F. Turner, Ben.H.A.

Action at Bulandshahr, September 28.—Seven days after he marched, Greathed was still within forty miles of Delhi. On the morning of the 28th his cavalry advanced guard arrived at cross roads, one of which led straight ahead to the town of Bulandshahr, a mile and a half distant, while another led to Malaghur, five miles away, where there was a fort. This fort had been held for the King of Delhi since the previous May by one Walidad Khan, and Greathed intended to reduce it. But when our cavalry reached the cross roads they discovered that Bulandshahr was occupied by the Jhansi and Nowgong mutineers, reinforced by local rebels; and Greathed at once attacked them, this being the only occasion on which he encountered the Delhi fugitives in noticeable strength. The rebels occupied a fortified serai and the jail, their left being extended in gard(n

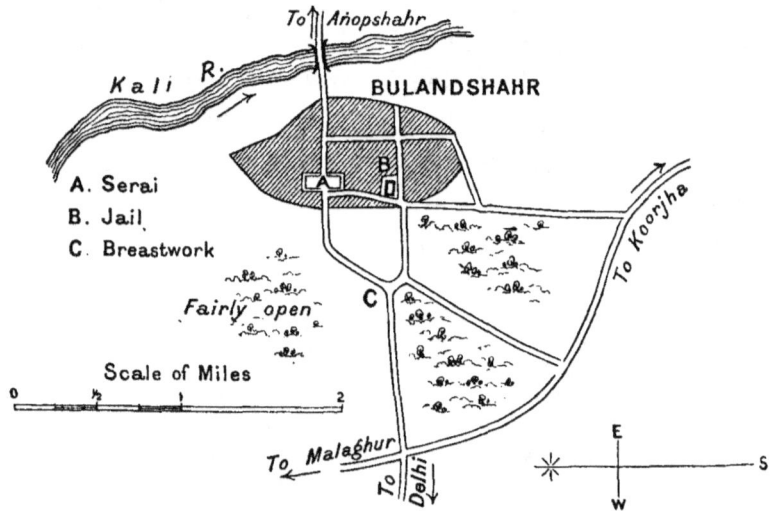

BULANDSHAHR, SEPTEMBER 28

enclosures. Some distance in their front was a breastwork armed with guns, and their cavalry hovered on either flank. Remmington's Troop and Bourchier's Battery, advancing down two roads which converged on the breastwork, speedily came into action and opened a cross fire which silenced the rebel guns. Our right wing, cavalry, guns, and infantry, then went forward, the breastwork was carried and three guns were captured; and, after a salvo or so of case had cleared the way, another rush was made and the jail was taken. Meanwhile the bulk of the cavalry under Major Ouvry, 9th Lancers, supported by two guns of 2/3 Ben.H.A. (Lieutenant G. Cracklow), advanced against the rebel right; but on approaching the houses one of the gun teams was disabled by musketry fire. Cracklow, however, got his other gun forward, and placing it so as to rake the courtyard of the serai, held his position in spite of loss for a quarter of an hour, when some infantry arrived in support; but only two men in the detachment, Sergeant A. Diamond and Gunner R. Fitzgerald, were left to work the gun.[1] Our cavalry then entered the town from both flanks, and after some hot encounters drove all the rebels out of Bulandshahr. Meanwhile a rear guard, including three guns of Blunt's Troop, left at the cross roads under Major F. Turner, Ben.H.A., was attacked by the enemy, who circled round our left flank with guns and cavalry; but were easily driven off. A reconnaissance pushed on to Malaghur found the place unoccupied. Our casualties during the day were 6 men killed and 6 officers and 35 men wounded, the rebel loss amounting to 300. The force halted four days at Bulandshahr and the fort at Malaghur was blown up, but most unfortunately Lieutenant Home, Ben.E., was accidentally killed during this operation.[2] Continuing his march to Cawnpore, Greathed reached Alighur on October 5 and dealt

---

[1] Diamond and Fitzgerald both received the Victoria Cross.

[2] Lieutenant D. C. Home, Ben.E., who commanded the party that blew in the Kashmir Gate at Delhi.

severely with some local rebels. By October 8 he was at Akbarabad, where he hoped to intercept the rebel Bareilly Brigade which was said to be in the neighbourhood; but here letters of the most pressing kind reached him from Agra, setting forth that the city was in great peril and imploring aid. It appeared to Greathed that he was bound to respond to this appeal, and setting all other duties on one side, he marched for the capital of the North-West.

Panic at Agra.—It will be remembered that the Indore mutineers reached Gwalior at the end of July. Being coldly received by Sindiah,[1] they crossed the Chumbal early in September and assembled at Dholpore, having inveigled 800 Mohammedan sowars from Sindiah's service. After the capture of Delhi they were reinforced by fugitives[2] from that city, and began to disturb the districts between the Jumna and Dholpore. They caused considerable apprehension to the Agra authorities, who, making light of their own strong and well-supplied fortress, with its European garrison and powerful guns, and listening only to their own selfish fears, wilfully diverted a force of British troops from important duties in order to soothe an altogether unjustifiable anxiety. When called upon to exert them, Greathed's troops showed splendid marching powers, and covering forty-four miles in twenty-six hours, they defiled early on October 10 across the bridge of boats into Agra.[3] On their way they

[1] See chap. ii. p. 38.
[2] Probably from the Neemuch Brigade (Mall., ii. p. 96).
[3] The European infantry were carried on elephants, carts, and camels.

passed through Hathras, where a few days previously the Bareilly Brigade had levied blackmail, and then marched unmolested to the Ganges.

Action at Agra, October 10.—On his arrival in Agra, Greathed was smilingly informed that the mere menace of his advance had sent the Dholpore rebels in haste across the Karee River, twelve miles away; and that there was no reason why his weary troops should not encamp amid the brushwood of the shady gardens lying to the west of the parade ground. He, however, chose to encamp on the parade ground, through which ran the Grand Trunk Road; but lulled into false security by the baseless reports he received from the Fort, he very unfortunately posted no pickets. Now, despite the astounding statement that there were no enemies north of the Karee, the rebels, with heavy guns,[1] were at this moment in a standing camp not four miles from the Fort—a camp actually on the Grand Trunk Road itself, though hidden by a village, where they were quietly and punctually supplied from the city with provisions and sweetmeats. Greathed's camp was pitched across the Grand Trunk Road. Many of the officers had gone to breakfast in the Fort, about one and a half miles off, the men were taking their ease in camp, whither huge crowds of sightseers had flocked from the city, the baggage was just arriving, and dhoolies with the sick were defiling down the

[1] These guns were taken in the impending action and are thus described by Bourchier: "Seven of the captured guns were of native manufacture.... Three were of enormous weight, though only 18- and 12-pounders" ("Eight Months' Campaign against the Sepoy Army," by Colonel G. Bourchier, C.B., Ben.H.A., p. 105).

road from the bridge of boats; when a storm of round shot from a battery of twelve guns on our right and front suddenly broke over the camp. An attack of cavalry was made on our left, and rebel infantry opened fire from the gardens on our right. For a short time there was a scene of wild confusion while a torrent of terrified humanity poured from the camp to the city, but when the sightseers disappeared discipline asserted itself. The gunners, dressed and undressed, rushed to their guns and brought them into action where they stood. The horses were soon hooked in by the drivers, the cavalry in their saddles and the infantry awaiting orders on parade.[1] Clearing a way through the crowds of fugitives, baggage carts, and dhoolis with great difficulty, Greathed and the officers who were in the Fort were quickly on the scene, and immediate measures were taken to resist the rebel attack. As their guns were too far off to be silenced by our lighter metal, the 8th Foot, the 4th Punjaub Infantry, and three squadrons of the 1st, 2nd, and 5th Punjaub Cavalry, supported by 1/1 Ben.H.A., were ordered to advance on the right, and the 75th Foot, the 2nd Punjaub Infantry, and the 9th Lancers, with 2/3 Ben.H.A. and three guns of 3/1 Ben.A., were sent to the left. The advance was covered by the other half of Bourchier's Battery, supported by infantry detachments. While the right

[1] See "Eight Months' Campaign against the Sepoy Army," pp. 101, 102. Describing the panic, Colonel Bourchier says : "Not satified with legitimate means of escape, the gun-horses in many cases were seized as they were being led to the guns, and were found next morning in the Fort."

column was driving the rebels from the garden enclosures, it was joined by No. 15 F.B. 2/5 Ben.A. (Lieutenant A. Pearson), which arrived from the

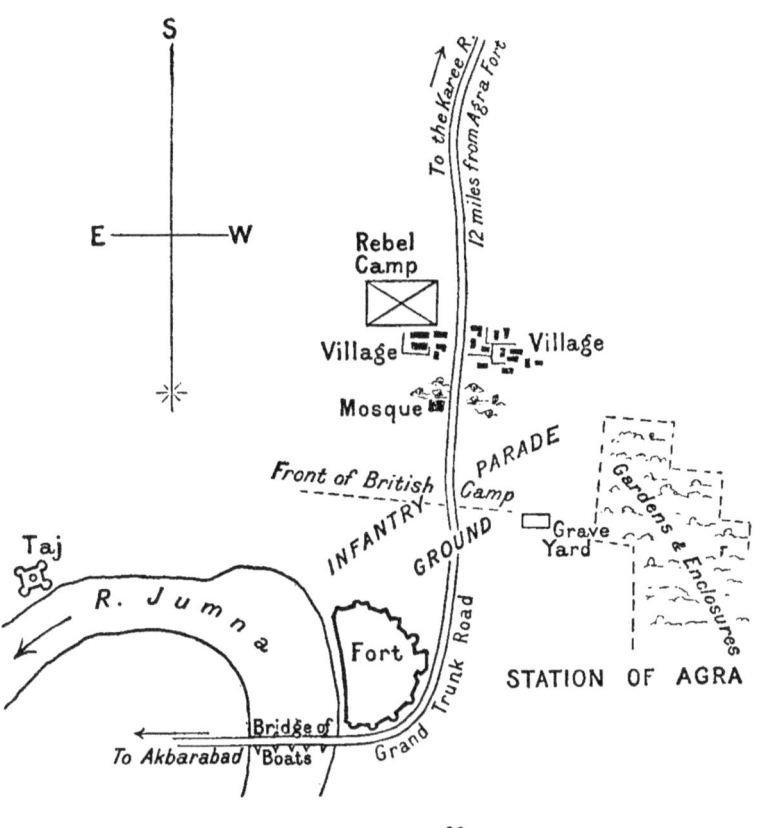

AGRA, OCTOBER 10

Scale: 1 mile = 1¼ inch (approximately)

Fort. Soon afterwards the Punjaubi squadrons, led by Captains Probyn, Watson, and Younghusband, defeated a large body of rebel horse, capturing three guns and five standards, while Remmington's fire exploded three ammunition waggons. But now a strong body of cavalry circling round our left flank threatened the camp. Blunt's Troop and a squadron of the 9th Lancers fell back for its protection, and presently a brilliant charge[1] led by Captain French, who was unfortunately killed, broke up the rebel horsemen, who were slain or put to flight. Our advanced troops now closed in and captured the guns, and Lieutenant-Colonel Cotton[2] came up with the 3rd Bengal European Regiment. As he was senior to Greathed he assumed the command, and the pursuit being pressed, the enemy's camp was soon reached. Leaving this to be disposed of by the infantry, the cavalry and artillery followed the flying foe to the Karee River, over which neither gun nor cart was allowed to escape. Thirteen pieces of ordnance and large quantities of ammunition were captured. The enemy's loss, which must have been considerable, is not recorded; our casualties amounted to 14 officers and men killed and 50 wounded.

Grant replaces Greathed in Command.—Greathed's column left Agra on October 15 and reached Manipuri on the 18th, where the command was assumed by Brigadier J. H. Grant, who arrived from Delhi. On the 21st the column was at Bewar, where Grant received a note from Lucknow pressing him to hurry to that city; and on the 26th it reached Cawnpore, after a skirmish near Kanoj, where the rebels aban-

[1] One of Blunt's guns was rescued by the 9th Lancers.
[2] Major and Brevet-Lieutenant-Colonel H. Cotton, 67th Ben.N.I., was in temporary command of the station. Mr. Colvin had died on September 9.

doned four guns and a party of sowars was driven into the Ganges.

The column from Delhi had now entered the Lucknow zone, and would receive its further orders from the Commander-in-Chief.

General Van Courtland and Brigadier Showers.—While Greathed was marching southwards, General Van Courtland,[1] who during the siege of Delhi had raised the Harriana Field Force, was reducing to submission the large Rohtak District, and on October 2 further measures of repression were undertaken by Brigadier Showers. This officer with a mixed force, including No. 14 F.B. 3/3 Ben.A. (Major Scott) and the Sikh Company Ben.A. (Lieutenant Sir W. Stirling-Hamilton), first cleared the suburbs of Delhi of all insurgents and then set out westwards to Dadri, where he captured 6 guns and punished some marauders. He then marched through the territories of two rebels, the petty chiefs of Jajjhar and Rewari, and returned to Delhi on November 10 with 14 captured guns and 5 lakhs of rupees.

Gerrard's Column.—The Rewari District, however, was once again disturbed by the arrival of the Jodhpore Legion from Rajputana. The Rajah of Jodhpore was loyal; but the Legion, as his contingent was called, had mutinied in the previous August, and after some adventures was marching towards Delhi.[2] Having defeated the troops of the Rajah of Jeypore, who opposed their advance, they

[1] General W. Van Courtland, who obtained his rank in the Sikh service, was an officer of foreign birth in civil employment. During the siege of Delhi he had assisted to keep the road to Umballa open.

[2] See chap. vi.

reached Nurnaul on November 16, and, being reinforced by the adherents of the Jajjhar and Rewari Rajahs, their numbers rose to 3,500 men, including a large force of cavalry, with 5 guns. A column under Brigadier Gerrard[1] was at once despatched from Delhi to oppose them. It was composed as follows :

6th D.Gs. (Carabineers) (1 squadron).	1st Ben.E.F. 7th Punjaub I.	3/1 Ben.H.A. (Lieutenant C. Cookworthy).
Guides Cavalry.	Punjaubi	Heavy Field Battery :
Mooltani Horse.	Sappers	Sikh Co. Ben.A. (Lieutenant A. Gillespie).

Note.—The Carabineers joined Gerrard after he had left Delhi ; they marched from Meerut. Later he was joined by a detachment of the Harriana Field Force, under Captain W. F. J. Stafford, numbering 2,500 of all ranks. The cavalry of the force was commanded by Lieutenant G. G. Pearse, Mad.A., who was in civil employment under Van Courtland.

Action at Nurnaul, November 16.—Early on the morning of the 16th Gerrard (who was at Kanund) set out for Nurnaul, 14 miles distant, but owing to the difficulties of the march he did not reach it until the afternoon. During the morning the rebels occupied the strong position offered by the outskirts of the town, but later, for some unknown reason, they withdrew to their camp, and in the subsequent engagement had no accident of ground in their favour. On approaching Nurnaul Gerrard deployed into line ; the Guides and Carabineers under Captain G. Wardlaw, with 3/1 Ben.H.A. in support, were on the right ; the infantry and heavy guns in the centre, and the Mooltani Horse and Sikh guns on

[1] Lieutenant-Colonel J. G. Gerrard, 14th Ben.N.I.

the left. The rebels advanced to meet him, and when their cavalry appeared on our right, their guns came into action; Gillespie's heavy guns at once replied, and Wardlaw charged the rebel sowars. They did not flinch, and the opposing horsemen met in mid career. Having driven back the cavalry, Wardlaw wheeled to his left, rode through the rebel guns, sabred some gunners, and brought his victorious troopers back to their place in the line. But the guns were not spiked, and after our cavalry had passed through them, they were once more brought into action. Then the Europeans, rushing forward, captured two of them, while the Mooltani Horse, after a successful skirmish on our left flank, possessed themselves of a third. The rebels then retreated, harassed by 3/1 Ben.H.A., which kept up a continuous fire of case and shrapnel; but they disputed the ground with great determination, and rallied for a moment when they reached some strong buildings in the vicinity of their camp. The Europeans and Sikhs now pressed forward, and the rebels, driven from their last defences, broke up and scattered with the loss of their two remaining guns.[1] About 350 rebels were killed, and the Jodhpore Legion broke up and was never heard of again. When their small numerical preponderance and their marked inferiority in artillery are taken into account, it must be admitted that the rebels on this occasion made a good fight. Brigadier Gerrard was unfortunately killed

[1] During the pursuit Cookworthy's Troop, advancing at a gallop, successfully passed over a mud wall fully 3 ft. high (*Blackwood's Magazine*, June, 1858).

during the pursuit, and Captain J. P. Caulfield, 3rd Ben.N.I., brought the victorious column back to Delhi. In all there were about 80 casualties, including Lieutenant G. G. Pearse, Mad.A., severely wounded.

THE KASHMIR GATE OF DELHI, SEPTEMBER 14, 1857

CHAPTER IV

THE LUCKNOW CAMPAIGN

I. CAWNPORE

Anxiety about Cawnpore.—When the mutineers were driven from Lucknow at the end of May, though confidence was felt in the safety of the city, no efforts were spared to complete its defences. During June large gangs of coolies worked in the Residency enclosure and the Machi Bhawan, while a continuous stream of elephants and carts brought in supplies from the districts. But as the month wore on, Cawnpore became a grave anxiety; the news of the mutiny that occurred there was rapidly followed by that of the hostility of the Nana Sahib and the ever-increasing peril of the Europeans. On June 25 the cheering intelligence reached Sir Henry Lawrence that a British force from Allahabad was expected to reach Cawnpore in from eight to ten days, but on the 27th Cawnpore capitulated.

Defence of the Entrenchment.—On the morning of June 7 a thousand souls were assembled in General Wheeler's entrenchment, the fighting men numbering some four hundred. They were composed of the British officers of the revolted regiments, of other officers detained in Cawnpore by various causes, and of a number of able-bodied civilians. There were about 70 men of the 32nd Foot (including invalids), 60 of the 84th, 15 of the 1st Madras European Fusiliers, and 40 Eurasian Drummers. Some 340 muskets in all might be counted on.[1] There were 60 trained gunners the men of 1/6 Ben.A., but the untrustworthy native artillerymen of the Oudh battery were dismissed, and Lieutenant Ashe was left with only his guns and horses and a volunteer *personnel*. The loyal remnant of the 53rd Ben.N.I., most of the native officers and about 100 men, were ordered to occupy the Artillery Hospital, 600 yards to the east of the entrenchment. The non-combatants under Wheeler's protection numbered about 80 Europeans or Eurasians incapable of bearing arms, 100 native servants, and over 500 women and children.[2]

As the General was prostrated through sickness during the greater part of the siege, and Major G. Larkins, the senior artillery officer, was also ill, the command was in the hands of Captain J. Moore, 32nd Foot. None of the officers of 1/6 Ben.A. were present, but in addition to Lieutenant Ashe there were six lieutenants of Bengal Artillery in the entrenchment.[3]

[1] Only the Madras Fusiliers were armed with the Enfield rifle.
[2] The exact numbers cannot be stated.
[3] Captain W. Olpherts, who commanded 1/6 Ben.A., was detached to Benares and the subaltern officers were employed in various ways. The artillery officers in the entrenchment were Major G. Larkins, 7th Batt., Lieutenant C. Dempster, Adjutant 7th Batt., Lieutenant J. A. H. Eckford, Quartermaster 7th Batt., Lieut. St. G. Ashe, 3rd Batt. Oudh Irr.A., Lieutenant B. Ashburner, and 2nd Lieutenants J. N. Martin, G. M. W. Sotheby, and F. W. Burney.

The crest line for musketry defence was about 900 yards in extent, and one man was allotted to every 15 yards. Even at the outset it was found impossible to allot more, but there was no lack of muskets and each defender of the parapet had from three to eight beside him, loaded and with bayonets fixed. The eight 9-pounder guns and one 24-pounder howitzer were placed in three groups under Lieutenants Ashe, Dempster, and Eckford, Ben.A.; they fired through intervals in the parapet and were without protection of any kind. A single 3-pounder [1] (Major W. R. Prout, 56th Ben.N.I.), flanked the main guard. Across the western corner of the entrenchment ran a row of half-finished barracks, of which Nos. 2 and 4 blocks were occupied by pickets, each of 16 men.[2]

About 9 a.m. on June 9 Lieutenant Ashe, who with 30 volunteers took out his guns to reconnoitre,[3] discovered the enemy crossing the canal, and shortly afterwards their field guns opened fire on the entrenchment.[4] Next day they brought up heavy

---

[1] This was a rifled gun, the first to be used by us in war.

[2] At the outset Block 2 was chiefly garrisoned by civilians, later it was placed in charge of Captain Mowbray Thomson, 53rd Ben.N.I. It had a temporary roof upon which a crow's-nest was raised. The other block was occupied after a few days; it was commanded for a time by Captain R. V. Jenkens, 2nd Bengal Light Cavalry.

[3] Lieutenant Ashburner, Ben.A., went out with Ashe, but was never seen again. It is supposed that his horse bolted and carried him among the rebels.

[4] The first casualty was Gunner P. Maguire, who was killed at his gun by a round shot. "Several of us," says Mowbray Thomson, "saw the ball bounding towards us, and he also evidently saw it, but, like many others whom I saw fall at different times, he seemed fascinated to the spot" (Mowbray Thomson, p. 66).

guns,[1] maintaining at the same time an all-round musketry fire at ranges from 300 to 800 yards. They also occupied the barrack blocks adjacent to those held by our pickets, and it was here that some

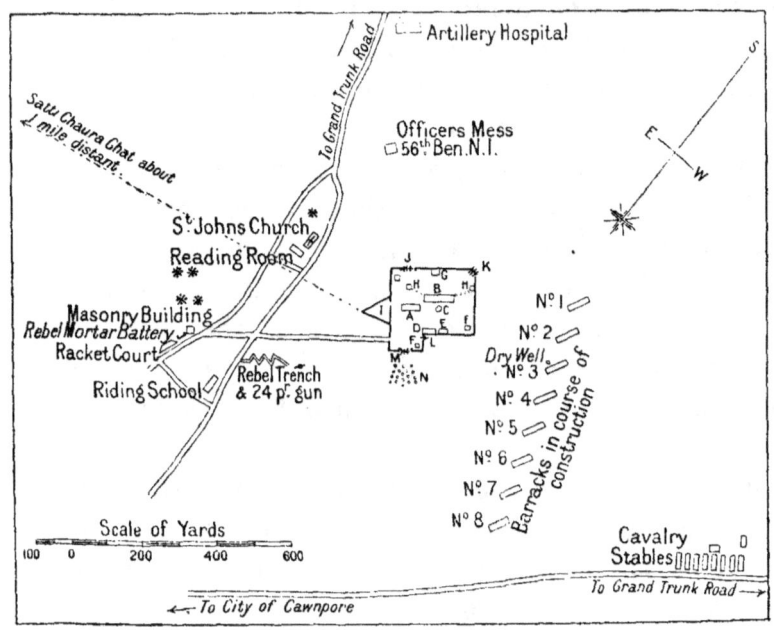

WHEELER'S ENTRENCHMENT

A. Masonry Barrack.
B. Masonry Barrack with thatched roof.
C. Well.
D. Provision Store.
E. Main Guard.
F F. Field Magazines.
G. Cook House.
H H. Outhouses.
I. Redan.
J. Lieut. Ashe's Battery.
K. Lieut. Eckford's Battery.
L. Rifled 3-pr. gun.
M. Lieut. Dempster's Battery.

N. Trous de loup.
No. 1. Block occupied by rebels in large numbers.
No. 2. Block occupied by British Picket of 16 men.
No. 3. Block unoccupied.
No. 4 Block occupied by British Picket of 16 men.
Nos. 5–8. Block unoccupied by British from lack of men.
* The rebels had 3 large guns at Church.
** These buildings commanded the Entrenchment.

of the closest and fiercest fighting took place. Otherwise they seldom ventured into the open except at night, when they allowed the garrison no rest.

[1] These were taken from the Arsenal.

The two large buildings in the entrenchment were given up to the women and children, and to the more serious cases of illness or infirmity; but after two or three days their doors and windows were all shot away, all internal partitions destroyed, and many casualties caused by shot and shell amongst the unfortunate inmates.

The 9-pounder ammunition was soon expended, and 6-pounder shot (with great consequent loss of accuracy) had to be used, so that after a few days the task of engaging the rebel guns was abandoned and the artillery devoted all their efforts to support the infantry defence.[1]

[1] As the siege progressed the trained gunners were all killed, and the guns were worked by volunteers. " Our 59 artillerymen had all been killed or wounded during the first week. With the exception of four of their number these fine fellows all perished at the batteries." The howitzers had been dismounted and many of the 9-pounders injured, the bores having been indented by the impact of round shot, and at length only two were available for firing case shot. Even in these the irregularity of the bores prevented the ordinary case shot being employed. Then " the women gave us their stockings; and having tapped the canisters, we charged these stockings with their contents" (Mowbray Thomson, p. 151).

Lieutenant H. G. Delafosse, 53rd Ben.N.I., had from the first been attached to Eckford's Battery, and survived that officer. Two of his exploits are recorded by Mowbray Thomson. He was annoyed by a 1-pounder gun the rebels had mounted in one of the barrack buildings outside the entrenchment. " As he was compelled to load his gun with 6-pounder shot, he could secure no regularity in his fire, so he gave his worn-out gun a monster charge of three 6-pounder shots and a stocking full of grape well rammed down. The faithful old weapon did not burst, and the troublesome little antagonist was never heard of again!" (p. 38). On another occasion the rebels exploded one of our limbers, which set fire to a gun carriage in dangerous proximity to a quantity of ammunition. There seemed to be no way of putting out the fire until Delafosse crept under the burning carriage, and lying on his back pulled away splinters of the wood, and scattered handfuls of earth on the flames, being subjected all the time to the fire of 18- and 24-pounders (pp. 138-9).

On the evening of June 13 the building with the thatched roof was set on fire by a red-hot shot, and, as a breeze was blowing, it was rapidly consumed. The sick and wounded were removed with the greatest difficulty—several were suffocated or burnt to death; all medical stores and surgical instruments were destroyed; and many women and children henceforth had no shelter. The flames lighted up the interior of our works, and the enemy's fire was incessant until past midnight, when, thinking our attention was fully occupied, they ventured an assault. A numerous body crept up towards Ashe's position in the hope that they were unseen; but when they got within 80 yards of the guns they were received with a salvo of case shot and volleys of musketry. Baffled and dismayed, they at once took to flight, leaving 100 dead upon the ground. The following night the garrison struck another blow. Captain Moore and 50 picked men stole out of the entrenchment, and, making for St. John's Church, they spiked 2 or 3 siege guns. After bayoneting some sleeping sepoys in the mess compound of the 56th Ben.N.I., they disabled 3 more guns and returned to the entrenchment with a loss of 1 man killed and 4 wounded.

But the rebels had no lack of men and were always able to replenish their artillery from the Cawnpore Arsenal, while day by day the garrison was dwindling away. On June 16 the Artillery Hospital was set on fire, and the loyal band of sepoys who held it applied for admission to the entrenchment; but the scarcity of food forbade any increase to the diminished garrison. Each sepoy was given a few rupees and a certificate of fidelity, and they were told to consult their own safety and depart. Indeed, after the first week the question of food became crucial, for the supplies that had been laid in were both unsuitable and inadequate, and a time came when a handful of split peas and a handful of flour was the daily ration for all alike. The only well was out in the open and exposed to fire; it was 70 feet deep, and the noise of the apparatus which raised the buckets drew fire even after darkness had set in. Eventually it was shot away, and then the process of raising the buckets by hand had to be resorted to, a method which, though slower, proved equally perilous.[1] In these circumstances the sufferings of the sick and wounded and of the women and children were terrible, and letters describing the awful state of affairs reached Lucknow, but Lawrence sorrowfully felt he was powerless to help, and the miserable days dragged on.

Treachery of the Nana Sahib.—On June 23 the Nana, who had been reinforced, wishing to celebrate the centenary of Plassey by the capture of the entrenchment, made an attack upon it with a large force of the three arms;[2] but the half-starved

[1] "Many lives were lost at the well. Water was as a rule only drawn at night, but when the rattling of the chains was heard a fire of grape was always opened by the rebels. After a while a Bengal Civil Servant, John McKillop, constituted himself captain of the well. He ran many narrow escapes, but after he had performed his duties for a week he was killed. It is needless to add that from the first day of the siege not a drop of water could be got for washing" (Mowbray Thomson, p. 87).

[2] "A party of rebel cavalry took part in this attack. They came

defenders gave the rebels so warm a reception that, after losing 200 men, they retired whence they came. Force having failed, the Nana now resorted to treachery.

The Massacre.—On the 25th honourable terms were offered and accepted. Of those who entered the entrenchment three weeks previously, nearly a third were dead,[1] many more were helpless from wounds and sickness, and the miserable food was almost expended.[2] Had it been a military garrison pure and simple, the men might have cut their way out and died fighting; but there was just a shred of hope that the women and children might be saved, and the gallant defenders clung to it. On the understanding that boats for transport to Allahabad would be provided, they quitted the entrenchment on the morning of the 27th, and the sad procession made its way to the river, while a body of sepoys, secretly appointed for the purpose, closed up in rear. At the Satti Chaura Ghat a number of boats were in readiness, and the embarkation was in progress, when the crews suddenly plunged into the river, having previously scuttled the boats or set fire to the thatched roofs with which they were provided; at the same time fire was opened on the British from each bank. In the midst of the slaughter carried out under the

from the neighbourhood of the Riding School, whence they started with great impetuosity. When they got near the entrenchment their horses were winded, and a round from the guns threw them into confusion" (Mowbray Thomson, p. 127).

[1] The common burying-place was a dry well close to one of the barracks held by our pickets. Over 250 bodies were placed in it.

[2] When the force capitulated there was three days' supply of starvation rations remaining.

eyes of Tantia Topi, who is said to have planned its details, an order arrived from the Nana to spare the lives of the women and children. Of the men who quitted the entrenchment that morning, but four [1] survived to tell the black story of Cawnpore; but 125 women and children were snatched from the merciful waters of the Ganges to undergo a further spell of torture and meet a more terrible fate. For the greater part of the day they were kept herded together on the blood-stained banks of the river without food, drink, or shelter from the sun, and as evening closed they were consigned to prison under conditions of the most revolting barbarity.[2]

II. THE DEFENCE OF THE RESIDENCY

Activity of the Rebels.—On June 27 the Nana Sahib was proclaimed Peshwa,[3] and on July 9 he despatched a force of 3,500 sepoys towards Allahabad, the advanced guard of the army which should conquer the south.

The events at Cawnpore also roused to action the Oudh mutineers, who concentrated at Newabgunge-Barabanki, and on June 29 pushed their advance guard to Chinhut, a few miles from Lucknow. Sir

---

[1] The four survivors were Lieutenants H. G. Delafosse (afterwards Major-General), and Mowbray Thomson, 53rd Ben.N.I., Gunner Sullivan, Ben.A. (who died shortly afterwards of cholera), and Private Murphy, 84th Foot.

[2] In the Beebeeghur.

[3] The mutineers present on this occasion, in addition to those of the Cawnpore garrison, were the 17th Ben.N.I. and 13th Irregular Cavalry from Azimghur and Benares, the Jhansi-Nowgong Brigade, a detachment of the 10th Ben.N.I. from Futtehghur, the 6th Ben.N.I. from Allahabad, and Oudh Irregular Regiments.

Henry Lawrence's information at this time was very inaccurate,[1] and under the impression that only a weak force was at hand he determined to crush it. With this intention he ordered Colonel Inglis, 32nd Foot, to march on the morning of the 30th to the Kokral Bridge with the following troops:

European Vol.C. (36 sabres) (Captain C. W. Radcliffe, 7th Ben.L.C.). Oudh Irr.C. (80 sabres) (Lieutenant G. N. Hardinge).	32nd Foot (300 men). 13th Ben.N.I. (150 men). 48th Ben.N.I. (50 men). 71st Ben.N.I. (20 men).	No. 9 F.B. 4/1 Ben.A. (4 guns) (2nd Lieutenant F. J. Cunliffe). No. 2 Battery Oudh Irr.Art. (2nd Lieutenant D. MacFarlan). No. 3 Battery Oudh Irr.Art. (2nd Lieutenant J. H. Bryce). 8-inch howitzer[2] with detachment from 4/1 Ben.A. (2nd Lieutenant J. Bonham).

O.C.A., Captain A. P. Simons.

The Battle of Chinhut, June 30.—There was some delay in starting, and the sun was well up when Inglis reached the bridge. Lawrence had gone on in front to reconnoitre, and seeing the rebel scouts falling back before him, he sent orders to Inglis to continue his advance. Beyond the Kokral bridge, four miles from Lucknow, the road to Chinhut was merely a newly raised embankment, made of loose

[1] On the evening of that day (June 29) several reports reached Sir Henry Lawrence that the rebel army, in no very considerable force, would march from Chinhut, etc.

[2] The 8-inch howitzer was found by Captain Fulton, Ben.E., lying unmounted at the Kotwali in the city. He brought it into the Residency, and Lieutenant Thomas, Mad.A. (Acting Commissary of Ordnance) mounted it on a howitzer carriage. It was intended for the fixed defences of the Residency, but on the day of Chinhut it was taken out with an 18-pounder limber (no proper limber being available) drawn by an elephant. It was worked by Sergeant Settle and a detachment of European gunners from 4/1 Ben.A. Sergeants Bewsey and Miller (who accompanied Lieutenant Bonham from Secrora) were also with it.

and sandy soil, with gaps here and there for future bridges. It was a fatiguing road to travel under favourable circumstances; under the sun of an Indian June it was intolerable, and by the unpardonable fault of some one the men were sent out to march and fight without having been supplied with their breakfasts. Two and a half miles east of the bridge the track ran between two villages. Bijaipur on the right was a small hamlet lying in the open; Ismailgunge on the left was larger and showed up against a background of trees extending to Chinhut, about a mile distant. As the British column advanced, the enemy's guns[1] in front of Chinhut opened at 1,400 yards range, and at the same time their infantry was seen in occupation of the thick mango groves surrounding the town. The native infantry supported by Cunliffe's and MacFarlan's guns took possession of Bijaipur, while the 32nd lay down some little distance from Ismailgunge, whence our vedettes had been fired upon as we advanced. Bonham's howitzer came into action on the road; Bryce's guns were in reserve. Bonham speedily silenced the rebels' artillery, and for a short time it was thought they were in full retreat; but presently a large force issuing from the woods about

---

[1] The rebel force consisted of 700 or 800 cavalry made up by the 15th Bengal Irregular Cavalry from Sultanpore and mutineers from the Oudh Irregular Cavalry. The infantry were the 22nd Ben.N.I. from Fyzabad, mutineers from the seven regiments of Oudh Irregular Infantry, and two regiments of Military Police. The artillery consisted of No. 13 F.B. 5/7 Ben.A. from Fyzabad, and No. 1 Batt. Oudh Irr.A. from Secrora ("Mutinies in Oudh," by M. R. Gubbins, p. 187). The rebels were also joined by several Talukdars who brought up a few (and probably useless) pieces of artillery.

Chinhut advanced with colours flying across the plain on our right, while another large force under cover of the trees threw themselves into Ismailgunge.

The native infantry in Bijaipur presented a firm front; but the advancing tide of rebels threatened not only to turn their flank, but to intercept their retreat, and our cavalry were ordered to charge.

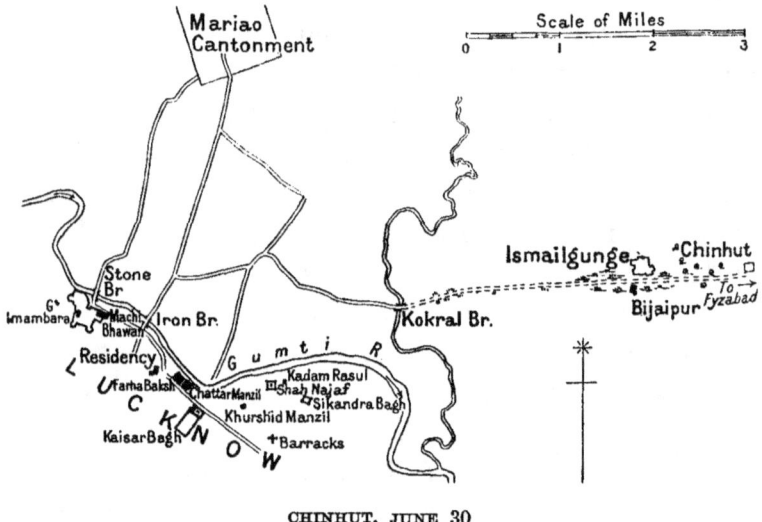

CHINHUT, JUNE 30

The Volunteers promptly obeyed this order, but the Irregulars refused to follow, and some of them fled the field and did not pull up till they reached Lucknow.[1] The plucky action of the Volunteers, however, checked only for a moment the rebel advance, and case shot from the field guns fired at a range of 500 yards was ineffective.

[1] Later in the day Hardinge succeeded in rallying a portion. The accounts of their subsequent behaviour in the action are conflicting. (Compare Mr. M. R. Gubbins's account with Colonel J. E. W. Inglis's Despatch.)

On our left we met with disaster. The 32nd Foot were ordered to take Ismailgunge, but were repulsed with heavy loss, including their commanding officer, Lieutenant-Colonel Case, and Captain Steevens who succeeded him. MacFarlan's Battery had been ordered to the left to support the 32nd, but on receipt of the order the Oudh gunners mutinied, driving off with some of the waggons, and cutting the traces and riding away with the teams of others. They attempted to remove the guns also, and were only prevented from doing so by MacFarlan, who mounted one of the horses of a team, pistol in hand, One gun limber actually did get away, and its gun had eventually to be abandoned. Bryce's men behaved in the same way, riding away on the teams when the advance of the rebels was observed. They were prevented from taking the guns, which in the end had to be abandoned.

Having silenced the guns in front of Chinhut, Bonham had ceased firing. From his position on the road he was unable to see the flanking movements of the rebels, and while the 32nd were attacking Ismailgunge he received orders from Lawrence to retire. But the elephant had bolted with the limber and the native drivers had allowed the bullocks, provided for emergencies, to stray away. Resolved to save the howitzer if possible, Bonham mounted one of the lead horses and assisted in bringing up a field limber. In vain; for the native drivers galloped off with it at once when he dismounted to limber up. At this moment the elephant was recaptured; but the animal's restlessness foiled every attempt to limber up the awkward short-trailed howitzer, and while efforts were being made to save the piece Bonham was badly wounded. The howitzer was then spiked and abandoned.[1]

On the repulse of the 32nd and the failure of the cavalry, Lawrence saw himself in danger of being turned on both flanks, and he ordered a retreat. Our famished troops were utterly exhausted; no water could be obtained, for the water carriers had all fled; many of the muskets had become foul and could not be fired; the European battery had no rounds left, and the retreat became a rout.

When the Kokral came in sight it was found that the enemy's cavalry had taken possession of the bridge, and the retreating column turned aside from the road and collected at a bend in the stream on our left. The moment was critical, for the cavalry blocked the road to Lucknow and the pursuing rebels pressed upon our rear. Though Cunliffe's guns had no ammunition left, Lawrence ordered him to come into action and light his port fires. This checked the rebel advance, while Radcliffe, by a

[1] When Bonham perceived that the limber was missing he sent Sergeant Miller in search of it and told Sergeant Bewsey to go and report matters to Sir Henry Lawrence. Neither non-commissioned officer had returned when Bonham was wounded and the howitzer abandoned. Bonham was carried off by Sergeant Settle and the gunners, and later was placed on one of Cunliffe's gun carriages; but when the bridge was cleared by Radcliffe's charge the guns limbered up and went· off, leaving him behind. He was discovered by Bewsey, who now came up and, placing Bonham on his horse, went along holding on by the stirrup. In this way they proceeded until overtaken by Miller, also mounted. Bewsey then committed his officer to Miller's care, and he himself walked into Lucknow. For this gallant conduct Bewsey was rewarded (some years later) by receiving a commission.

JUNE 30, 1857] BATTLE OF CHINHUT

brilliant charge at the head of his 36 Volunteers, drove off 500 rebel horsemen and cleared the bridge. The guns then limbered up and our troops made their way into Lucknow, pressed by the enemy as far as the Iron Bridge, where they came under fire from the Redan Battery at the Residency.

On this disastrous day we lost 292 officers and men killed, and 78 wounded.[1] To find a parallel to it we must look back to the battle of Adrianople, August 9, 378, in which the Emperor Valens was defeated and lost his life. " His men reached the battlefield after a twelve-miles march under a burning sun, exhausted by thirst, hunger, and intolerable fatigue,"[2] half beaten before the battle began like our own unfortunate soldiers at Chinhut.[3]

The Residency invested.—The pursuing sepoys crossed the river lower down, and on being joined by the two regiments of Oudh Irregular Infantry quartered in the city, and by the Military Police, they closely invested the Residency. They did not, how-

---

[1] The proportion of killed to wounded is extraordinary. It was caused by the fact that wounded men were often of necessity left where they fell, and no quarter was given. The following heroic episode is told by Gubbins in his "Mutinies in Oudh." "Captain Bassano, 32nd Foot, who had been looking for his commander, Colonel Case, discovered that officer lying wounded on the field. The men had passed on, but Captain Bassano desired to bring some back to carry off the colonel. The enemy were, however, close at hand, and Case would not suffer him. Finding ordinary remonstrance unavailing, the gallant soldier turned to his subordinate : ' Leave me, sir,' said he, ' and rejoin your company.' The order was reluctantly obeyed." (*Note.*—Gubbins was in the Residency during the siege.)

[2] Gibbon's " Decline and Fall of the Roman Empire," vol. iii. p. 111 (Bury's ed.).

[3] The author is indebted to Colonel H. W. L. Hime, (late) R.A., for this historical parallel.

ever, molest the Machi Bhawan, the garrison of which was safely transferred to the Residency on the night of July 1. Lieutenant L. F. C. Thomas, Mad.A., was the last to leave, after lighting a twenty-minutes time fuse; and as the retreating garrison entered the Residency the Machi Bhawan magazine blew up.

On July 2, 3,000 souls were assembled within the Residency enclosure. There were:

British officers . .	133	Christian non-combatants,
British N.C.O. and men .	671	women, and children . 600
Eurasian drummers .	51	Native non-combatants,
Volunteers . . .	153	women, and children . 680
Total	1,008	Total 1,280
Native troops . .	712	
Total combatants . .	1,720	

The Defences of the Residency.—But for a projecting spur, the Residency enclosure was a rough rectangle containing some 33 acres, of varying elevation. Its northern half commanded all the adjacent ground, while the southern was level with the neighbouring roads and buildings. The northern face ran along the edge of a high bank, sloping somewhat sharply down to the Gumti, with a free field of fire; but elsewhere buildings and ruins were close to the parapet, and on the south only ten yards separated the defenders from houses held by the rebels. The defences at this moment were by no means completed; both here and at the Machi Bhawan hundreds of coolies were at work, but when they caught sight of our retreating troops they disappeared as if by magic.

The garrison of the Residency were all told off

to various posts as indicated on the plan. Each post was self-contained and none of its allotted detail was allowed to leave its precincts. The dis-

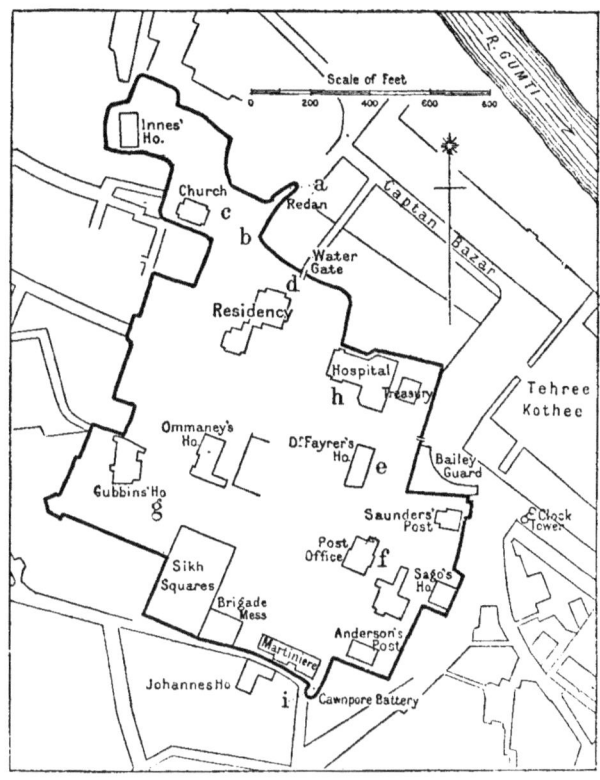

THE DEFENCES OF THE RESIDENCY

DISPOSITIONS OF GUNS AT BEGINNING OF SIEGE

a. Lieut. Cunliffe's post.
b. Lieut. MacFarlan's post.
c. Capt. Evans' * post.
d. Lieut. J. Alexander's post.
e. Lieut. Brice's post.
f. Lieut. Bonham's post.
g. Lieut. Thomas' post.
h. Lieut D. C. Alexander's post.
i Lieut. Lewin's † post.

* Capt. H. L. Evans, 17th Bom. N.I., Deputy Commissioner, Oudh.
† When Lewin was killed, July 15, these guns were placed under Bonham in addition to those at the Post Office.

position of the ordnance was permanent, and in order to supply the want of a heavy mobile piece,

2nd Lieutenant J. Bonham devised a triangular wheeled platform for an 8-inch mortar and bed. The usual coin giving a permanent elevation of 45° was replaced by small wedges, and this novel mounting, nicknamed the "ship," proved a most efficient howitzer which could be fired at from 1° to 15° elevation.

Repulse of Rebel Attacks.—When the rebels first took possession of the city they appear to have quarrelled about who was to rule, but at length a son of the deposed sovereign, whose energetic mother will be henceforth known as the Begum, was proclaimed King on July 9, and a somewhat burlesque government was set up. At the same time futile attempts at military organisation were made, but no spark of military capacity was shown, and there was little discipline in the heterogeneous crowd that formed the rebel army.[1] However, after certain orgies had marked their entry into Lucknow, they pressed on the siege of the Residency to the best of their ability. They brought guns close up to our boundary, which, being silenced at once, were speedily removed to other sites, whence their fire was spasmodic, and as a rule ineffective.[2] Most fortunately

---

[1] Included in the rebel army were a number of Passees, an indigenous race of Oudh. They were not a warlike tribe, but were accustomed to mining operations. On the other hand the 32nd Foot, a Cornish regiment, numbered in their ranks many a man who had been employed in the tin mines of Cornwall.

[2] "They kept their guns in isolated and sheltered nooks and corners, and fired them at random, hitting mainly the upper parts of the buildings and defences, their shots going in large numbers clear over the entrenchments, and plunging into their own posts beyond. . . . Besides guns more or less worthless, the enemy had two good light field

they possessed no mortars, but their musketry fire was a great scourge, and, directed as it was at close quarters against our loopholes and embrasures, it caused us daily loss. The existence of a building, Johannes House, over against the Cawnpore Battery was a coign of vantage the rebels utilised to the utmost, and for nine days our battery was reduced to silence by its fire; but at the end of that time the house was captured and blown up. During the progress of the siege thirty-seven attempts at mining were made by the enemy.[1] On August 10 a mine intended to form a breach was placed short of its proper position, but succeeded in destroying a portion of the stockade in front of the Martinière; and on August 16 a breach 30 feet long was made at the corner of the Sikh Square; but on both these occasions the subsequent attacks were repulsed and the damages repaired. All other attempts failed. Lodgments were effectually dealt with by artillery fire, hand grenades or sorties. The rebel galleries, with the two exceptions above noted, were either wrongly directed, checked by counter-mines, or (being made too near the surface) collapsed of themselves. On one occasion this collapse was assisted by a shell cleverly lobbed from an 8-inch mortar.[2]

batteries, and also the 8-inch howitzer which they captured at Chinhut. These they moved about from place to place as the desire seized them. The store of ammunition for these guns was small " ("Lucknow and Oudh," p. 118).

[1] An excellent account of the mining operations, and indeed of the siege generally, will be found in the book quoted in the previous note.

[2] Lieutenant J. Bonham was in command of 8-inch mortars near the Cawnpore Battery when Captain Fulton, Ben. E., came up with an

Three general attacks were made on the entrenchment, namely, on July 20, August 10, and September 5; but the intention of the enemy being always known from the look-out on the Residency roof, these attacks were all repulsed with heavy loss, and the casualties of the garrison were comparatively small.

At the end of August the rebels displayed more enterprise in their use of artillery, and made a battery of two siege guns at the Clock Tower. A counter-battery was at once placed at the Treasury, and armed with 18-pounder guns, and when the rebels opened fire from their new battery on September 5, in support of a general attack, it was almost immediately silenced.[1] A little later an attempt made to bring a cross-fire of artillery on the Cawnpore Battery was foiled by the " ship," and two or three other mortars mounted in a similar way.

The Relief.—But death and disease were busy among the inmates of the enclosure. Cholera and scurvy were rife, and amputations were usually fatal. On July 2 Sir Henry Lawrence was mortally wounded

empty pillow-case which he wished to have filled with powder in order to destroy a rebel gallery he had discovered close to our defences. The danger of the operation was obvious, and Bonham asked Fulton to wait and see what could be done with a mortar shell before risking his life. Fulton agreed, and a mortar was lined on the place he pointed out, which was only about 5 yards from the wall behind which the mortar stood at some 15 yards' distance. In order to get a range of only 20 yards, the mortar had to be elevated almost vertically, and it was a chance which side of the wall the shell would fall. After three or four trials, however, the powder charge being gradually reduced, a shell got home and the gallery was destroyed.

[1] The battery was built and worked by the 13th Ben.N.I. under Captain R. H. M. Aitken, who received the Victoria Cross for various acts of gallantry during the siege.

by the splinter of a shell, and was succeeded by Major Banks in his civil, and by Brigadier Inglis[1] in his military capacity. A few weeks later Banks was killed, and Inglis added his duties to his own. As the enemy were within speaking distance, bad news, generally false, but occasionally true, was continually shouted to the garrison, but no friendly communication was received until July 22—three weeks after the beginning of the siege—when a spy brought in a note from Havelock, and from that date onwards hopes of relief were often raised, doomed only to be disappointed. The garrison also were ever haunted by the fear that provisions would fail.[2] Days passed and the garrison dwindled. Of the Europeans 140 had been killed and 190 wounded; of the natives 72 and 131.[3] Captain G. W. W. Fulton, Ben.E., the soul of the engineer defence, was dead, and all the artillery officers present when the siege began were killed or wounded except one.[4] But on September 22 authentic news arrived

[1] Captain and Brevet-Major J. S. Banks, 33rd Ben.N.I., and Lieutenant-Colonel and Brevet-Colonel J. E. W. Inglis, 32nd Foot.

[2] The chief Commissariat officer was wounded at Chinhut, and a subsequent error in stock-taking led to the belief that provisions were running short. This error was subsequently discovered.

[3] Eleven women and 54 children died. Three women were killed.

[4] Captain A. P. Simons died in hospital on September 7 from the effect of his wounds at Chinhut; Lieutenant E. P. Lewin was killed on July 15 in the Cawnpore Battery; Lieutenant J. H. Bryce, wounded on July 16, died of cholera on the 18th;-Lieutenant F. J. Cunliffe was slightly wounded, but died of fever September 22; Lieutenant D. MacFarlan was severely wounded on July 20; Lieutenant D. C. Alexander was severely burnt on July 17 by the premature explosion of a mortar charge, and was killed by a round shot on the day after the arrival of Havelock's force; Lieutenant J. Alexander was shot through the arm while laying an 18-pounder in the Hospital Battery; Lieutenant J. Bonham after receiving two wounds was wounded a third time severely. Lieutenant L. F. C. Thomas. Mad.A., was alone untouched.

that General Havelock had crossed the Ganges. After three days of suspense it was reported on the afternoon of the 25th that officers dressed in shooting-coats and solar topees, a regiment of Europeans in blue pantaloons and shirts, and a bullock battery were seen near the Moti Mahal. At 5 p.m. volleys of musketry rapidly growing louder were heard in the city, and soon the whiz of a Minié rifle bullet was recognised by the delighted garrison. Five minutes later the relieving force was seen fighting its way through the streets.[1]

III. HAVELOCK'S EFFORTS TO RELIEVE LUCKNOW

Lines of Communication.—Troops could move from Calcutta to Allahabad by road or by river. The course of the Ganges between the two cities was over 800 miles long, and passed through the military stations of Bhagulpore, Dinapore, Gazipore, and Benares. At certain seasons of the year steamers towing flats could perform this journey in from twenty to thirty days.[2] The Grand Trunk Road from Calcutta to Ranigunge ran along the 125 miles along which the E.I. Railway was already working. It then went through Hazaribagh to Benares, crossed the Ganges and ran along the left bank to Allahabad, the journey by rail and route march occupying five weeks. By utilising bullock carts, etc., the time could be reduced to a fortnight, but no more than 200 men a day were ever carried by this mode of

---
[1] "The Defence of Lucknow," by a Staff Officer, pp. 173, 174.

[2] During June, July, and August the river was generally made use of, but towards the end of the rains in September the water subsided.

transport.[1] The electric telegraph ran along the Grand Trunk Road.

Formation of Relief Columns at Allahabad.—It has already been told how white troops, at first in very small bodies, began to move northwards along these lines; how the 84th Foot and 1st Madras European Fusiliers reached Allahabad; and how by the end of June the energetic Neill despatched a small force towards Cawnpore. It set out on the 30th, under Major S. G. C. Renaud, Madras Fusiliers, and consisted of 400 European infantry (84th Foot and Madras Fusiliers), 400 Sikhs (Ferozepore Regiment), and 20 invalids of the Bengal Artillery, with two bullock-drawn 9-pounder guns (Lieutenant T. N. Harward). On July 1, Neill sent Captain J. B. Spurgin up the Ganges in a small steamer, the *Brahmapootra*, with 100 Madras Fusiliers, a few artillery invalids, with two 6-pounder guns and supplies of food, etc., intended for the Cawnpore garrison.

On the day of Spurgin's departure Major-General Havelock arrived at Allahabad and approved the measures that had been taken. He was shortly followed by detachments of the 64th Foot and 78th Highlanders, and by 3/8 R.A. (2nd Captain F. C. Maude).

Maude had only one subaltern, Lieutenant E. Maitland, and his effective strength was reduced by

---

[1] This mode of transport was called a Bullock Train. Relays of oxen were placed along the road, and waggons which started from the rail head at Ranigunge, filled with soldiers, went forward without any halt during the night and the cool hours. In the daytime the men rested. It was not until after the arrival of Sir Colin Campbell that these trains got into full working order.

sickness to a total of 50; Lieutenant C. W. Crump, Mad.A., and 30 infantry volunteers were therefore attached to him, and he took over two 6-pounder guns, two 9-pounder guns, and two 24-pounder howitzers, with bullock draught and native drivers. This somewhat composite command was for a time the only representative of the Royal Artillery, which now, after a lapse of fifty years, again appeared on the mainland of India.

Havelock at Futtehpore.—Difficulties of transport and the absence of necessary stores delayed Havelock's further progress till July 7, when, at the head of 1,100 men, he set out for Futtehpore by forced marches. There was good reason for his haste. Renaud had started in ignorance of the destruction of General Wheeler's garrison, and of the subsequent movements of the rebels, but both these things were known to Havelock before he left Allahabad. On the night of July 11 by great efforts he overtook Renaud, within striking distance of the rebel force 3,500 strong. At daybreak Havelock halted 5 miles from Futtehpore, the united forces being thus composed:

European Vol.C. (20 sabres).	64th Foot (435 men).	3/8 R.A. (76 men).	⎫
Detachments 13th Ben.Irr.C.	78th Highlanders (284 men).	Ben.A. Invalids (22 men).	⎬ Composite Battery
3rd Oudh Irr.C. (95 sabres).	84th Foot (190 men).	Golandauz (18 men).	
	1st Mad.E.F. (376 men).	8 guns [1]	
	Ferozepore Regt. (448 men).	(2nd Capt. F. C. Maude, R.A.).	⎭

TOTAL { Europeans . . . . . 1,403 } 1,964
       { Natives . . . . . .   561 }

[1] Lieutenant Harward's command was absorbed in Maude's Battery.

Action at Futtehpore, July 12.—The town of Futtehpore, surrounded by walled enclosures and possessing many well-built houses, was a position of considerable strength. The Trunk Road ran through it, and, owing to the flooded state of the country, was the only means of convenient access to it. In front, but separated from the town by swamps, were hamlets, hillocks, and mango groves. On the morning of the 12th the Nana's force of 3,000 infantry, 500 cavalry, and 12 guns occupied Futtehpore. The breakfast bugle was sounding in Havelock's camp when it became apparent that the enemy were advancing to the attack, and, from the determination of their onset and their subsequent behaviour, there is reason to believe they imagined they had only Renaud's force in front of them. Havelock at first merely occupied a copse in his immediate front with 300 men of the 64th Foot, but on the enemy persisting in their attack, he formed the bulk of his infantry into line of columns with skirmishers in front. Maude, with two 9-pounders, advanced along the road, and his three lieutenants, each with two guns, conformed to the movement of the infantry on either flank. Maude came into action at 800 yards, and with his second shot disabled one of the attacking guns. The enemy then began to fall back, for the fire from the Enfield rifles reached them from an unexpected distance, and Havelock's mode of advance menaced their flanks. But when our guns limbered up to follow, a large body of rebel cavalry advanced down the road, and Maude again came into action at 650 yards. The cavalry at once opened out and unmasked two heavy guns behind which large masses of infantry were visible, with whom were seen two elephants richly caparisoned. But a concentrated fire of our guns from the centre and flanks, aided by volleys from the Enfield rifles, overpowered the rebel battery, and when one of the elephants was drilled through and through by a round shot, and his rider, evidently a person of distinction, brought heavily to the ground, the enemy broke and fled. Their own guns, found to be loaded, were turned upon them, and our infantry, following them up, drove them through and from the town. Our native cavalry, ordered to pursue, behaved with treacherous cowardice; but some guns were brought through the town, and under their fire the infantry put the rebels to final and irretrievable flight. The victory was won by rifles and cannon. The fire of the enemy never reached our men, and the few casualties that occurred were among our native cavalry, but 10 European soldiers died of sunstroke. The enemy lost about 150 men and 12 guns, as well as stores, baggage, and money. With the aid of the captured ordnance Maude organised a battery of five 9-pounder guns and one 24-pounder howitzer, with stores and carriages complete. The Irregular Cavalry, who had shown themselves to be worthless, were dismounted and 40 of their horses were given to infantry soldiers selected from the British regiments. On the 13th the force halted, and Futtehpore was sacked and burnt. Next day the advance was continued.

Action at Aong, July 15.—On July 15 the enemy

were again encountered in front of the village of Aong, situated on the Grand Trunk Road in the midst of a thickly wooded district. Our small body of improvised cavalry was in advance with Lieut.-Colonel Tytler,[1] Havelock's Adjutant-General, when it was found that the enemy had dug an entrenchment and had posted two guns to bar the road. Maude's Battery and a detachment of the 64th Foot were at once sent forward in support of Tytler, but the enclosed nature of the ground enabled the rebels to hold our troops in check for some time during which large bodies of cavalry threatened both our flanks, and twice Maude had to resort to case shot to clear his front. After two hours' fighting, however, the sepoys began to yield ground, and then a bayonet charge drove them out of Aong. Beyond the village was a large mango grove, and our troops were resting in its grateful shade when the General called upon them for fresh exertions; for on reaching Aong he had learnt from his spies that the bridge over the Pandoo River was still intact. The march was at once resumed.

Action at the Pandoo River, July 15.—When our troops approached the river, now much swollen by the recent rains, the enemy opened fire from a low ridge in rear of it, and several 24-pounder shot reached the column. The bridge which carried the Trunk Road was at a salient bend of the river in our direction, and in order to envelope it with artillery fire Maude with Lieutenant Crump ad-

[1] Captain and Brevet-Lieutenant-Colonel J. M. B. F. Tytler, 37th Ben.N.I.

vanced along the road, while Lieutenants Maitland and Harward were sent to the right and left. On the flanks the ground was heavy, and the centre guns getting in advance came into action at 600 yards. After three rounds of shrapnel shell the enemy's sponge staves were broken, and his 24-pounders were put out of action. Then, after a futile attempt to blow up the bridge, the rebels fled before a bayonet charge delivered by the Madras Fusiliers, and Havelock's force bivouacked two miles beyond the river, after marching for twelve hours and fighting two actions under the burning sun of July. The enemy lost 200 men and 4 guns. Our force had 26 casualties in the two combats, including Major Renaud mortally wounded.[1]

Battle of Cawnpore, July 16.—Next day, July 16, Havelock reached Maharajpore, a village six miles from Cawnpore, where he left his camp and baggage under escort, and then pushed on to the junction of the Grand Trunk Road with that leading direct to the Cawnpore cantonments. Here the enemy, having been strongly reinforced, were in position. Their right rested on a wooded village at A, their left on an elevated hamlet at B, and at both villages two 24-pounder guns were posted. Their centre was supported by a third village at C, and a small entrenchment was made across the junction of the two roads.[2] When Havelock arrived within half a

---

[1] The Royal Artillery lost one man, Bombardier Harding. He was laying his gun when a round shot shattered his arm from the fingers to the elbow. "If you please, sir, may I fall out?" said he to Maude, saluting, and then pointing to his mangled arm. He died that evening from loss of blood ("Memoirs of the Mutiny," i. p. 49).

[2] See p. 174.

1ST BATTLE OF CAWNPORE

mile of this entrenchment his column turned off the road and marched round the left of the rebels, who opened fire with some effect from the guns at B. Having passed the woods on the east of the Cantonment Road, Havelock deployed and formed line in such a way that his front now faced the left rear of the rebels. The Madras Fusiliers covered the deployment, and the guns came into action at 900 yards against the 24-pounders at B; but these were well posted, and Maude's lighter pieces made no impression on them. Havelock, however, sent the 78th Highlanders forward with the bayonet, and after a brilliant charge they captured the guns, rolled up the rebels' left and centre, and made themselves masters of the two villages they had occupied.

Making his way with difficulty across the sodden ground, Maude came again into action in the fork of the two roads, and having silenced the guns at A, the village speedily fell before the attack of the 64th Foot. He himself personally accompanied their advance, and finding the rebel guns loaded, turned them, with the assistance of some infantry soldiers, on their late owners, and afterwards spiked them when his own men came up.

He then was ordered to halt on the Grand Trunk Road a little in front of the village C, whither it was understood the infantry would return and encamp as soon as they had cleared the front. In this position, however, he was unsupported and found himself in constant danger from bodies of hostile cavalry and from skirmishers advancing from the west of the railway embankment, and when eventually he got a pressing order to advance, the capture of his guns could have been easily effected if the rebels had made the attempt. Havelock was again heavily engaged.

On being driven from the line of villages, the enemy's infantry appeared to be everywhere in full retreat, and our small body of volunteer cavalry made a successful charge on a party of horse that

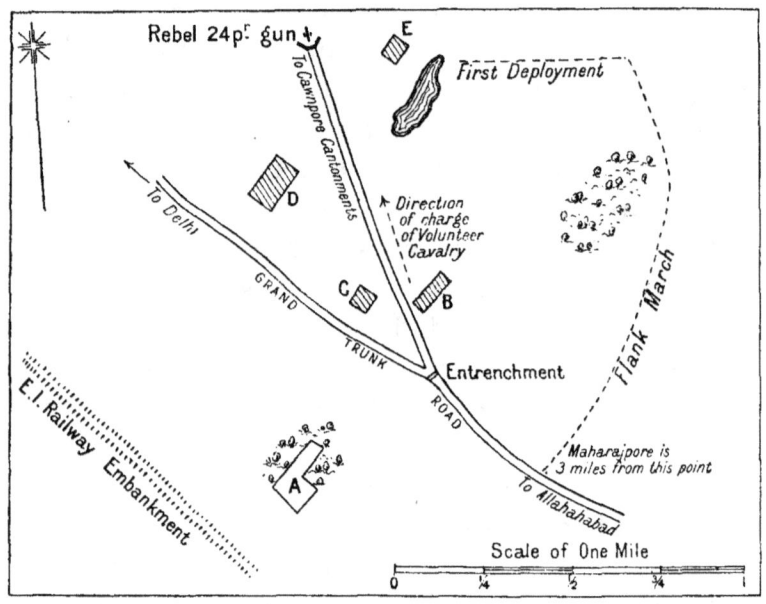

CAWNPORE, JULY 16

came within striking distance; but when Havelock moved down the road leading to cantonments, a 24-pounder gun was suddenly unmasked and round shot struck the head of our column. It was then realised that a large force of all arms, which, as later became known, was under the immediate leadership of the Nana Sahib, intervened between us and Cawnpore, and was occupying the villages D and E.

With much noise of drum and trumpet the rebels advanced, their infantry threatening our flanks round which their cavalry circled, and cut up our wounded and stragglers. When their 24-pounder gun opened fire our men lay down, but now, under the menace of the rebel movements, Havelock, whose horse had been shot, mounted a pony, and placing himself in front of the recumbent infantry gave the order, " The line will advance." At this moment four of Maude's guns made their appearance and were received with ringing cheers, and in obedience to the General's command the whole line moved forward. Maude was unable to get to the front in time to support the attack, but our infantry, though outnumbered ten to one, were not to be denied, and their adversaries broke and fled. They were in full retreat when Maude brought his four guns into action and turned retreat into a rout. The enemy lost about 250 men and 8 guns.[1] Our casualties reached a total of 108.

Occupation of Cawnpore.—Without food or baggage the victorious force bivouacked two miles from cantonments—too late, alas! to save the imprisoned women and children who were barbarously murdered by order of the Nana Sahib when it was known that our troops had crossed the Pandoo. On July 17 Havelock marched into Cawnpore. After blowing up the magazine in the Arsenal the Nana fled and never again personally encountered the British. On the 19th his palace at Bithur was pillaged and 20 guns brought into Cawnpore.

[1] These consisted of five 24-pounder guns and three 6-pounder guns.

Havelock crosses Ganges.—On July 20 the *Brahmapootra*, after an adventurous voyage, reached Cawnpore, and the two guns and most of the gunners were transferred to Maude's Battery. The same day Neill marched in with a small reinforcement, and, being appointed a Brigadier, was left in command at Cawnpore with 300 men, while Havelock with 1,500 men and 10 guns [1] marched to Mangalwar, six miles on the Lucknow side of the Ganges. He halted there for a few days, and resuming his march on the 29th, almost at once encountered the enemy at Unao.

Action of Unao, July 29.—The straggling houses of this village spread out for some distance to the right of the main road, while on the left of and behind the village were extensive swamps. After a frontal attack the enemy was driven from Unao, but when our troops debouched on the far side it was found that the rebels had rallied and that reinforcements were coming up on their right and left. These reinforcements were advancing in close formation on both sides of the main road and their movements were much hampered by the swamps. Unlimbering at 700 yards, Maude soon silenced the guns that were supporting their advance, and then, moving along the road to shorter range, and coming into action right and left, he threw some effective shells into the dense masses of their infantry, whose flanks were at the same time threatened by our skirmishers wading over the flooded ground. Leav-

---

[1] The extra guns were doubtless obtained at Cawnpore and manned as best could be arranged.

JULY 29, 1857] 1ST ACTION AT BASSIRATGUNGE 177

ing some 500 dead and wounded and 15 guns [1] behind them, the enemy retreated to Bassiratgunge, eight miles nearer Lucknow, whither, after a halt of two hours, Havelock followed them.

Capture of Bassiratgunge, July 29.—Bassiratgunge was a walled town surrounded by wet ditches;

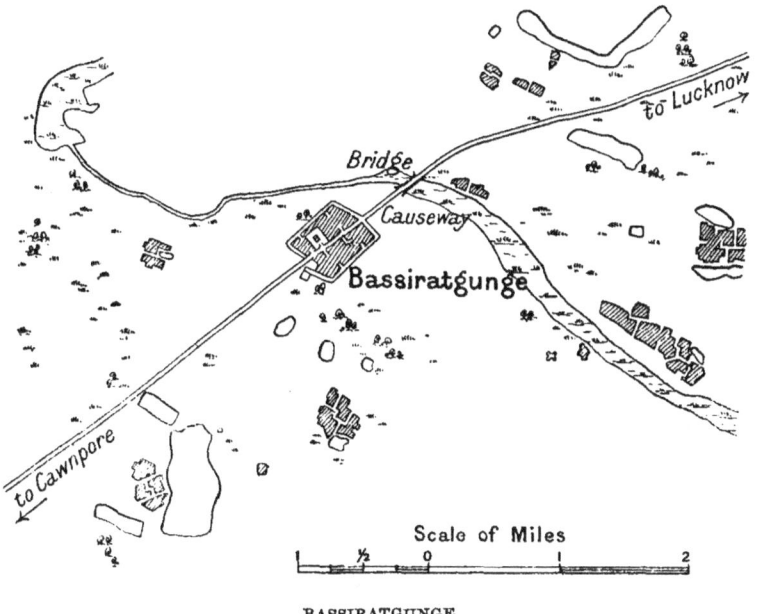

BASSIRATGUNGE

the gate was defended by a round tower with four guns, while the adjacent buildings were loopholed. Behind the town lay a wide and deep inundation crossed by a narrow causeway and bridge. Our guns were pushed to the front, supported by the infantry, and, in spite of a heavy cannonade directed against them, continued to gain ground. Then the 64th

[1] These included the guns of No. 13 F.B. 5/7 Ben.A., which mutinied at Fyzabad and fought at Chinhut.

Foot, advancing from our left and circling round the town, penetrated between it and the inundation, thus menacing the enemy's line of retreat, while the Madras Fusiliers and 78th Highlanders delivered a frontal attack, and Bassiratgunge fell into our hands. The rebels lost 400 men and 4 guns. During the 29th our casualties amounted to 12 killed and 76 wounded.

Retirement to Mangalwar.—Though always victorious in his conflicts with the mutineers, the wastage of battle and disease bore heavily on Havelock's forces, and now he had lost one-sixth of his small army. While his troops were resting on July 30 after their labours of the previous day, the news reached him that no reinforcements could at present be expected from Calcutta. He therefore retired to Mangalwar and sent for a company of the 84th Foot and two 24-pounder guns[1] from Cawnpore, while half of No. 12 F.B. 2/3 Ben.A. (Lieutenant W. Smithett) reached him from Benares.

2nd Action at Bassiratgunge, August 5.—Hearing on August 5 that the enemy had returned to Bassiratgunge, Havelock again moved against that place. A purely artillery combat ensued in which the guns drove the enemy out of their positions, which were then occupied by the infantry. The rebels were pursued beyond the causeway and bridge with a loss of some 300 men, but their guns were kept at such a distance and withdrawn so rapidly that they escaped capture. Our total loss was 2 killed and 23 wounded.

[1] This involved no increase of *personnel*. The guns were given to Maude.

Havelock then returned to Mangalwar, where he was joined by Major G. L. Cooper, Bengal Artillery, who assumed command of the artillery of the force.

Action at Boorhya-ki-Chowki, August 12.—On August 12 Havelock learnt that the enemy had returned to Bassiratgunge once more, and, on proceeding to attack them, he found them at Boorhya-ki-Chowki, a village on the main road, one and a half miles on the Cawnpore side of Bassiratgunge. They were soon driven back by combined gun and rifle fire and were pursued through and beyond Bassiratgunge. Their loss was about 300 men, ours 4 killed and 21 wounded. Two guns were captured. After the combat Havelock marched back to Mangalwar. His strength now amounted to only 1,415, with 335 sick. The relief of Lucknow had for the moment to be abandoned; but weak as he was, he still could strike.

Action at Bithur, August 16.—Learning that followers of the Nana Sahib were lingering around Bithur and that mutineers from various stations[1] had begun to swell their ranks, on August 16 Havelock determined to attack them. They numbered about 4,000, and occupied a position in front of the town, with the Ganges on their left, a marshy rivulet in their front, and on their right an unfordable watercourse which wound round their rear. The road to Bithur ran parallel to the Ganges at half a mile's distance, with bridges across the rivulet

---

[1] These consisted of mutineers of the 31st and 42nd Ben.N.I. from Saugor, the 22nd from Fyzabad, the 2nd Bengal Light Cavalry from Cawnpore, and the 3rd Bengal Irregular Cavalry from Saugor.

in front of, and the watercourse in rear of, the rebel position. A frontal attack was therefore a necessity. The two 24-pounder guns advanced parallel to the road, the remainder of Maude's Battery on the right,

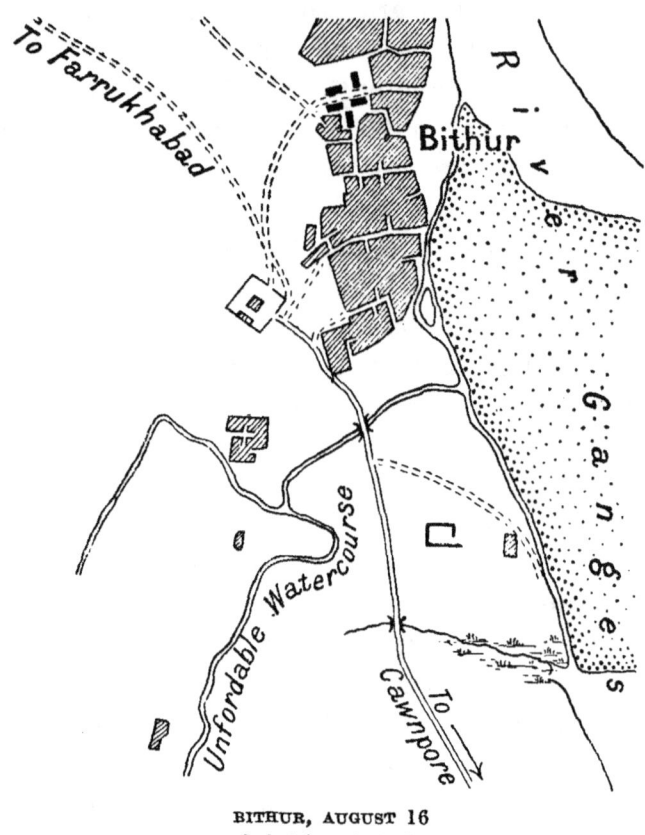

BITHUR, AUGUST 16
Scale 1 in. = 1·15 mile

2/3 Ben.A. on the left, and a brisk cannonade was opened on the enemy's position. The rebels had held their ground for an hour when the Madras Fusiliers, supported by the 78th Highlanders, penetrated their left. Their line was then speedily rolled up and they retreated across the watercourse bridge into Bithur. Had Havelock possessed but a squadron of cavalry, few rebels would have escaped. Our infantry following them up drove them through Bithur, inflicting a loss of 250 men and capturing 2 guns. Our casualties amounted to 8 killed and 38 wounded.

Havelock's Dangerous Position.—This was the last effort of Havelock's gallant force, which since it left Allahabad was never out of touch with the enemy and had fought nine engagements. On August 20, exclusive of detachments guarding the bridge of boats and keeping open communications with Allahabad, there were only 700 men fit for duty.

His position was one of great danger. At any moment the veil which hid the right bank of the Jumna might be lifted, and an army issue from Kalpi ; on the north-west the Nawab of Furruckabad waited with his levies, 30,000 strong, to take advantage of any difficulty which might beset him ; and it was in the power of rebels from Oudh to cross the Ganges below that city at any moment and threaten his communications with Allahabad. He was in pressing need of reinforcements, which would have reached him already had it not been for the untoward events on the lines of communication about to be related.

IV. THE LINES OF COMMUNICATION

Mutiny at Dinapore.—While Havelock with his little band of heroes was struggling with the impossible, the lines of communications were seriously threatened and the slender stream of reinforcements moving northwards was brought to a standstill. The menace of the three sepoy regiments at Dinapore had from the first been a source of great perplexity to the Governor-General; for though their disarmament was certainly desirable, it was thought it might provoke outbreaks in other places in the Bengal Presidency which were absolutely without protection. After delaying for the arrival of oversea reinforcements, the Government shuffled off the responsibility upon the shoulders of Major-General G. W. A. Lloyd, commanding the Dinapore Division, a meritorious officer who had become lethargic and infirm, and who, like many others, held an infatuated belief in sepoy loyalty. Although informed about the middle of July that he might take advantage of the passing of European troops through the station to disarm the sepoys if he thought it desirable,[1] and urged thereto by the stout-hearted Commissioner of Patna, Mr. W. Taylor,[2] for a while the General did nothing; but on July 25 he decided on a step which in his view would meet the necessities of the case.[3]

[1] On July 15 Sir P. Grant wrote to Lloyd as follows: "... If when the regiment (the 5th Fusiliers) reaches Dinapore you see reason to distrust the native troops, and you think it is desirable to disarm them, you are at liberty to disembark the 5th Fusiliers to assist you," etc. Letter is given *in extenso*, Kaye, iii. p. 90.

[2] Mr. William Taylor was a strong man, not afraid of responsibility, who did not scruple to strain the law of the land under the pressure of necessity. His actions, which have shocked certain writers, led the Government to remove him from his post. Nevertheless he saved Patna in 1857, and the exposure of the Wahabee conspiracy some years later fully vindicated him (Malleson, i. p. 121).

[3] The following is mainly based on the General's own letter to his brother given *in extenso* by Ball, ii. p. 121, etc.

On that day he ordered a parade of the European troops, including two companies of the 37th Foot, and whilst they stood to arms he sent some carts to the other end of the station (where the sepoys were quartered) to collect the percussion caps from the magazines in their lines; thus hoping to render them harmless without degrading them by removing their weapons. The sepoy regiments, however, when they saw the carts removing the caps, assembled in excited groups and broke out into mutinous cries.[1] On this, the senior officers of the European troops strongly advised the General to disarm the sepoys there and then; but Lloyd contented himself by ordering their native officers to collect the 15 caps which each man carried in his pouch. He then dismissed the European parade. Early in the afternoon the British officers who had gone to the native lines to see that the General's orders were obeyed found themselves surrounded by mutineers and had to fly for their lives. A parade was instantly ordered, but it was too late. While the " assembly " was sounding at the eastern end of the cantonments the sepoys slipped away from the western. The British infantry fell in and the bullocks were yoked to the guns,[2] but no pursuit worthy of the name was undertaken[3] and the rebels, avoiding the direct

---

[1] The 7th and 8th Ben.N.I. were by far the more demonstrative. The 40th remained steady, and, it is said, only joined in the subsequent flight when their lines were fired on from the hospital, which was situated between the European barracks and native huts.

[2] No. 11 F.B. 4/5 Ben.A. was a bullock field battery.

[3] There seems to have been some confusion as to the seniority of Major and Brevet-Lieutenant-Colonel W. Fenwick, 10th Foot, and Captain and Brevet-Lieutenant-Colonel A. Huyshe, Ben.A. Directions had

road to Arrah which ran along the Ganges, spread themselves out of range over the swampy ground lying to the south. Lloyd himself could not walk, and his horse (on which he had to be lifted) was not at hand; he therefore went on board a steamer with some riflemen, as he was fully possessed with the idea that the rebels would take the Arrah Road which ran within rifle range along the river. But they did not take the Arrah Road, and except that a few stray mutineers were shot and a boat or two run down the steamer effected nothing, and the European troops, having burnt the natives lines, returned to their quarters.

Kunwar Sing.—Meanwhile the mutineers made for the River Sone, and as it was in flood but few sepoys would have escaped if they had been followed up and caught between that river and the Ganges. By the evening of the 26th, however, they were safe in Shahabad,[1] where they were joined by Kunwar Sing, a Rajput noble of commanding influence, and they lost no time in marching for Arrah, the chief station of the district.

Mutiny at Segowlie.—The very day of the Dinapore mutiny the 12th Bengal Irregular Cavalry, in which the greatest trust was reposed, suddenly rose at Segowlie, murdered Major J. G. Holmes and other

been given to Fenwick by the General as to his course of action, should the native regiments at any time attempt to escape; but Huyshe, who had lately come to the station, was senior to Fenwick, and this was not realised by Lloyd until some days after the mutiny. See his letter to his brother above quoted.

[1] The magistrates of Chuprah had ordered all the boats to be collected and removed, but the subordinate to whom this task was confided fled without attempting to carry out the duty (Ball, ii. p. 122).

Europeans, and then dispersed throughout the country. Serious trouble now menaced the Government on both banks of the Ganges, and every prospective rebel in Behar was a-tiptoe with expectation. Patna, indeed, was firmly held in hand by the Commissioner, but General Lloyd spoke of entrenching himself,[1] and Dinapore might have shared the plight of Agra. Worse, our lines of communication might have been cut, and Havelock's gallant band might have been isolated, had it not been for a series of incidents initiated by the forethought of one man, and dominated by the vigour and energy of another.

Mr. Vicars Boyle.—Mr. Vicars Boyle, a civil engineer in the railway works at Arrah, began to put his house in a state of defence towards the end of May, and despite some ridicule provisioned it for a siege. As times grew more critical, Mr. Taylor sent 50 of Rattray's Sikhs[2] to the station, where 15 Europeans and Eurasians of the Civil Service and E.I. Railway remained at their duty. But resolute as these men were, and splendid fighters as they proved themselves to be, they all must have perished if the " little house at Arrah "[3] had not been prepared for defence. On July 27 the mutineers, having plundered the treasury and opened the gates of the jail, made for this same little house to murder its inmates,

[1] See the General's letter to Mr. W. Taylor quoted by Malleson, i. 76.

[2] These were a body of Sikh policemen raised in 1856 by Captain T. Rattray, 64th Ben.N.I.

[3] There were two houses in Boyle's compound. The smaller one was the better suited for defence. Among the defenders was one Mohammedan (a subordinate official).

but were received in a way that speedily taught them caution. Then began the siege, which kept the mutineers occupied during a critical week and prevented them spreading over Behar and raising fresh rebellion.

The Siege of Arrah.—Rattray's Sikhs proved staunch and loyal despite the repeated temptations and threats of their co-religionists in the rebel ranks; many of the civilians were excellent rifle shots, and the little garrison under the command of Mr. H. Wake, the magistrate, defended itself with the greatest spirit and success. The rebels surged round the little fort in overwhelming numbers, for reinforcements of all kinds joined them;[1] they rained bullets on its loopholes, they brought up guns to batter its walls, they sought to overcome the garrison by lighting fires on which chillies were thrown, and by leaving carcasses of animals to rot beside the house; but Boyle's defences defied their efforts, and the wind blew the stifling smoke and poisonous fumes into their own camp. Then they tried mining; and although the garrison at once replied by countermines, success might have at last been achieved by them[2] if on August 2 Arrah had not been relieved.

Defeat of Dunbar's Column.—The news of the attack on Arrah reached Dinapore the same day,

[1] A great many sepoys on leave or on pension joined Kunwar Sing. In the battle subsequently fought on August 2 the sepoys of nine regiments were found amongst the killed. Two guns were brought from Jugdispore by Kunwar Sing, and one of them was mounted on the roof of the larger house in Boyle's compound.

[2] When Arrah was relieved it was found that the enemy's mine had been pushed forward close to the house, and a canvas tube full of gunpowder was found lying ready beside it.

and, after some abortive attempts had been made, Captain C. Dunbar, 10th Foot, landed on July 29 with a force of 450 men in the Sone Delta. Two miles on the road to Arrah a branch of the Sone had to be crossed, and while a detachment of Sikhs went to secure the necessary boats the dinners of the troops were cooked. The Sikhs met with some opposition and were in need of slight reinforcements, but at the sound of the firing Dunbar's whole force was ordered forward, leaving the dinners behind. Resistance being brushed aside, the march was resumed, and by 10 p.m. Arrah was three miles distant still; yet Dunbar determined to push on, though the moon was setting, because his men were footsore and starving and carried no food with them. Sikh skirmishers had hitherto led the way, but they were now drawn in and marched in close formation 200 yards in advance of the main body. Presently, as the troops in their conspicuous white clothing were passing through a wood, a withering fire was opened on them in the darkness, and for a time the utmost confusion ensued. Dunbar was among the first to fall, and at least 50 of our men were killed before the rest collected together under the shelter of a high bank, where they passed the night. At dawn it was apparent that they were in the presence of 5,000 rebels, and a disastrous retreat of twelve miles was begun. When at last the boats were reached it was discovered that hostile hands had been busy; oars had been removed, the rudders were tied up, and the boats themselves hauled up out of the water. Again a scene of great confusion arose, and when the British reached the steamer there were but 250 survivors, of whom nearly a hundred speedily died of their wounds and exposure.[1]

Vincent Eyre.—But Arrah was not destined to fall; and even as the remnant of the stricken force were steaming down to Dinapore help was coming from another direction. It so happened that Major V. Eyre with the men and guns of No. 3 F.B. 1/5 Ben.A. was on the afternoon of July 25 passing by river through Dinapore on his way to Allahabad, and, at General Lloyd's order, he landed 3 guns until 4/5 Ben.A., gone in pursuit of the rebels, should return. He was therefore aware of the mutiny of the three regiments, and when he reached Buxar on July 28 a rumour reached him that they were marching towards that station by way of Arrah. Eyre therefore ordered his steamer to halt, in the hope that some infantry would soon overtake him. But as none came, he next day steamed up to Ghazipore, where a detachment of the 78th Highlanders were detained to watch the 65th Ben.N.I. Having landed 2nd Lieutenant D. W. Gordon and 2 guns, and having obtained a reinforcement of 25 Highlanders, Eyre steamed back to Buxar, where Captain F. W. L'Estrange and a detachment of the 5th Fusiliers had meantime arrived. On his own responsibility Eyre took L'Estrange under his command, and, ordering the Highlanders to return to Ghazipore on the first opportunity, he set out on the

[1] Messrs. R. L. Ross Mangles, and W. F. McDonnell of the Bengal Civil Service, who accompanied Dunbar as volunteers, and Private D. Dempsey of the 10th Foot all received the Victoria Cross for gallantry during the retreat.

30th towards Arrah. As the drivers and horses of his battery were on their way by route march from Dum-Dum, he obtained plough oxen for two 9-pounder guns and one 24-pounder howitzer, carrying his ammunition in country carts. Having been joined by 13 mounted volunteers, his total force was 220.

Action at Bibigunge, August 2.—When the news reached him of Dunbar's disaster and of the plight of Arrah, he pushed on with all speed, and on the morning of August 2 was confronted by the rebels a mile beyond Goojeragegunge, where they occupied the woods in front and on both flanks. But the guns and the Enfield rifles of the 5th Fusiliers speedily drove them back to Bibigunge, where they had broken the bridge over a branch of the Bunas River and protected the approaches with earthworks. Covering the movement by the fire of his guns, Eyre then wheeled to his right and made for the railway embankment which at about a mile's distance ran parallel to the road and clear of the branch of the Bunas which passed through Bibigunge. But the enemy conformed to this movement, and when our line was formed to the original front, the right resting on the embankment and the left occupying a small wood 300 yards from it, Eyre was hard put to it to hold his own. On two occasions determined attempts to rush the guns were only repelled by case shot; but after an hour's fighting our skirmishers turned the right flank of the enemy, and Eyre, having prepared the way with a rapid fire of case shot and shrapnel, ordered the Fusiliers forward with the bayonet and the rebels broke and fled.

Relief of Arrah.—The majority made for Arrah, but during the night their flight was continued and on August 2 none remained in that station. Wake's gallant garrison was thus relieved, and on the 3rd was joined by Eyre, who was delayed by the necessity of bridging the main stream of the Bunas.

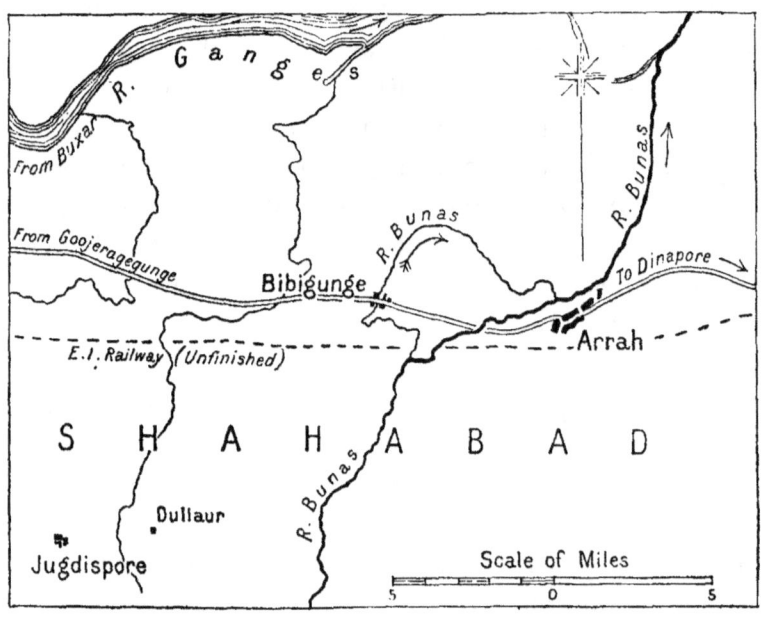

BIBIGUNGE AND ARRAH

The brilliant victory of the previous day had been gained at a cost of only 16 casualties over a force containing at least 4,000 trained soldiers, and our prestige, lost by Dunbar's disaster, was completely restored.

Capture of Jugdispore.—Kunwar Sing retreated to Jugdispore, where some 3,000 sepoys in addition to his own retainers collected. Eyre remained at Arrah until August 11, and then, having been reinforced by a detachment of the 10th Foot, and the Sikhs of the garrison, he marched on Jugdispore with a force of 522. Next day he early encountered the enemy, and after skirmishing for some hours pressed them back to Dullaur, an entrenched village where they attempted to make a stand. Driven from this, they fell back on Jugdispore, which, after a running fight through a mile and a half of thick jungle, was occupied and two guns captured. Kunwar Sing and the sepoys fled southwards, and after some wanderings made for Banda, where they were made welcome by the Nawab. They had lost during the fighting 300 killed; the British casualties were 6 men wounded. Shortly afterwards the field force was broken up and Major Eyre with 1/5 Ben.A. proceeded on his way to Allahabad.

Help from Nepaul.—When the mutiny at Segowlie occurred, a division of Gurkhas was close to the frontier; for the Governor-General had accepted the offer of active assistance from Nepaul, tendered to him by Jung Bahadur[1] shortly after the Meerut outbreak. On July 30 the Gurkhas entered Gorruckpore, and the detachments of the 17th Ben.N.I. and 12th Irregular Cavalry still at that station were disarmed.[2] They then went on to Azimghur, where they halted, Gorruckpore being temporarily abandoned.

Trouble on the Grand Trunk Road.—Road and

---

[1] Jung Bahadur was the all-powerful Minister of the Ruler of Nepaul.
[2] When the main body of the 12th Bengal Irregular Cavalry mutinied at Segowlie the detachment at Gorruckpore remained quiet. Eighty-three of them eventually joined Outram on his way to Cawnpore, and did good service. Certain officers of disarmed or mutinous regiments of Ben.N.I. were attached to the Gurkha force.

river were now both fairly secure between Allahabad and Dinapore, but trouble arose eastward as a sequel to the mutiny at the latter station. About the end of July the detachment of the 8th Ben.N.I. at Hazaribagh followed the example of its head-quarters and was soon joined in mutiny by the Ramgurh Battalion;[1] the electric wires were cut on the Grand Trunk Road and the whole of Chota Nagpore was in a ferment. Hazaribagh remained in the hands of the rebels till the middle of August, when the Commissioner of Chota Nagpore, Captain Dalton, with the aid of the Rajah of Ramgurh and Rattray's Sikhs, restored order in the station and reopened the road. But the mutineers were not followed up and punished until they suffered a severe defeat at Chuttra on October 2.[2] On that date Major English with 180 men of the 53rd Foot and 200 of Rattray's Sikhs encountered over 3,000 rebels, and after an hour's fighting captured 4 guns and a quantity of stores.[3] All serious danger from Chota Nagpore thereafter disappeared, but the district remained in a disturbed state for many months.[4] The Grand

[1] The Battalion was maintained by the loyal feudatory Rajah of Ramgurh.
[2] At this time Carthew's Brigade of Madras troops arrived by sea at Calcutta, and Lieutenant-Colonel Fischer was marching with others by land from Cuttack. It was at first proposed to confide the pacification of Chota Nagpore to them, but eventually a wing of the 53rd, under Major F. English, was sent there from Fort William.
[3] English had 5 killed and 37 wounded. Captain J. H. Smyth, Ben.H.A., acted as his staff officer; he was absent from his troop (5/1 Ben.H.A.) when the Mutiny broke out.
[4] Rattray's Sikhs took the greater part of the duty of keeping order; they were helped at times by the Shekwattee Battalion from Midnapore, and by a detachment of the 13th Foot from Benares.

Trunk Road was also threatened from the north when the 5th Bengal Irregular Cavalry mutinied at Rohni, and, after a vain attempt to raise the 32nd Ben.N.I. at Baosi, set off to join Kunwar Sing.

Action of the Government.—But the Government was beginning to appreciate sepoy loyalty at its true value, and its hands were now strengthened by the arrival of reinforcements. The Hindustani regiments were all disarmed throughout the Bengal Presidency before the close of the year,[1] with the exception of the 73rd Ben.N.I. at Jalpagorie, under the steadfast Colonel G. M. Sherer.

Major-General Lloyd was removed from the command of the Dinapore Division, which, combined with that of Cawnpore, was placed under Major-General Sir James Outram, appointed Chief Commissioner of Oudh.

Sir James Outram.—Outram arrived in Calcutta on August 1, reached Allahabad on September 2, and marched for Cawnpore on September 5 with the 5th Northumberland Fusiliers, detachments of the 64th Foot and 1st Madras Fusiliers, and No. 3 F.B. 1/5 Ben.A. (Major V. Eyre). He was joined on the way by the loyal remnant of the 12th Irregular

[1] The Governor-General's Bodyguard was disarmed at Calcutta; the 63rd Ben.N.I. and the 11th Bengal Irregular Cavalry at Berhampore; the 32nd Ben.N.I. at Baosi; the 65th Ben.N.I. at Ghazipore, and a certain number of the golandauz of the 9th Batt. Ben.A. There were also two mutinies not mentioned in the text; namely, that of the detachment of the 34th Ben.N.I. at Chittagong, and that of the golandauz and a detachment of the 73rd Ben.N.I. at Dacca. The mutineers, with the exception of a portion who managed to reach the north-eastern districts of Oudh, were driven into Nepaul. At Dacca Lieutenant W. Dowell, Ben.A., was in charge of two post guns, and did good service.

Cavalry from Azimghur. Learning, as he approached Futtehpore, that rebels from Oudh had crossed the Ganges with intent to sever his communications, he sent Major V. Eyre to operate on his right flank with 150 infantry, 2 guns of 1/5 Ben.A. (Lieutenant D. W. Gordon), and a detachment of the 12th Irregular Cavalry. On the 11th Eyre reached Koondum Puttee and came up with the rebels, some 300 strong, who at once fled to their boats, where they defended themselves till Gordon opened fire with his two guns. Then, in a panic, they threw themselves into the river and were nearly all killed by case shot and musketry fire. These 300 were apparently the advanced guard of a larger force, whose object was to overrun the Doab, but this sharp defeat put an end to the movement and Outram's march was henceforth uninterrupted and his communications unmolested.

V. THE ADVANCE ON LUCKNOW

Outram's Self-negation.—Outram, having left a small garrison at Futtehpore, reached Cawnpore on September 15, and naturally General Havelock and the Field Force fell under his command. The bulk of the troops at Cawnpore were about to march on Lucknow and Outram meant to accompany them, but by a noble act of self-negation—cordially approved and sanctioned by the Commander-in-Chief and the Governor-General—he waived his military rank, allowed Havelock to command the advance, and determined to serve as a volunteer until Lucknow should be reached.

Havelock's Force.—The force now at Havelock's disposal was as follows :

| European Vol.C. 12th Ben.Irr.C. | 1st Infantry Brigade (Brigadier J. G. Neill): 5th Foot (Northumberland F.). 64th Foot (2 Companies). 84th Foot. 1st Mad.E.F. 2nd Infantry Brigade (Brigadier W. Hamilton)[1]: 78th Highlanders. 90th Foot. Ferozepore Regiment. | Artillery Brigade (Brigadier G. L. Cooper, Ben.A.): 3/8 R.A. (2nd Captain F. C. Maude). No. 12 F.B. 2/3 Ben.A. (Captain W. Olpherts). No. 3 H.F.B. 1/5 Ben.A. (Captain and Brevet-Major V. Eyre).[2] |

	Cavalry	
European	109	
Native	59	
		168
	Infantry	
European	2,388	
Native	341	
		2,729
Artillery		282
Total		3,179

Capture of the Alam Bagh, September 23.—Leaving Colonel N. Wilson[3] with the head-quarters of the 64th Foot, and Lieutenant Harward with some invalid gunners at Cawnpore, Havelock crossed the Ganges on September 21, by a newly constructed bridge of boats, and then, sweeping the rebels from his path at Mangalwar, he reached the River Sye on September 22. Happily the Bunnee Bridge was found intact and a royal salute was fired to warn

[1] Major and Brevet-Lieutenant-Colonel W. Hamilton, 78th Highlanders.

[2] Eyre's Battery was now armed with 18-pounder guns obtained at Cawnpore.

[3] Lieutenant-Colonel and Brevet-Colonel 64th Foot.

the beleaguered garrison that relief was at hand. Next day the enemy were encountered in a strong position across the road to Lucknow. Their left occupied the Alam Bagh and their centre and right held a line of hillocks on the west of the road. Movement off this road was for a time found impossible on account of the swampy nature of the ground, but as our troops pressed onwards to the attack, drier ground was met with. Then a wide outflanking movement to our left was carried out under the protection of Olpherts' Battery and the Volunteer Cavalry, while Maude and Eyre, whose heavy guns were especially useful, came into action on the right. Driven from the Alam Bagh by a charge of the 5th Fusiliers and 78th Highlanders, the enemy were soon in full flight over the Char Bridge, whither they were pursued by Olpherts' guns and the cavalry. On September 24 Havelock halted at the Alam Bagh, where his field hospital, baggage, and stores were to remain protected by Lieutenant-Colonel C. C. M'Intyre,[1] 78th Highlanders, with 300 men of his regiment and 4 guns detached from Olpherts' and Eyre's Batteries.

Capture of the Char Bridge, September 25.—Next day at 8 a.m. the advance on the Residency began. Neill's Brigade, headed by two companies of the 5th Fusiliers followed by Maude's Battery, led the advance along the direct but narrow road to the Char Bridge, Outram riding with Maude alongside the leading gun. The enemy occupied a position about a mile from the Alam Bagh, their centre

[1] Major and Brevet-Lieutenant-Colonel.

posted in some walled enclosures and their right and left occupying the fields on each side of the road. Our column had hardly passed through its advanced pickets, when it was assailed by the close fire of musketry and of two guns well placed a little in rear of the Yellow House. Then, for some reason

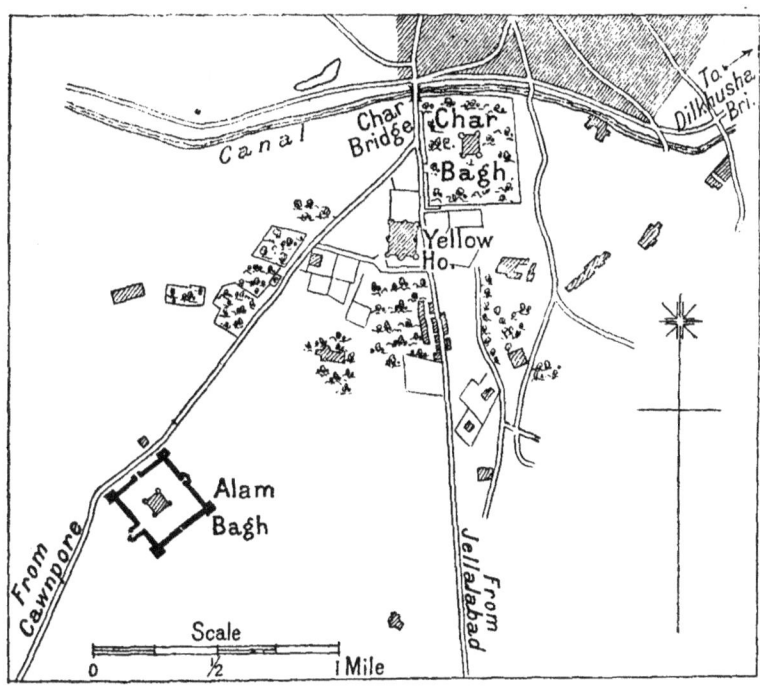

THE CHAR BRIDGE, SEPTEMBER 25

that has never been fully explained, it halted, the infantry lay down on each side of the road, and our two leading guns were at once brought into action upon it. Maude's men immediately began to fall. In a few minutes he lost his sergeant-major and senior sergeant, but as the men in the leading detachments were struck down they were replaced by volunteers from the other guns. Several times he appealed to Outram, who himself was wounded in the hand, to counsel a forward movement, but, as the order to halt had come from Havelock, he refused to interfere. About 9 a.m. a staff officer brought the welcome order to advance, and as our infantry pressed forward the enemy fell back and the guns at the Yellow House, which had been admirably served, were moved to the enemy's left flank. Then the column advanced and reached a point where the road turned sharply to the left. Two hundred yards in front was the Char Bridge, behind which guns were entrenched, flanked on either side by loop-holed houses. The Madras Fusiliers were sent to the left to occupy the banks of the canal, while the 5th Fusiliers wheeled to the right to drive the rebels from the Char Bagh. Our leading guns came into action at the angle of the road—2 9-pounders in the open—and at once encountered the full force of the rebel battery, which consisted of 5 guns, 2 being 24-pounders. Maude had already lost 21 men, and now the first rebel salvo disabled a gun detachment. But eager volunteers from the 84th Foot came forward, and the two 9-pounders, although they did not silence the enemy's guns, succeeded in keeping their fire down.[1] A hot combat raged for half an hour and then the 5th Fusiliers and the 84th Foot carried the bridge with the bayonet and stormed the entrenchment, the rebel gunners sticking to their guns to the last and dying beside them. The bridge, however, had hardly been won when the

---

[1] Maude received the Victoria Cross.

British were attacked by artillery fire from their right rear—the two guns from the Yellow House were again in action. Thereupon the 90th Light Infantry, taking ground to their right from their place in the column, charged with the bayonet and captured both, and Olpherts, who accompanied the attack, having galloped back to his battery and brought forward a couple of spare limbers under a heavy fire, carried them off in triumph.[1]

Advance to the Moti Mahal.—A further advance was now possible. The direct road to the Residency by Ganeshgunge was known to have been broken up and barricaded, and the houses on either side of the street loopholed. Havelock therefore adopted the more circuitous route along the left bank of the canal, while the 78th Highlanders and the Sikhs were ordered to hold the Char Bridge until all the columns had passed by. Some difficulty was met with in getting Eyre's heavy guns along, but no opposition was encountered, as the march was continued to a point opposite the Dilkhusha Park wall and then directed on the Sekundra Bagh. At this place the column turned sharply to the left and, moving across the plain, made for the Moti Mahal, but until its shelter was reached the troops were under a heavy fire. The Khurshid Manzil[2] and an entrenchment made in front of Sadat Ali's Mosque were occupied by the rebels, and as our troops moved across the plain they came under a heavy but not very effective fire;

---

[1] Olpherts received the Victoria Cross.

[2] The Khurshid Manzil was used by the officers of the 32nd Foot as their mess. It is often referred to as the Mess House.

while our batteries, unlimbering in turn, fired on guns posted in the entrenchment and on the left bank of the Gumti. Considerable anxiety existed as to the fate of the rear guard until about 3 o'clock, when it was seen to be fighting its way along Hazratgunje.

Action of the Rear Guard.—Immediately after the main column had passed by, the Sikhs and Highlanders at the Char Bridge who formed the rear guard were sharply attacked by the enemy advancing down Ganeshgunge and the adjacent streets. After a combat of several hours' duration the rebels were repulsed and their guns thrown into the canal; but touch with the main body was lost, and when the rear guard followed it up, instead of marching on the Sekundra Bagh, it turned up Hazratgunje and engaged in a strenuous street fight.

The Advance to the Bailey Guard.—Having halted at the Moti Mahal to allow the main body to recover its formation, Havelock, now that the rear guard was in sight, determined to push on to the Chattar Manzil. Lieutenant-Colonel R. P. Campbell, with 100 men of the 90th Light Infantry, and two of Eyre's guns, under Lieutenant J. McK. Fraser, was left behind in charge of the wounded and baggage in the Moti Mahal, and the main body continued its march across fairly open ground. At the same time the rear guard was close to Sadat Ali's Mosque, where a rebel battery was in action. This battery was at once attacked, a gun spiked, and the rebels driven from that particular spot; but they still occupied the buildings of the Kaisar Bagh and kept up a heavy

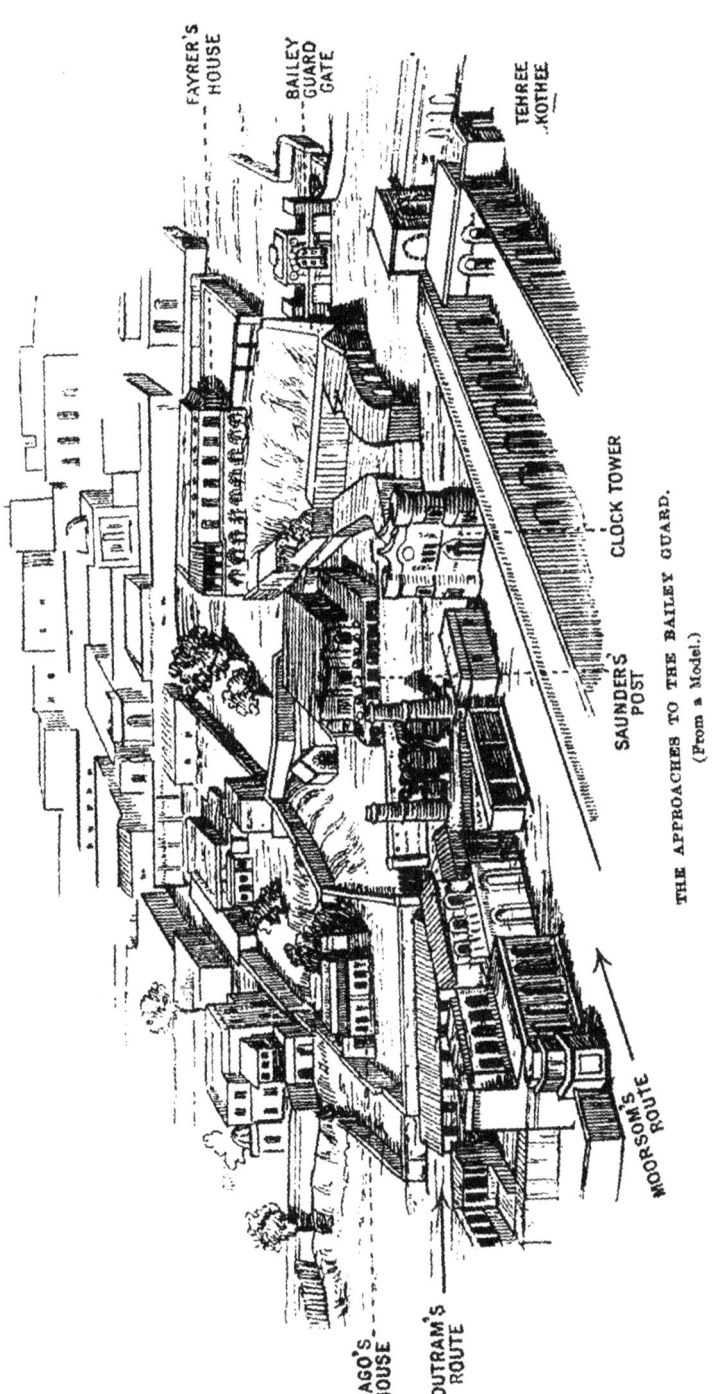

THE APPROACHES TO THE BAILEY GUARD.
(From a Model.)

fire of guns and musketry on both the rear guard and main body.

A short and narrow lane, running along the south wall of the Chattar Manzil Garden led to a large courtyard since known as Dhooly Square. Once within its precincts the main body was sheltered from fire, for the Chattar Manzil was unoccupied by the rebels; and now another pause was made while a staff officer, Lieutenant Moorsom,[1] attempted to find a way through the palaces—the Chattar Manzil, Farha Baksh, etc.—which would avoid the open streets. Meanwhile the rear guard preserved its original line of advance, and the Highlanders, arriving opposite the south entrance of Dhooly Square, became the head of the column. But the halt in Dhooly Square was short. Outram, indeed, wished to wait for Moorsom's report, but Havelock determined to go on at once, and he and Outram, placing themselves at the head of the 78th Highlanders, led the way to the top of the Khas Bazaar. Neill followed with the main body, but was killed close to the archway spanning the southern entrance of the square, where Olpherts had already brought a gun into action to fire on the Kaisar Bagh. So long as they were in the street running past the south of Dhooly Square, the British were assailed in rear by the guns and muskets of the Kaisar Bagh; but when they turned to the left down the Khas Bazaar they were met with both flanking and frontal fire. Highlanders and Sikhs, however, dashed impetuously forward for 200 yards down this street, and then,

[1] Lieutenant W. R. Moorsom, 52nd Foot.

guided by Outram up the long curved road that led direct to the Bailey Guard, they fought their way to the Residency and the two Generals rode in to Aitken's post. Meanwhile Moorsom, who had been unsuccessful in his search for a road through the palaces, overtook the column and guided the main body across the Paeen Bagh through the Clock Tower gate to the Bailey Guard. The rebel battery at the

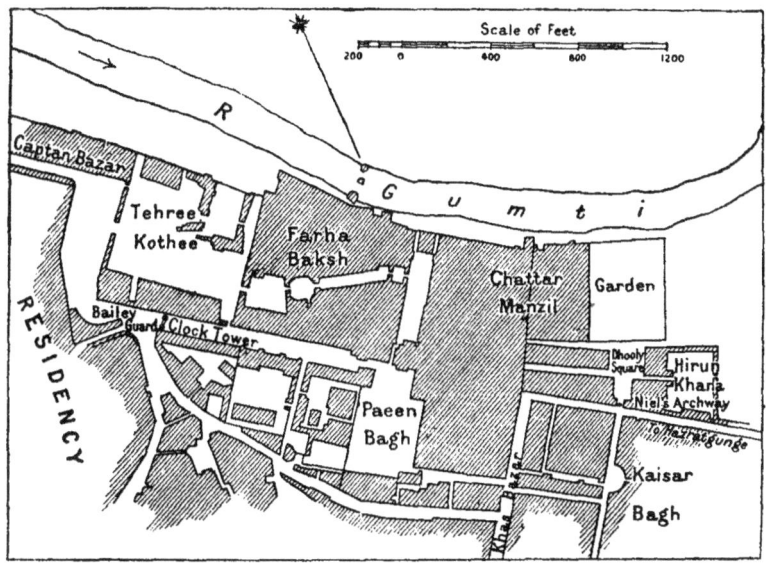

THE APPROACHES TO THE BAILEY GUARD.

Clock Tower was thus taken in reverse, and the sepoys swung round one gun and fired an ineffective shot before they fled. That night the whole of Havelock's force entered the Residency with the exception of Colonel Campbell's party, Fraser's two guns, the sick and wounded, and the stores and baggage.

Final Movements. — Early next day, September 26, a party of infantry under Lieutenant E. R.

Simmons, 5th Foot, occupied Martin's House, to give support to Colonel Campbell, whose position in the Moti Mahal was full of danger. But a strong body of the enemy held the Khurshid Manzil and it was necessary to reinforce Simmons. Later, Colonel R. Napier, Ben.E., made his way with a detachment of infantry along the banks of the river by a sheltered route which, when darkness came, would serve for Campbell's withdrawal to the Residency. Meanwhile one of Fraser's guns was out in the open where it had been in action and whence it had been impossible to withdraw it on account of the musketry fire from the Khurshid Manzil. Anxious for its safety, Brigadier Cooper and Lieutenant C. W. Crump, his adjutant, accompanied Napier and Captain W. Olpherts, who had a plan for its recovery. Private Duffy, Madras Fusiliers, crept out at dusk and attached a drag rope to the trail of the gun, a 24-pounder, and the piece was pulled into safety by a team of gun bullocks which Olpherts had brought with him, but unfortunately both Cooper and Crump were killed while the operation was being carried out.[1]

During the night Campbell's detachment, the guns, and the stores were all safely transferred to the Residency; but the sick and wounded, in a long string of dhoolies, were brought by mistake into the square where Neill had been killed, and falling into the hands of the rebels were ruthlessly put to death.

The Relief.—The Residency was now reinforced by about 2,000 men, and Outram, resuming his military rank, took supreme command, Havelock

[1] Private T. Duffy received the Victoria Cross.

commanding the relieving field force and Inglis the original garrison. Active operations were at once undertaken. The Captan Bazaar, which harboured the enemy in vexatious proximity to the Residency, was cleared, and the position extended to include the Farha Baksh and Chattar Manzil, while successful sorties destroyed some of the nearest rebel batteries. At the same time Colonel M'Intyre maintained his position at the Alam Bagh, and preserved touch with Cawnpore. A careful survey of the garrison's food supply revealed the fact that Havelock's reinforcement would cause no present anxiety on that score, while it removed the dangers of being overwhelmed by superior numbers. The gallant defenders of the Residency could show a bold front to the enemy and calmly await the relief that now would make no tarrying.

DHOOLY SQUARE, WHERE NEILL FELL, SEPTEMBER 25, 1857

CHAPTER V

THE LUCKNOW CAMPAIGN—*continued*

I. ARRIVAL OF SIR COLIN CAMPBELL

The New Commander-in-Chief.—When the extent of the Mutiny was realised in England, Lieutenant-General Sir Colin Campbell, G.C.B., was appointed Commander-in-Chief in India and arrived at Calcutta on August 13. It was a time of deep anxiety. Delhi still defied us; Lucknow was closely besieged; Agra was isolated; anarchy raged in Central India and Rohilkhund; danger was deepening in the Punjaub and Rajputana; Havelock's relieving force lay exhausted at Cawnpore, and revolt in Behar threatened to cut off Calcutta from Allahabad. But it was the darkest hour before the dawn. In a few weeks the British emerged victorious from the crisis; for though much labour and peril were yet to be encountered, the vital struggle of the Indian Mutiny was over when Delhi had fallen and Lucknow was relieved.

Preparations for an Advance on Lucknow.—For more than two months Sir Colin was detained in Calcutta by the general dearth of all the munitions of war. Horses were purchased, guns were cast, Enfield rifle ammunition was accumulated, stores collected, and the bullock train service along the

Grand Trunk Road was improved and developed. As these works proceeded reinforcements poured into India. The Naval Brigades from H.M.S. *Pearl* and *Shannon* were landed at Calcutta in August, and following on their heels came the troops diverted from China and others from the Cape of Good Hope. Slowly but surely they all were moving to the front.

Menace to the Grand Trunk Road.—It has been told how Eyre's successful operations cleared the way for Outram's advance, and how a Gurkha force had been established at Azimghur; yet there still were dangers on both sides of the Grand Trunk Road. When Gorruckpore was evacuated in August it became a rallying-place for rebels from Oudh ; and though on three occasions [1] they were driven back over the frontier by the Gurkhas, the country lying between the lower waters of the Gumti and the Gundak was open to invasion until Jaunpore and Gorruckpore had been occupied in force by British troops [2] and Jung Bahadur's reinforcement had arrived from Nepaul. But it was on the south side of the Grand Trunk Road that the real peril lay. The Nawab of Banda was now able to maintain the independence he had assumed, and mutineers flocked to his territory. During September the Dinapore Brigade, the 5th Bengal Irregular Cavalry, and Kunwar Sing and his followers arrived at his captial and at any moment an attack might be made on our lines of communication. Such an attack was actually

[1] At Mandori, ten miles from Azimghur, September 19 ; Kudua, near the Oudh frontier, October 19 ; Chanda, October 30.
[2] The field forces under Brigadier Franks and Brigadier Rowcroft (see p. 247).

carried out at the end of October when the 7th and 8th Ben.N.I. and other rebels crossed the Jumna and advanced on Futtehpore.

Action at Khujwah, November 1.—On the afternoon of October 31 Lieutenant-Colonel T. S. Powell, 53rd Foot (escorting the heavy guns of the Naval Brigade) was halted eleven miles from Futtehpore, when news of the rebel advance reached him. Though he had just made a twelve-miles march, he at once pressed on to Futtehpore; and having been reinforced by troops at that station, he set out at dawn on November 1 for Khujwah, about twenty-four miles distant, where the rebels were reported to be. His force now consisted of the *Shannon* Naval Brigade [1] (Captain W. Peel, R.N.), with two 9-pounder guns, a wing of the 53rd Foot, detachments of the Royal Engineers, 64th Foot and 93rd Highlanders, and two guns of No. 20 F.B. 3/5 Ben.A. (Lieutenant P. C. Anderson), in all some 500 men. Early in the afternoon Powell came in contact with 4,000 rebels, half of them sepoys in their uniform, with two 16-pounder guns and one 1½-pounder gun.[2] They were in position in front of Khujwah, across the Futtehpore Road, down which they opened fire with round shot when the British appeared. Their right occupied a long line of sand hills, and their left, somewhat refused, was on higher ground on the other side of the road. When the rebel guns opened fire, skir-

[1] There were six 64-pounder guns, two 24-pounder howitzers, and two 9-pounder guns with the Naval Brigade. Only the 9-pounders accompanied the march to Khujwah.

[2] It was fortunate that this force was not accompanied by the Irregular Cavalry, which would have proved very useful to the rebels.

mishers from the Royal Engineers and 93rd Highlanders advanced through the cornfields which lay on either side of the road; and the 53rd, edging off to their left and pressing forward in column with their Colonel at their head, charged and captured the 16-pounders, Powell being killed in the moment of victory.[1] At the same time Lieutenant E. Hay, R.N., with a party of seamen and marines, and the two naval 9-pounders, established himself on the enemy's left. The rebels now threw their right forward, threatening to take us in rear; and Captain Peel, who succeeded Powell in the command, realising that the British (who had covered seventy-two miles in three days) were unequal to a protracted fight, resolved to bring matters to a conclusion. Hastily collecting fresh troops from the supports, he led them against the enemy's centre, and turning the flank of the sandhills he rolled up their right wing, while Hay steadfastly maintained his position on their left. The rebels were completely broken, and, leaving 300 dead on the field, they fled back to Banda. We had about 90 casualties, including Colonel Powell and 22 N.C. officers and men killed.

Hopes of the Mahrattas.—But more serious danger than that arising from wandering bands of sepoys was known to lurk behind the Jumna, which screened all rebel movements from our eyes. The first result of the capture of Delhi was to increase the number of our mobile enemies, for the sepoys who escaped

[1] One of these guns was taken by Powell's adjutant, Lieutenant J. M. Macneill, who with his own hand killed several men who were working it.

from that city reinforced the rebels on both sides of the Doab. Moreover, when the Imperial City fell the star of the Moghuls set for ever, and the Mahrattas saw their opportunity in the south, while Oudh and Rohilkund occupied our attention in the north. It has been told how the Gwalior Contingent, restless and discontented, still remained in the neighbourhood of Sindiah's capital; but now, roused by the fall of Delhi, they yielded to the reiterated persuasions of the Begum of Jhansi and rallied to the standard of the Peshwa, which she had raised in company with the Rao Sahib, the Nana's nephew. Numerous adherents joined their ranks, for the Peshwa was still a name to conjure with. An acceptable General was found in Tantia Topi, who was soon at the head of an army containing a large proportion of trained troops, and the possession of the bridge at Kalpi gave him a strategical position which was a standing menace to our lines of communication.

Sir Colin goes to the Front.—But approaching reinforcements would increase Sir Colin's active army to a strength of over 6,000 men, and the relief of Lucknow was imperative both for military and humanitarian reasons. He therefore determined to march at once upon the capital of Oudh, leaving a containing force behind him at Cawnpore under Major-General C. A. Windham.[1] Meanwhile Brigadier Grant, reaching Cawnpore, October 26, and being reinforced by troops from the south,[2] crossed the

---

[1] The officer who led the stormers at the Redan, September 8, 1855.
[2] Among these was 4/5 R.A. (Captain W. N. Hardy), which arrived at Cawnpore on October 28, and on the 31st marched out in company

Ganges on the 30th, and after a successful skirmish with the enemy at Bantara he brought away all the sick and wounded from the Alam Bagh. On October 27 Sir Colin[1] left Calcutta, and on November 9 rode into the camp at Bantara.

II. THE BRITISH FORCES IN OUDH AND AT CAWNPORE

The British forces in Oudh and at Cawnpore as they existed in November 1857 were as follows:

1. CAWNPORE GARRISON (Major-General C. A. Windham)

                 Detachment, *Shannon* Naval Brigade
64th Foot.   Bullock Field Battery.[2]
Details of other Regiments.
                 Total: 500 men and 4 guns.

Note.—Reinforcements of 1,200 men were close at hand.

2. ALAM BAGH GARRISON (Major and Brevet-Lieutenant-Colonel C. C. McIntyre)

75th Foot.[3]   Detachment 1/5 Ben.A. (Lieutenant D. W. Gordon).
                 No. 20 F.B. 3/5 Ben.A. (Captain G. Moir).

with a small mixed force to join the camp at Bantara, with two 24-pounder guns drawn by elephants. There were one or two skirmishes on the road and the 24-pounders were useful in battering down village towers, etc.

[1] Sir Colin and his staff travelled in *dak gharris* (post carriages) to Allahabad. Twelve miles beyond Shergati, fourteen elephants laden with sepoys and a party of sowars were seen crossing the Grand Trunk Road. They were men of two companies of the 32nd Ben.N.I. on detachment at Deoghur, who mutinied on October 9, stole some elephants and made off westwards. Luckily Sir Colin's carriages were not perceived, and as a bullock train with British infantry was close at hand, the Commander-in-Chief escaped a very great danger.

[2] This was an extemporised battery. It consisted of four 9-pounder guns manned by Sikhs and invalids Ben.A.

[3] See note p. 213.

3. LUCKNOW GARRISON (Major-General Sir James Outram)

Residency Garrison
(Brigadier J. E. W. Inglis).

32nd Foot.
Detachments:
13th, 41st,
71st Ben.N.I.
4/1 Ben.A.(Lieut.
J. Alexander).

Vol. C.; 12th Irr. Ben.C.
1st Inf. Brig.:
5th, 64th (2 Comp.),
84th Foot, 1st Mad. E.F.
2nd Inf. Brig.:
78th, 90th Foot, Ferozepore Regt.

Field Force
(Major-General H. Havelock).

3/8 R.A. (2nd Captain F. C. Maude).
2/3 Ben.A. (Captain W. Olpherts).
1/5 Ben. A. (Captain and Brevet-Major V. Eyre)

4. RELIEF FORCE (General Sir Colin Campbell with Brig. J. H. Grant second in command)

Cavalry Brigade (Brigadier A. Little):[1]
9th Lancers (2 squadrons).
Military Train (2 squadrons).
Detachments:
1st Punjaub C.
2nd Punjaub C.
5th Punjaub C.
Hodson's Horse.

Royal Engineers:
(One company)
Detachments:
Ben. Sappers and Miners
Mad. Sappers and Miners
Sikh Pioneers.

O.C.R.A., Captain F. J. Travers.

O.C.Ben. and Mad. A., Captain and Brevet-Major F. Turner.

Naval Brigade { 6 24-pounder guns
2 8-inch howitzers
2 rocket tubes
(Captain W. Peel, R.N.).

Artillery Brigade (Brigadier W. T. Crawford, R.A.)[2]:
4/5 R.A. { 2 18-pounder guns
1 8-inch howitzer
(Captain W. N. Hardy).
5/13 R.A. { 6 8-inch-mortars
10 5½-inch mortars
(Captain C. S. Longden).
6/13 R.A. { 4 9-pounder guns
2 24-pounder howitzers
(Captain W. A. Middleton).
1/1 Ben.H.A., 5 6-pounder guns
(Captain F. F. Remmington).
2/3 Ben.H.A., 5 6-pounder guns
(Captain C. H. Blunt).
E. Troop Mad.H.A.,[3] 2 6-pounder guns
(Lieutenant C. Bridge).
No. 17 F.B. 3/1 Ben.A. { 4 6-pounder guns
2 24-pounder howitzers
(Captain G. Bourchier).
Detachment 1/5 Ben. A., 3 9-pounder guns (Lieutenant E. W. E. Walker).

---

[1] Lieutenant-Colonel and Brevet-Colonel A. Little, 9th Lancers.

[2] Lieutenant-Colonel W. T. Crawford, R.A., arrived at Calcutta from Hong Kong August 31, and was followed by 5/13 and 6/13 R.A.

[3] E. Troop. Mad. H.A., reached Calcutta in July.

3rd Infantry Brigade (Brigadier E. H. Greathed):	4th Infantry Brigade (Brigadier The Hon. A. Hope):[1]	5th Infantry Brigade (Brigadier D. Russell):[2]
8th Foot. Detachments:[3] 5th Foot. 64th Foot. 78th Highlanders. 90th Foot. 2nd Punjaub I.	53rd Foot. 93rd Highlanders. 4th Punjaub I.	23rd Foot. 82nd Foot.

Note.—The military train, which acted as cavalry, contained many ex-cavalry soldiers.

III. THE ADVANCE ON LUCKNOW

Sir Colin's Plans.—On November 12, after a slight skirmish, Sir Colin encamped at the Alam Bagh, which was cleared of cattle and lumber to receive the tents and baggage, ammunition and stores, and on the 14th he set out for Lucknow. A few days previously Mr. T. H. Kavanagh, a Government clerk, succeeded in escaping from the Residency disguised as a native. He carried plans and despatches which informed the Commander-in-Chief that the rebels had extended their defences to the east of Hazratgunje, that a line from the European Barracks to the Sekundra Bagh was occupied, and that strong buildings lay between this line and the Residency. Sir Colin thereupon determined to advance across country to the Dilkhusha, and then, hugging the right bank of the Gumti, to attack the Sekundra Bagh.

The Advance on the Dilkhusha.—At 9 a.m. on

---
[1] Lieutenant-Colonel the Hon. A. Hope, 93rd Highlanders.
[2] Lieutenant-Colonel and Brevet-Colonel D. Russell, 84th Foot.
[3] Sir Colin took these detachments from the Alam Bagh and replaced them by the 75th Foot, which had marched from Delhi under Greathed.

November 14 the force was in motion. Hope and Russell's Brigades, covered by the cavalry, formed the main column, and Greathed protected the left rear. 1/1 Ben.H.A. was with the cavalry, and 4/5 R.A. and 3/1 Ben.A. were near the head of the main column; the remainder of the guns were distributed in rear. Every man carried three days' provision in his haversack; tents and baggage were left behind. The line of march was roughly parallel to, and a mile and a half from, the canal, and ran through well-cultivated country, dotted with clumps of trees and villages. The enemy were first encountered near the park wall of the Dilkhusha, and were driven back by the cavalry and light guns to the Martinière. The Dilkhusha stood on a small, elevated plateau, and when our troops reached the forward brow they came under fire of guns in the Martinière and musketry from the broken ground in front. The Martinière was about a thousand yards distant, and Remmington and Bourchier came into action. They were soon joined by Hardy, who brought up first his 8-inch howitzer and later the two 18-pounder guns. After a few rounds from the heavy ordnance, the enemy were seen removing their artillery with teams of oxen, and our infantry captured the Martinière with little or no loss. The rebels were then pursued and driven across the canal by the cavalry.

It was now about noon. Hope's Brigade occupied the grounds of the Martinière, Little, with the cavalry, being on the right front. Russell's Brigade was ordered to take up a position to cover the left and keep the road open for the troops still in rear. To do this the more effectually he placed two companies in Rampur, a village on the canal close to the Dilkhusha park wall; but its possession was at once challenged by the enemy, who occupied in force a belt of trees situated on the left bank of the canal between Rampur and Banks' house. Hardy at once opened fire from the Dilkhusha plateau, and Little, accompanied by Bourchier's Battery, reinforced Russell. The rebels were then speedily driven back into the city. But hardly had Little withdrawn to his old position when the enemy made a second attack, advancing with great determination by the Hazratgunje bridge. On this Hope's Brigade was sent to the front, and Bourchier's Battery and two of Peel's 24-pounders were placed on a small hillock on the left rear of the brigade, from which they were able to rake the bridge. Remmington came into action on the right flank, while the 93rd Highlanders deployed into line. Their fire crushed the attack, the enemy fell back, and the 53rd and the 4th Punjaubis pursued them across the bridge and set fire to the suburb on the other side. Sir Colin's force then remained concentrated round the Martinière with pickets on the canal.

Halt at the Dilkhusha.—The Commander-in-Chief, intending to make an advanced depôt at the Dilkhusha, halted during the 15th to allow a convoy to come up from the Alam Bagh. It arrived safely on that day, but its march was not unmolested. Several attacks were made upon it, but they were frustrated mainly by 2/3 Ben.H.A. (Captain Blunt), and

6/13 R.A. (Captain W. A. Middleton), which reached the Alam Bagh on the 14th in time to accompany the rear guard. The rebels, however, did not let the day pass undisturbed. About noon our right flank was threatened by a strong demonstration, but the pickets were on the alert, and the two guns of E Troop Mad.H.A. did good service. The enemy then became quiet, but in order to disguise his intentions Sir Colin Campbell made a reconnaissance on the left, on which flank his guns were massed. He withdrew his pickets from the right and ordered Captain Longden and Captain Peel to bombard the Barracks and adjacent palaces during the night. As evening fell a large bonfire was lighted in the Dilkhusha plateau, and a semaphore erected on the Martinière sent the welcome message to the Residency, "Advance to-morrow."

Preparations in the Residency.—Meanwhile the approach of the relieving force was anxiously watched from Lucknow, and it was determined to begin active operations as soon as Sir Colin Campbell reached the Sekundra Bagh. With this view the outer boundary wall of the Chattar Manzil Palace was mined, and behind it were placed 4 18-pounder guns, 1 8-inch howitzer, 4 9-pounder guns, and 2 24-pounder howitzers, under Captain W. Olpherts, and in an adjacent quadrangle 6 8-inch mortars under 2nd Captain F. C. Maude.[1] The

[1] Major Eyre (acting Brigadier) himself took charge of half of this extensive gun and howitzer battery. Maude had only one artillery subaltern—Lieutenant E. Maitland. Lieutenant E. H. C. Simpson, 39th Ben.N.I., and Lieutenant E. B. Ward, 48th Ben.N.I., were attached to him.

Hirun Khana and Engine House were also mined, so that when the moment arrived a sortie would find free egress.

November 16.—On the 16th Sir Colin proposed to advance from his right flank, leaving a force to secure the Dilkhusha. The protection of this important post was confided to Brigadier Little with a squadron of the 9th Lancers, the Military Train, a squadron of Punjaub Cavalry, 300 men of the 8th Regiment, the 3 guns of 1/5 Ben.A., and the 2 guns of E. Troop Mad.A. The remainder of the force set out at 8 a.m. The advance guard was composed of a squadron of Hodson's Horse, a company of the 53rd Foot, and 2/3 Ben.H.A. It was followed by Hope's and Russell's Brigades, Greathed's Brigade and the remainder of the cavalry being in rear. Of the artillery, 4/5 R.A. followed the advance guard, farther to the rear was 6/13 R.A., then came the Naval Brigade, followed by Longden's mortars (5/13 R.A.); 1/1 Ben.H.A. and 3/1 Ben.A. were with Greathed.

The Sekundra Bagh.—The canal was crossed without difficulty near its junction with the Gumti along the right bank of which the march was continued till, after circling round to the westward for a mile and a half, the column turned due south down a narrow lane flanked by mud huts and gardens and with steep banks on each side. Four hundred yards in front was a walled enclosure—the Sekundra Bagh —situated 100 yards west of the lane, and 200 yards further on was a large brick serai surrounded by mud cottages. When the advance guard ventured into this dangerous pass they were at once assailed in front and on the flanks by a sharp musketry fire, and the lane was found to be barricaded. For a time confusion reigned; but the cavalry were got out of the way into some side paths, the 53rd occupied the enclosures on the right, and Lieutenant A. Ford, 4/5 R.A., brought the 8-inch howitzer into action and fired some rounds of shell and shrapnel into the Sekundra Bagh. Then Blunt, forcing his way through the obstacles with which the lane was encumbered, led 2/3 Ben.H.A. forward till he got clear of the houses, and, taking ground to his right, brought his guns up over the seemingly impracticable bank which intervened between him and the open ground. Galloping forward he came into action between the serai and the Sekundra Bagh, against both of which he opened fire as well as in the direction of the Kaisar Bagh, the three directions from which he himself was assailed. His position was precarious until the 93rd Highlanders, the leading regiment of Hope's Brigade, pressing eagerly forward and clearing all the enclosures as they advanced down the lane, captured the serai and the surrounding buildings. When the remainder of the brigade came up in support, the sappers cut down a portion of the bank, and, aided by the infantry, the two 18-pounders of 4/5 R.A. were dragged into action 50 yards from the south-east corner of the Sekundra Bagh. The building was 120 yards square, with round bastions at the corners; it was well loopholed, and garrisoned by 2,000 men, nevertheless the breaching fire of the 18-pounders was carried out with little interference.

Capture of the Barracks.—Meanwhile on the left front the combat was continued, and pressing on from the serai detachments of the 93rd and 53rd, aided by two of Blunt's guns, fought their way to the Barracks, which the Highlanders converted into a military post, while the 53rd, in skirmishing order, kept up connection with the main body of the brigade at the Sekundra Bagh.

Capture of the Sekundra Bagh.—Seven rounds from the 18-pounders sufficed to make a breach in the wall of the Sekundra Bagh, which the sappers rushed forward to enlarge. At the same time small bodies of Highlanders and Punjaubis forced their way through it into the enclosure. Others, impatient of delay, assailed the main doorway in the south front, and after a gallant fight burst it open, while almost simultaneously a third means of ingress—a weakly barred window—was found by the 53rd Foot. Then Hope's Brigade poured into the Sekundra Bagh. Round the walls were isolated pavilions and mural chambers of the usual oriental construction, and each held its garrison. The sepoys were like rats in a trap. There was no escape, quarter was neither given nor expected. Room after room was searched out by bullet, bayonet, and tulwar, and though the sepoys fought with the courage of despair they were all doomed men. Two guns took part in the contest. A heavy door barred the access to one of the bastions, and a gun of Blunt's Troop, under Lieutenant C. Hunter, was sent in to blow it open. In a pavilion in the east wall of the enclosure a party of rebels were making a stout resistance, and a gun, under Lieutenant W. Smith, detached from Middleton's Battery, was dragged in through piles of dead and wounded to bear upon them. Smith fired until he had expended all his case shot; he then loaded with unplugged common shell, the fuse hole being next the charge. When the pavilion was carried, Smith rejoined his battery, but Hunter's gun was kept in the enclosure till late in the afternoon, for even after all real resistance was beaten down isolated combats continued, and when evening fell over 1,850 dead sepoys lay upon the ground.

The Advance continued.—About half-past 12 Hope withdrew the main body of his brigade from the bloodstained building. The position, extending from the Barracks on the left to the Sekundra Bagh on the right, was virtually won, and a further advance was possible. The way to the Residency lay to the westward over a fairly open plain, traversed by a road. Eight hundred yards from the Sekundra Bagh and 200 yards north of the road was the Shah Najaf, a large mosque standing in a garden 200 yards square, bounded by a strong loopholed wall, 20 feet high. This enclosure was almost hidden by numerous mud cottages and a mass of brushwood which formed a fringe of jungle round it. Close by to the eastward was the Kadam Rasul a substantial building standing on a high mound. The Kadam Rasul and the Shah Najaf were the next objects of attack.

Between the Sekundra Bagh and the Shah Najaf lay a village from which our infantry drove the rebels, and then Hardy's Battery advanced 300 yards to the westward. It came into action under a heavy cannonade of shot, shell, and grape, and Hardy was killed when laying his second round by a shot which carried away part of his head and neck. Lieutenant A. Ford then assumed command of the battery. About 2 p.m. the 8-inch howitzer was disabled and Ford was wounded, but he was able to see the howitzer safely removed before handing over the command to Lieutenant W. S. Brown.[1]

Soon after Hardy's death Middleton brought 6/13 R.A. into action near the Sekundra Bagh, where also 2nd Captain H. L. Talbot, detached from Longden's Battery, arrived with two 10-inch mortars which had been pushed on to reinforce Hardy's 18-pounders; then Peel came up and brought his guns into action on the left of 4/5 R.A.

Capture of the Shah Najaf.—The enemy's cannonade began to fail, and about 2.30 p.m. our infantry advanced on the Shah Najaf and Kadam Rasul. Blunt and Middleton were sent in support while Peel, Brown, and Talbot kept up a vertical fire, assisted by Longden, who was now in action on the north of the Sekundra Bagh. Soon after 3 p.m. the Punjaubis captured the Kadam Rasul. Our infantry were just able to hold their footing on the east side of the jungle surrounding the Shah Najaf; but in the mosque itself the sepoy garrison behind their strong walls were immovable, and the heavy vertical

---

[1] Lieutenant W. S. Brown was posted to No. 3 Company, 8th Batt., but was attached to Captain Hardy's Battery at Allahabad in the previous October.

fire to which they were subjected was powerless to turn them out. Sir Colin determined, therefore, on a bayonet attack supported by the close fire of artillery, and the 93rd Highlanders and Middleton's Battery were ordered forward. As he passed to the front at a gallop along the ranks of that gallant regiment, Middleton was greeted with loud cheers, which his men returned, and then, taking ground to his right, he brought the guns into action within 60 yards of the south wall of the Shah Najaf, and poured case shot upon the mud huts and brushwood all around. The 93rd rushed forward through the cheering battery and fought their way forward, but the twenty-foot wall was an insuperable obstacle.[1]

While the infantry replied as best they could to the musketry fire of the rebels, two of Peel's 24-pounder guns were brought up, and—soldiers and sailors together manning the drag ropes—came into action close to the walls. But the masonry resisted the impact of the heavy projectiles and it seemed that failure was inevitable. Under a well-directed fire of his rockets Peel withdrew his guns, and Hope, collecting his wounded, was about to retire, when Sergeant J. Paton of the 93rd Highlanders reported that he had discovered a breach in the north wall. Hope at once was on the spot, and some sappers having speedily got to work, the Brigadier himself led some fifty men into the enclosure. No opposition was met with, and the main gate was thrown

---

[1] The Despatches are silent on the subject of Middleton's gallant advance, but it was brought to notice eventually in *Blackwood's Magazine*, October, 1858, p. 494, by Major A. Alison (afterwards Major-General Sir A. Alison, Bt.), who served on the Staff at Lucknow.

open; for strange to say, in the moment of their victory the rebels fled. The Shah Najaf was then occupied by the 93rd Highlanders. Middleton was sent forward to clear the front, and came into action in a position south of the road, and about 400 yards from the the Khurshid Manzil, which was still occupied by the enemy. But in its gallant action at the Shah Najaf 6/13 R.A. had suffered so severely that one gun had to be driven into its new position by Lieutenants W. Smith and W. D. Milman, mounted as lead and wheel drivers; and if the battery officers had not helped to serve the guns—in which duty they were assisted by Major W. W. Barry,[1] R.A.—they could not all have been brought into action. After being joined by one of the naval guns, Middleton continued his fire till nightfall, and the army bivouacked in the position it had won.

Casualties on November 16.—November 16 was the critical day in the Relief of Lucknow, and the brunt of the fighting fell upon Hope's Brigade and the guns. The Naval Brigade had 17 casualties, there were 14 in Blunt's Troop, and 41 in the Royal Artillery, of which 20 occurred in Middleton's Battery.

Action taken by the Residency Garrison.—From early morning Outram and Havelock watched Sir Colin's advance from the roof of the Chattar Manzil, and when the Sekundra Bagh was attacked the mines were exploded which were designed to blow down the boundary wall and unmask Olpherts' guns.

---

[1] This officer came from Hong Kong with Lieutenant-Colonel W. T. Crawford, R.A., as his adjutant.

The action of the mines was feeble, but the guns themselves soon completed the work. The infantry were then assembled in the great square of the Chattar Manzil, and the two mines at the Hirun Khana and the Engine House were fired. Olpherts' heavy guns and Maude's mortars opened on the enemy, and our infantry, rushing out of their cover, were speedily masters of all the buildings between the Chattar Manzil and Moti Mahal. During the night some of the heavy guns were placed in position to breach the Kaisar Bagh, while Maude, with howitzers and mortars prepared to support Sir Colin's attack on the Moti Mahal and Khurshid Manzil, which alone intervened between the relieving forces and the besieged garrison.

November 17.—Next day, with the object of protecting his left flank, which was only secured by the military post at the Barracks, Sir Colin ordered Brigadier Russell to occupy the Bungalows in Hazratgunje, situated about 600 yards from the bridge over the canal. This undertaking was successfully carried out with the support of Longden's mortars, which were brought into action at daybreak. In the centre, the Khurshid Manzil and surrounding buildings were kept under a hot fire by Peel, Olpherts, and Maude, and about noon Longden's mortars were moved forward from the left, and joined in this bombardment. Fire from the left bank of the Gumti was replied to by the 18-pounders [1]

---

[1] The author is unable to state who commanded 4/5 R.A. at this time; it was either Lieutenant W. S. Brown or Lieutenant H. C. S. Dyer (see note, p. 227).

which were placed in a small breastwork constructed by the engineers.

Capture of Khurshid Manzil and Moti Mahal.— About 3 p.m. our infantry attacked the Khurshid Manzil. They met with little resistance, and pursuing their advantage, captured the Moti Mahal also, and now only a small open space intervened between them and the Residency. Although this space was under fire from the Kaisar Bagh and the left bank of the Gumti, Outram and Havelock, accompanied by some other officers, crossed it at once to greet the Commander-in-Chief at the Khurshid Manzil, and loud rang the cheers when the news spread along from post to post that the three generals had met.

Skirmishes on the Left and Centre.—Early next day, November 18, Bourchier with 2 guns was sent to reinforce our left flank, where our hold on the Bungalows was somewhat precarious. Talbot had been left in charge of the $5\frac{1}{2}$-inch mortars of Longden's Battery, and Lieutenant E. S. Burnett was now detached with four of them to reinforce Bourchier and to act under his orders. The remaining two, under Talbot, were brought into action near the Barracks. This artillery support enabled our infantry to advance and clear the ground in front. Bank's House and the Hospital were captured, but the last named caught fire and our main body retired to the Bungalows, though Bank's House was held as a detached post for some days. This retirement appeared to encourage the enemy, for they at once made a brisk attack on the pickets in front of our centre. Two companies were sent forward in support, and, no more infantry being at the moment available, Remmington's Troop dashed forward into line with the leading skirmishers and opened fire with extraordinary force and precision. The enemy were at once driven back.

Bombardment continued, November 17 and 18.—In the centre the Kaisar Bagh was a target for the guns of the relieving force, aided by those in the Residency; and during the afternoon Lieutenant F. G. E. Warren was detached from Longden's battery with two 8-inch mortars to a position near the Kadam Rasul, to fire on a battery of the enemy established on the left bank of the Gumti. Here he was joined next day by the 18-pounder guns, now under Lieutenant Dyer,[1] and this position was maintained until the relief was completed.

The Exodus of November 19.—On the 19th the fire on the Kaisar Bagh was continued, and the exodus from the Residency began. A flying sap was made to defilade the open space before referred to from the fire of the Kaisar Bagh, the mortars and 18-pounders dealt effectively with the guns on the left bank of the Gumti, and the long column of sick and wounded, women and children was conducted without mishap to the Dilkhusha, following the road by which the relieving column had advanced.

Bombardment continued, November 21-24.—Peel's 24-pounders were now placed in battery close to

---

[1] Lieutenant H. C. S. Dyer, R.A., came with Lieutenant-Colonel W. T. Crawford from Hong Kong, as his Quartermaster. He was temporarily attached to 4/5 R.A.

Martin's House, and in company with the Residency guns began to breach the Kaisar Bagh, to divert the attention of the enemy from the Residency by the fear of an assault.

Relief of the Residency.—On the 21st Middleton had special charge of a convoy of food, treasure, and such guns as could be removed. Under his command were his own battery, Blunt's Troop, the 18-pounders from the Kadam Rasul, and 3/8 R.A., who once more found themselves with mobile guns. The convoy reached the Dilkhusha in safety, and Middleton under orders of Brigadier Little placed his guns in position to resist a threatened attack. Next morning the enemy advanced with the intention of cutting the communications between the Dilkhusha and the Alam Bagh, but they were driven off without difficulty.

Evacuation of the Residency.—On the night of the 22nd all the guns not to be removed were burst, the Residency outposts were called in, and the garrison marched out of the Bailey Guard Gate.[1] By 4 a.m. on the 23rd the whole force reached the Dilkhusha, while the sepoys, in complete ignorance of the evacuation, continued their fire on our old positions. On the 24th the march on the Alam Bagh began, and by the evening of the 26th was completed the reorganisation of the army which

[1] "The illustrious garrison marched past Outram and Inglis, who stood at the Bailey Guard Gate. 'All have passed, sir,' reported an A.D.C. Outram waved his hand to Inglis to precede him. But Inglis at once said, 'You will allow me, Sir James, to be the last and to shut the gate of my old garrison.' Outram at once yielded, and Inglis closed the gates" ("State Papers," Introduction, ii. p. 315).

was designed to carry out the course of action proposed by Sir Colin Campbell and approved by the Governor-General.

IV. THE DEFEAT OF TANTIA TOPI

Sir Colin's Plans.—Briefly summarised, the course of action proposed by the Commander-in-Chief consisted in (1) the temporary evacuation of the city of Lucknow, (2) the holding of the Alam Bagh as an advanced post for future operations, and (3) immediate action against such mutineers as should cross the Jumna and threaten Cawnpore. Outram's Division,[1] reinforced to a strength of 4,000, was to garrison the Alam Bagh, the rest of the army was to accompany the Commander-in-Chief with his huge convoy to Cawnpore. As regards the artillery, 3/8 R.A., 4/1, 2/3 and 1/5 Ben.A. remained under Outram's command, Lieutenants Ford and Brown and the men of No. 4 Comp. 5 Batt. R.A. being attached to Maude and thenceforward forming part of 3/8 R.A. The remainder of the artillery accompanied the Commander-in-Chief.

Death of Havelock.—On November 24 a great sorrow fell upon the army and the nation: the gallant Havelock, who had long been ailing, died.[2] He was buried on November 26 in the Alam Bagh, where a cross carved by Outram marked his grave.

Sir Colin marches to Cawnpore.—On November 27 Sir Colin began his march to Cawnpore with 3,000 fighting men, 2,000 sick, wounded women and chil-

---
[1] The 1st and 2nd Brigades (Lucknow Garrison).
[2] He had just been promoted K.C.B.

dren, and the treasure that the forethought of Sir Henry Lawrence had saved. Since November 19 he had received no news from Windham, and on the evening of his first day's march, when he encamped at the Bunnee Bridge, he could learn nothing beyond the fact that for twenty-four hours heavy firing had been heard from the west. Next day he himself could hear the sound of a distant cannonade which became louder as the troops pressed forward, and about noon he received definite intelligence that Windham was sorely in need of help. Leaving the cavalry and horse artillery, with which he had rapidly advanced to encamp at Mangalwar, Sir Colin, accompanied only by his staff, galloped towards Cawnpore, and as the sun was setting the city came in view. A general action was evidently in progress, and the smoke and flames of incendiary fires seemed to show all too plainly that some disaster had taken place; but the bridge upon which so much depended, the bridge of boats across the Ganges, was still intact.

Windham at Cawnpore.—When General Windham was left to protect Cawnpore on November 9 the following orders were given to him. He was to show as bold a front as possible without inviting attack, and was only to move out against the enemy to save from bombardment the entrenchment that protected the west end of the bridge of boats across the Ganges; he was to keep open communications with Sir Colin's force, and to pass on all reinforcements arriving from the south; but he was permitted to retain the services of Carthew's Brigade of Madras Troops, which was marching up from Allahabad,[1] until the movements of Tantia Topi were known. It soon, however, became evident that the rebel leader meant to seize the opportunity offered by Sir Colin's absence of falling on the small force of British at Cawnpore, and about the middle of the month his troops began to cross the Jumna. On November 15 Windham got definite instructions from Sir Colin to detain all detachments until further orders. On the 14th Carthew had arrived with the 27th Mad.N.I. and four 9-pounder guns of C/5 Mad.A. (Lieutenant S. H. E. Chamier), the 17th Mad.N.I. having been left at Futtehpore. During the next few days two battalions of the Rifle Brigade and detachments and drafts of the 34th, 82nd, and 88th Foot brought the effective forces at Cawnpore up to a strength of 1,700 men.

His Plans.—When the rebels first crossed the Jumna their force was scattered, and Windham determined to crush the various detachments before they could unite; but as this course was forbidden by his orders he was obliged to submit his plan for the approval of his chief. A letter was despatched on November 15 to Lucknow, but no answer was received. By the 19th it was evident that all communication with Sir Colin was suspended, and on the 23rd Windham learnt that the post at the bridge over the Sye at Bunnee had been captured by a party of raiders. He at once despatched a wing of

[1] Carthew's Brigade consisted of the 17th and 27th Mad.N.I., the Madras Rifles (a battalion composed of the rifle companies of several regiments), a detachment of Madras Sappers, half E Troop Mad.H.A. (which had already joined Sir Colin), and C/5 Mad.A. In his rear was the 18th Mad.N.I. under Colonel Fischer and A/3 Mad.A.

the 27th Mad.N.I. and two of Chamier's guns to restore the lost communication, but, by a letter which arrived in camp the same day, he was led to believe that the Commander-in-Chief was himself hemmed in and would not be able to return, at least for a time.

Combats round Cawnpore, November 26.—Believing that safety lay in audacity, Windham determined to attack the rebels, and on November 26 he fell upon their leading division on the Pandoo River and drove them in confusion across that stream, capturing three guns.[1] But the time for striking at a scattered enemy had passed. Undismayed by the defeat of his advance guard and confident in his overwhelming numbers, Tantia Topi pressed boldly forward, and even in the moment of victory Windham saw that he must retire. He fell back to a position covering Cawnpore, pitching his camp across the Kalpi road close to some brick kilns on the western margin of the city.

Tantia Topi's Force.—Tantia Topi had now under his hand 14,000 disciplined men, sepoys of the line and the Gwalior Contingent and 11,000 irregulars. also some 60 guns. After following up the British with great boldness, this force on the night of November 26 occupied a quadrant of a circle which swept round from the Ganges canal on the west to the Bithur road on the north.

November 27.—On November 27 Windham posted

[1] Lieutenant J. R. Oliver, R.A., whose company (3/14 R.A.) had arrived at Calcutta, was sent up to Cawnpore to instruct infantry in gun drill. He was at this time in command of the extemporised field battery.

Brigadier R. Walpole [1] at the junction of the Kalpi and Grand Trunk roads with the Rifle Brigade and 88th Connaught Rangers, the bullock field battery under 2nd Captain D. S. Green, R.A.,[2] and two 24-pounder guns of the Naval Brigade, which had just arrived under Lieutenant E. Hay, R.N. This force was to protect the front of the camp. Brigadier Carthew on the right with detachments of the 34th and 82nd Foot and two guns under Lieutenant Chamier guarded the Bithur Road, while the main body of the 82nd occupied a wood west of the village of Seesambow in the centre. Colonel Wilson with the 64th Foot was in the entrenchment at the bridge-head. In this position Windham hoped to save the city from pillage and at the same time cover the bridge-head.

About noon the rebels opened a heavy cannonade against our front and right flank, and an enveloping attack followed. Carthew successfully held his ground, but Walpole was overpowered by superior artillery fire, and Windham brought the 34th from the right flank to reinforce him. Then drawing back his left to the brick kilns, Windham held the general position until 4 o'clock. At this hour news reached him that bodies of the enemy had penetrated Cawnpore, round Carthew's right flank, between the Bithur Road and the river, and, anxious for the safety of the entrenchment at the bridge, he galloped

---

[1] Lieutenant-Colonel R. Walpole came to Cawnpore in command of the 2nd Batt. Rifle Brigade.

[2] Major-General Dupuis (see next note), considering Lieutenant Oliver too young for an independent command, placed his A.D.C., Captain D. S. Greene, R.A., in charge of the guns.

off to his right flank, leaving Major-General Dupuis [1] in command at the front. About the same time our centre gave way, and the troops occupying the wood fell back.[2] Dupuis was thus in danger of being cut off, and retreat was inevitable. It was a work of no small difficulty to transport the wounded from the camp to the bridge-head through the narrow streets of the town, but this was safely accomplished. Our camp, however, was abandoned, and that night our tents were burnt by the rebels. The withdrawal from the wood was a disaster, and in plain words the retreat of our left wing was a rout. One of the naval 24-pounder guns was overturned and temporarily abandoned,[3] and by nightfall the bulk of the British troops were in and about the bridge-head with the exception of Carthew's command. When Windham reached his right flank he found that the intruding rebels had been driven back by a detachment of the 2nd Batt. Rifle Brigade, which had opportunely arrived from Futtehpore; he then ordered Carthew to hold the position he had defended in the morning, and after some successful skirmishes, in which the two guns of C/5 Mad.A. did good

[1] Major-General J. E. Dupuis and his staff arrived from England to command the Royal Artillery. They were at this time in Cawnpore.

[2] The officer in command of the centre was severely censured.

[3] General Dupuis states in his despatch that Lieutenant-Colonel J. M. Adye, R.A., his Assistant Adjutant-General, volunteered on the night of the 27th, and brought from the centre of the town this 24-pounder gun—assisted by Captain and Brevet-Major A. G. Austen, Ben.A., and Lieutenant W. J. Bradford, Mad.A., and that on the 29th he took command of the same gun (manned by a few sailors and Madras gunners) and placed it in position to protect the canal, where it proved of good service. Lieutenant E. Hay, R.N., was twice wounded during the fighting of November 27.

service, the Brigadier bivouacked near the Assembly Rooms.

November 28.—The bulk of the city, however, was in the hands of the rebels, and early next morning they pressed the advantage they had gained. Windham made a gallant stand, but failed to appreciate the vital importance of his right flank. It is true that Walpole, who was sent to operate on the south of the canal, and hold the rebels in check on that side of Cawnpore, performed his task with great success;[1] for he drove the mutineers back and captured two 18-pounder guns; but Carthew with the force at his disposal was unable to maintain his ground. Colonel Wilson and the 64th made a gallant effort to assist him, but in vain. Wilson was killed, and Carthew was obliged to fall back after a fight which lasted all day. This retreat gave the enemy the command of the river-side, put the bridge of boats in the greatest jeopardy, and enabled them to seize and burn the Assembly Rooms, where a mass of regimental property was accumulated, while they destroyed at their leisure stores and necessaries prepared for the use of the rescued garrison of Lucknow. If another day had been allowed him, it is probable that Tantia Topi would have triumphed, but Sir Colin arrived in the nick of time, and at dawn on the 29th his heavy guns were in action on the left bank of the Ganges. In the three days' fighting Windham's casualties

[1] Walpole's force consisted of 5 companies of the Rifle Brigade, 2 companies of the 82nd, and the extemporised battery under 2nd Captain D. S. Greene, R.A.

amounted to 36 officers and 306 men killed and wounded.

Sir Colin crosses the Ganges.—When the rebels' guns attempted, on the morning of the 29th, to injure the bridge of boats they were at once opposed by Peel, Longden, and the guns mounted in the bridge-head; and when their fire had been in part subdued, Hope's Brigade, accompanied by the field guns and cavalry, and, followed by Greathed's, crossed the river, while Brigadier Inglis [1] protected the rear of the convoy which stretched over six miles of road. By the evening of the 30th all were safely across, and a camp was formed for the women and children on the south side of Cawnpore, under the protection of the army, whose position stretched from the bridge-head on the right to the plain beyond the Grand Trunk road on the left. Greathed's Brigade occupied an advanced position in the General-gunge Bazaar, being supported by Peel's heavy and Bourchier's field guns. The rebels occupied the city on the north side of the canal, and kept up an incessant fire during the hours of daylight, but till Sir Colin could dispose of his convoy he was powerless to act against them. On December 3 he was able to despatch the women and children and half the wounded to Allahabad,[2] and on the 5th he learnt that they had arrived there in safety. He then determined to attack the rebels next day.

His Available Force.—His troops as reinforced

---

[1] Brigadier D. Russell was severely wounded on November 17, and Inglis was appointed to succeed him.

[2] Under escort of Lieutenant-Colonel R. D. Kelly, 34th Foot.

now amounted to 5,000 infantry, 600 cavalry, and 35 guns, organised as follows:

SIR COLIN CAMPBELL'S ARMY, DECEMBER 6

Brigadier J. H. Grant, Second-in-Command.
Major-General W. R. Mansfield,[1] Chief of the Staff.

Naval Brigade (H.M.S. *Shannon*, Captain W. Peel, R.N.).

Cavalry (Brigadier A. Little).
9th Lancers.
Detachments:
  1st, 2nd, and 5th Punjaub Cav.
Hodson's Horse.

Royal Engineers (Lieutenant-Colonel H. D. Harness).

Artillery (Major-General J. E. Dupuis).
1/1 Ben.H.A. (Captain F. F. Remmington).
2/3 Ben.H.A. (Captain C. H. Blunt).
E. Troop Mad.H.A. (2 guns) (Lieutenant C. Bridge).
5/13 R.A. (Bullock Draught) (Captain C. S. Longden).
6/13 R.A. (Captain W. A. Middleton).
7/14 R.A. (Captain and Brevet-Major C. H. Smith).[2]
3/1 Ben.A. (Captain G. Bourchier).

Infantry

Third Brigade (Brigadier E. H. Greathed):
  8th Foot.
  64th Foot.
  2nd Punjaub I.

Fourth Brigade (Brigadier the Hon. A. Hope):
  53rd Foot.
  42nd Highlanders.
  93rd Highlanders.
  4th Punjaub I.

Fifth Brigade (Brigadier J. E. W. Inglis):
  23rd Foot.
  32nd Foot.
  82nd Foot.

Sixth Brigade (Brigadier R. Walpole):
  2nd Batt. Rifle Brigade.
  3rd Batt. Rifle Brigade.
  Detachment 38th Foot.

Bridge-head Entrenchment (Major-General C. A. Windham).
Infantry Detachments.   Detachment 4/14 R.A.
  38th Foot, 88th Foot, etc. 2 8-inch howitzers (Captain C. Johnston).
                      Details Ben.A. (Captain J. H. Smyth).

---

[1] Lieutenant-Colonel (local Major-General) W. R. Mansfield, unattached.

[2] 7/14 and 4/14 R.A. arrived in India from England in October. The latter was at Allahabad in November and a detachment was sent on to Cawnpore,

Battle of Cawnpore, December 6.—On the morning of December 6 the rebels occupied the neighbourhood of the Assembly Rooms with their left; their centre faced Greathed's position and their right extended beyond the bridge, carrying the Grand Trunk Road across the canal. Two miles to the north of this bridge was the camp of the Gwalior Contingent and South Jumna rebels, whose line of retreat was on Kalpi, that of the Nana Sahib's followers being by the Bithur Road.

The Commander-in-Chief determined, while holding the enemy on their centre and left, to fall with the bulk of his force on their right flank. With this object in view a heavy cannonade was opened by Windham from the bridge-head, and briskly maintained for two hours, during which time our troops got into positions of readiness sheltered from the view of the rebels by the old cavalry lines. At 11 o'clock Greathed opened fire to his immediate front, and the horse artillery and cavalry were sent to make a wide detour to the left and cross the canal by a bridge about two and a half miles above that carrying the Grand Trunk Road. Walpole, supported by Smith's Battery, crossed by the Generalgunge Bridge, and, advancing along the western side of the city in the direction of the brick kilns, interposed between the rebel centre and right. Meanwhile Hope's Brigade, supported by Inglis's in second line, prepared to carry the Grand Trunk Road Bridge. In front of it on each flank were great mounds of broken bricks and clay which were strongly occupied by the enemy. Peel, Middleton, Longden, and Bourchier opened a brisk cannonade, and Hope's skirmishers rushing forward drove the rebels from the mounds, but they rallied on the bridge, and for a short time held our men at bay. Then the sailors gallantly brought a 24-pounder gun into the skirmishing line and in a few rounds broke down all resistance. The bridge was carried, and Hope and Inglis gained the enemy's side, Walpole's Rifles reinforced them on their right, and the right wing of the rebels were soon in full flight along the Kalpi Road.

Pressing on in advance of the infantry, Bourchier came within close range of the camp of the Gwalior Contingent, and fired round after round of shrapnel into it, until the 53rd Foot and 4th Punjaub Infantry marched up and took possession of it. There was now a splendid opportunity for cavalry and horse artillery to inflict serious loss on the rebels, but though it was nearly 2 o'clock nothing could be seen or heard of Little's Brigade. As an infantry pursuit might fail to capture the guns, Bourchier, without escort, was ordered to pursue with his battery along the Kalpi Road, the infantry following in his rear. The battery rapidly followed up the rebels for two miles, passing by many abandoned guns and ammunition waggons. Four times in this distance they came into action to clear their front and flanks, and then were halted by order of Brigadier Grant, who accompanied the advance. The cavalry column now at last appeared: it had been delayed by bad ground and a mistake of a guide. The pursuit was at once renewed, and, led by Grant himself, continued for fourteen miles from Cawnpore. When the horsemen drew bridle the enemy were scattered in all directions and every gun was captured.

As soon as the camp of the Gwalior contingent fell into our hands, Sir Colin Campbell ordered the 23rd Royal Welsh Fusiliers and the 32nd Foot, supported by 7/14 R.A., to act as its guard,[1] and, placing a strong force under General Mansfield, directed him to advance towards the Soubadar's Tank and threaten the left rear of the rebels and their line of retreat by the Bithur road. Mansfield's force consisted of Middleton and Longden's Batteries, the two battalions of the Rifle Brigade and the 93rd Highlanders. In the successful advance against the enemy's defeated wing the last-named regiment was on the British right, and when Mansfield received his orders was on the Grand Trunk Road close to where another road branches in a southerly direction to the civil station of Cawnpore. Here the Highlanders began to suffer from the enemy's guns posted about the village of Gutaiha, and Mansfield threw them forward to better cover, while Middleton answered the hostile fire. So matters continued for half an hour until the Rifle Brigade, which had advanced beyond and to the left of the Gwalior Contingent's camp, joined General Mansfield, and were extended in skirmishing order 300 yards on each side of the road leading to the civil station. Close on their heels came Longden's battery, the subalterns running along beside the teams of bullocks and

---

[1] During the afternoon, when a part of Hope's and Inglis's Brigade were pursuing the rebels on the Kalpi Road, the camp was attacked, but the rebels were repulsed with a loss of two guns.

urging them forward. The Rifles pressed on, the 93rd being held in reserve, and Middleton and Longden taking up successive positions and maintaining a rapid fire, the enemy fell back. So Bengalipore was reached, but enclosures were still held to a certain extent by the rebels. Dashing forward in advance of the infantry, Middleton led his battery at a gallop through the village, and coming into action in the open beyond it, fired on the enemy's guns as they disappeared along the Bithur Road, whilst the Rifles were still running up to his support. The 93rd then advanced and a position was occupied to the west of the Soubadar's Tank. Later in the day on two occasions the rebel guns fired on Mansfield, but were easily silenced and did little harm; indeed, the General could have inflicted very serious loss on the defeated enemy had he thought proper to advance.[1]

The Naval Brigade guns and 7/14 R.A. remained in support of our centre and left,[2] which were but weakly held. The enemy, however, made no attempt to molest them, and when early next morning two squadrons of cavalry patrolled the city no rebels could be seen.

Affair of Serai Ghat, December 9.—On December 8 Brigadier Grant was sent in pursuit in the direction of Bithur with 2,000 infantry, 500 cavalry, Remmington's Troop, and Middleton's Battery. By nightfall he reached Sherajpore, three miles from

[1] Malleson, ii. p. 275.
[2] This is the author's supposition only. The further doings of these guns are not mentioned in the Despatches.

Serai Ghat on the Ganges; at daylight on the 9th he was reconnoitring the ferry at that place. A large force of rebels were then discovered, about to embark their guns. Grant hastily ordered up the guns and cavalry, but the ground was difficult, dangerous quicksands were encountered, and the rebels turning at bay brought 13 guns into action. Lieutenant W. D. Milman (6/13 R.A.), however, found sound ground and opened fire, although quite unsupported. A little later Lieutenant C. H. Pickering joined him with his division[1] and 1/1 Ben.H.A. opened a flanking fire; then the mutineers fled, abandoning their guns and mounting the team horses. Presently some of the enemy's horsemen made an attack on our guns, but were at once driven back by the cavalry, who cut down some 80 or 90 stragglers. The bulk of the rebels escaped, but all their guns and some excellent gun bullocks were captured.

[1] Lieutenant W. Smith's Division remained with a small rear guard of the three arms which Grant left at Sherajpore.

THE SHAH NAJAF, NOVEMBER 16, 1857

CHAPTER VI

THE MILITARY SITUATION AT THE END OF 1857

I. Rajputana and Central India

The Jodhpore Legion.—While the siege of Delhi and the relief of Lucknow were in progress, important events had taken place south of the Jumna. The reader will remember that early in June small reinforcements from Deesa reached Brigadier G. St. P. Lawrence in Rajputana. The Rajah of Bikanir prepared to give him active support, and native levies were organised at Deoli and elsewhere; but notwithstanding, he could do little more than maintain small garrisons in Neemuch, Nusseerabad, Ajmere, and Mount Abu. On August 21 a portion of the Jodhpore Legion [1] mutinied and attacked Mount Abu. Beaten off by the British, they proceeded to Erinpura, the head-quarters of the Legion, and persuaded the main body to join the revolt. The Maharajah of Jodhpore was loyal and despatched his state troops to deal with the rebels; but the state troops proved utterly unreliable, and the mutineers, making common cause with the revolted Thakhur of Awah, established themselves at that

---

[1] The Jodhpore Legion consisted of 3 troops cavalry, 11 companies of infantry, and 2 9-pounder guns.

place, and successfully resisted all efforts to dislodge them. The Legion remained in Awah until the end of October, when it marched towards Delhi as already related.

Kotah.—Furthermore the state troops at Kotah on October 15 murdered the British Resident, possessed themselves of the town, and laid siege to the palace in which the Maharao sought refuge. But this was not all.

Prince Feroz Shah.—A man in the guise of a Mecca pilgrim appeared in Mandesar, and, styling himself the Shahzada Humayun, or Prince Feroz Shah of the Imperial House of Delhi, he raised the green flag of Islam and was soon at the head of a motley gathering that grew daily in strength. On August 31 he occupied Dhar, and, becoming bolder from the absence of opposition, he early in October seized the town of Jiran, ten miles south of Neemuch. The garrison of Neemuch, consisting of 2 squadrons of the 2nd Bombay Light Cavalry, detachments of the 83rd Foot and the 12th Bom.N.I., some native levies, and 4 guns manned by volunteers under Lieutenant C. H. Barnes, Ben.H.A., marched out on October 23 to attack the rebels. Jiran, however, proved too strong for the small British force, and a disaster was only saved by the dash and self-sacrifice of the officers.[1] Feroz Shah then advanced and laid siege to Neemuch. It was relieved at the end of November, as will presently be related; but

---

[1] Captain N. B. Tucker, 2nd Bombay Cavalry, who was in command, was killed, as also was Captain S. Read, 83rd Foot. Five officers were wounded.

Awah remained in rebel hands till the end of January and Kotah was not taken until March.

The Malwa Field Force.—Meanwhile the column organised at Poona by the Bombay Government crossed the Nurbudda on June 27 under Brigadier C. S. Stuart,[1] and on August 2 marched into Mhow. As soon as the rains were over and the country passable, Stuart set out to capture Dhar and relieve Neemuch, his column, henceforth known as the Malwa Field Force, being composed as follows:

14th L.Ds. (1 wing).	86th Foot (wing).	4/2 Bom.A. (Captain J. D. Woollcombe).
3rd Cav. Hyderabad Contingent.	25th Bom.N.I. Mad. Sappers and Miners.	2/6 Ben.A. (3 guns) (Captain T. J. W. Hungerford). Small Siege Train.

Action at Dhar, October 22.—On October 22 Stuart was attacked near Dhar by the rebels, who covered their movements with 4 guns. One gun was speedily put out of action by the fire of 4/2 Bom.A., and the other three were captured by the 14th Light Dragoons, and turned upon their late owner by the 25th Bom.N.I., who followed close upon the heels of the cavalry. The latter, pursuing their advantage, broke the left flank of the rebels, who fled in disorder.

Capture of Dhar, October 31.—Stuart then invested the fort of Dhar and established a practical breach by October 31, when a reconnaissance carried out previous to assault disclosed the fact that the fort was evacuated. Having been joined by a part of the Hyderabad Contingent, Stuart then marched to the

---

[1] Lieutenant-Colonel and Brevet-Colonel C. S. Stuart, 1st Bombay European Fusiliers.

relief of Neemuch, while the defeated rebels fell back and reinforced the besiegers.

Skirmish at Rawal, November 8.—On November 8 Feroz Shah detached a force which suddenly attacked Mehidpore, the head-quarters of the Malwa Contingent, who not only offered a feeble resistance, but supplied many deserters to the enemy before they dispersed. The triumphant rebels with three captured guns made for Neemuch, but were overtaken and defeated at Rawal by the Hyderabad Cavalry, who recaptured the guns.

Relief of Neemuch.—On November 21 Stuart was four miles south of Mandesar and easily repelled an attack made on his position. Continuing his advance, he routed the rebels at Mandesar and Goravia on the 22nd, and raised the siege of Neemuch. Leaving Hungerford with the three guns of 2/6 Ben.A. in Rajputana and the Hyderabad Contingent at Mandesar, Stuart marched back to Mhow, and on December 15 disarmed Holkar's troops at Indore. After his defeat at Neemuch, Feroz Shah drifted eastwards towards Saugor and he did not again take the field until he appeared in Rohilkhund at a later period.

The Nagpore Column.—The Nagpore movable column, consisting of detachments of the 4th Madras Cavalry, 33rd Mad.N.I., the Nagpore Rifles, and D/3 Mad.A., reached Jubbulpore on August 1 under Brigadier Millar.[1] Taking with him in addition two companies of the 52nd Ben.N.I., Millar marched to Damoh, freeing it from rebel pressure, and carried

---

[1] Lieutenant-Colonel J. Millar, 33rd Mad.N.I.

out some small operations; but as disquieting rumours came from Jubbulpore, he, on September 7, sent back some Madras troops to that station. Next day he set out for Saugor, but hearing that the Dinapore mutineers were marching south, he returned to Damoh. On the 19th news arrived not only that the 50th Ben.N.I. had mutinied at Nagode, but also that the 52nd had risen at Jubbulpore. Millar then taking its garrison with him left Damoh to its fate;[1] he disarmed the two companies of the 52nd, and brushing aside their mutinous comrades who tried to bar his passage, he on October 1 reached Jubbulpore, which the Madras troops had saved. Thenceforward the column remained at Jubbulpore till the end of January, 1858, when the advance of a fresh British force cleared the district.

II. Events North of the Jumna

Benares District.—During the time which elapsed between the departure of the Commander-in-Chief from Allahabad and the end of the year troops from England began to arrive, and the orders he gave before going to the front were carried into effect. Lieut.-Colonel H. E. Longden with 500 infantry (10th Foot and 17th Mad.N.I.), and 2 guns[2] reached Jaunpore from Benares on November 2. The Oudh rebels having again crossed the frontier and seized the fort of Atraulia, Longden, supported by the Gurkha force from Azimghur, bombarded the fort and drove the enemy back to Oudh. He then re-

[1] Damoh was held for a time by the loyal Rajah of Panna, but was eventually sacked by the rebels.

[2] Unidentified.

turned to Jaunpore, which shortly afterwards was converted into a brigade command under Colonel T. H. Franks,[1] 10th Foot, who was strongly reinforced. At the end of the year Franks was holding Azimghur with his right; his centre was a few miles in front of Jaunpore and his left was at Badlapore. Another field force under Brigadier Rowcroft[2] had assembled in Behar, to the north of Chuprah. It consisted of the Naval Brigade of H.M.S. *Pearl* (Captain Sotheby),[3] Bengal Police, and Gurkhas. On December 26 Rowcroft attacked a body of rebels on the right bank of the Chota Gundak, drove them across that river with the loss of a gun, and then marched towards Burhat Ghat on the Gogra. Meanwhile Jung Bahadur with a Gurkha force of infantry and artillery, 10,000 strong, entered British territory, and on January 5 was in the vicinity of Gorruckpore.

Allahabad.—Brigadier W. Campbell,[4] 2nd Dragoon Guards, commanding at Allahabad, from time to time sent out small columns to patrol the neighbourhood, which was in a disturbed state, but it was not until January 5 that any considerable body of rebels were brought to an encounter. On that date marauders numbering some hundreds were known to be in position near the village of Pandora, and Campbell marched to attack them with detachments

[1] Lieutenant-Colonel and Brevet-Colonel T. H. Franks, 10th Foot.
[2] Lieutenant-Colonel and Brevet-Colonel F. Rowcroft, 62nd Ben.N.I.
[3] Captain E. S. Sotheby had 5 guns with teams trained by himself. Lieutenant N. E. R. Turnour, R.N., was in charge of his artillery.
[4] Lieutenant-Colonel and Brevet-Colonel W. Campbell, 2nd Dragoon Guards.

of the 79th Highlanders and Rifle Brigade and a division of 4/14 R.A. (Lieutenant-Colonel S. E. Gordon). He had no cavalry, but it happened that E Troop R.H.A. (Major J. Anderson) was approaching Allahabad from Benares, and was only one march away. Orders were despatched to him to send his guns to Allahabad, but to march his detachments direct to Campbell and act as cavalry. The rebels offered but a feeble resistance, and after a few rounds from Gordon's guns Anderson charged and cut down two or three hundred of them.

Futtehpore.—Early in December a column was formed at Allahabad under Lieutenant-Colonel G. R. Barker, R.A., consisting of detachments of the 13th Light Infantry, the 53rd Foot, Madras Rifles, and 3/14 R.A. (Major W. G. Le Mesurier), in order to clear the Futtehpore district of rebels, who, well supplied with boats, made incursions from the right bank of the Jumna. Barker successfully performed this duty, although unable to bring the rebels to bay, and joined the Commander-in-Chief about the middle of the month. Immediately after his departure Futtehpore was occupied by the Madras troops under Brigadier Carthew, who during January made a march that cleared for a time the south-east Doab; but the summer of 1858 was well advanced before rebels on the right bank of the Jumna ceased to be a source of annoyance.

North-West Provinces and the Punjaub.—In Oudh the Begum, aided by the Moulvie of Fyzabad,[1] were busy organising their forces and add-

---

[1] See p. 47, n. 2.

ARTILLERY CROSSING A NULLAH

ing to the defences of Lucknow. The beleaguered British garrison had been rescued, it is true; but when our flag disappeared from the Residency many of the Oudh Talukdars, hitherto neutral or even friendly, threw in their lot with the rebels, and the task before the British included not only the recapture of the capital, but also the subjugation of the province. Although Rohilkhund was still in open revolt, all danger from the Punjaub had disappeared; for when the Sikhs saw their brethren returning laden with the spoils of Delhi, they believed that the old prophecy had been indeed fulfilled, and that the "fair people from the west" had stamped out for ever the power of the Moghuls.[1]

III. PLANS OF THE GOVERNMENT

The reader is now in a position to appreciate the military situation at the end of 1858. The crisis of the Mutiny being over and reinforcements crowding the harbours of Calcutta, Bombay, and Karachi, the Government were able to take a comprehensive survey of the whole theatre of war, and to lay down the outlines of a far-reaching strategy. The general plan was as follows: A force known as the Central India Field Force (with that of Malwa as a nucleus) was to assemble on the Nurbudda under the command of Major-General Sir Hugh Rose, and, having

[1] In the time of Aurangzíb the chief priest of the Sikhs was put to death in Delhi, and he was said to have declared that fair people from the west would sweep away the Moghul power and that the Sikhs would carry the spoils of Delhi back to the Punjaub.

captured Chanderi and Jhansi, was to advance on Kalpi and establish communications with the Bengal Army. Another force, the Rajputana Field Force under Major-General H. G. Roberts, was to advance from the Bombay Presidency on Nusseerabad, and after driving the rebels from Awah and Kotah was to operate on Sir Hugh Rose's left. A third force, the Saugor Field Force, consisting of Madras troops under Brigadier G. C. Whitlock,[1] was to assemble at Kampti and, after relieving Saugor, to sweep through Bundelkhund on Sir Hugh Rose's right. Meanwhile the Commander-in-Chief would subjugate Oudh and Rohilkhund, and for this purpose would draw to himself reinforcements both from the north and south. First the Doab was to be cleared of rebels, and a concentration effected at Futtehghur. From this point both Oudh and Rohilkhund were threatened, but every effort was to be made to induce the rebels to believe that Bareilly was the point selected for our attack. When siege guns had been collected and his reinforcements were at hand, the Commander-in-Chief would lay siege to Lucknow, and its capture would, it was hoped, leave him free to deal with such opposition as might still be met with north of the Jumna and in the Ganges basin.

[1] Lieutenant-Colonel and Brevet-Colonel G. C. Whitlock, 3rd Madras European Regiment.

CHAPTER VII

THE LUCKNOW CAMPAIGN—*continued*

I. THE ADVANCE INTO OUDH

Seaton at Gangaree, December 15.—In order to carry out the concentration at Futtehghur various columns were set in motion. On December 6 Brigadier T. Seaton [1] left Delhi for Futtehghur with a large convoy of stores, stretching over twenty miles of road, and the following troops:

Detachments:	1st Ben.E.F.	3/3 Ben.H.A. (Lieutenant H. P. Bishop).
6th D.Gs.	7th Punjaub I.	
9th Lancers.	Ben. Sappers and Miners.	5/1 Ben. H.A. (2 guns) (Lieutenant E. C. Griffin).
Hodson's Horse.		
		H.F.B. Sikh Co. Ben.A. (Lieutenant A. Gillespie).
		O.C.A., Lieutenant-Colonel R. R. Kinleside, Ben.A.

Having reached Alighur, he found Lieutenant-Colonel Farquhar [2] with a small mixed force encamped near Gangaree beyond the Kali River. A large body of rebels from Rohilkhund crossed the Ganges about December 10 with the evident intention of falling upon Farquhar; but their approach was discovered, and Seaton, leaving his convoy under the protection of the fort at Alighur, was waiting for them at Gangaree. When on December 15 the enemy made

[1] Lieutenant-Colonel and Brevet-Colonel T. Seaton, 60th Ben.N.I.
[2] Major and Brevet-Lieutenant-Colonel R. Farquhar, Beluchi Batt.

their appearance they were greeted by a brisk fire from the horse artillery, to which their 3 guns made but an ineffective reply, and realising that they had made a mistake they attempted to withdraw from the combat. The Carabineers and Lancers then made a dash for their guns, and were received by a steady and effective fire of case shot; but the charge was pushed home and the guns were taken. Simultaneously Hodson delivered a heavy blow on their left flank, and the withdrawal quickly became a rout. Their loss was 300 men, that of the British 48.

Action at Patiala, December 18.—Continuing his march, Seaton on the 18th found Patiala occupied by the rebels in force, and his advanced guard (Hodson's Horse) came under fire from the enclosures in front of the town. Sending two guns of 3/3 Ben.H.A. (Lieutenant T. A. Dirom) to reinforce Hodson, he massed the bulk of his cavalry on his right and formed up his infantry and heavy guns under cover of a slight undulation. Then Griffin with the 6-pounders of the native troop reinforced Dirom, and Bishop, making a détour to the right with his 4 remaining guns, came into action on the flank of the enemy, who now opened fire from 11 guns posted in front of Patiala. But the artillery duel was of short duration, for as our infantry and the heavy battery appeared on the crest of the undulation, our full strength was visible to the rebels, who at once began a retreat through the town. From their position on the extreme right our cavalry were able to gain the open country beyond the town without making their way through narrow streets, and they fell upon the flying rebels, and aided by the guns inflicted severe loss upon them. At least 700 were killed, and their 11 guns were taken. Seaton lost 1 man killed and 4 wounded. The effect of these two victories was drastic and the district was freed from organised bands of rebels, who all fled into Oudh or Rohilkhund.

Seaton's Junction with Walpole.—Seaton now ordered his convoy to continue its march from Alighur, escorted by Wale's Horse,[1] while he himself set out for Mainpuri, and on his way brushed aside an armed rabble which the Rajah with inconceivable fatuity had assembled to oppose his progress. Having taken 6 guns and killed 250 rebels with the loss of 2 men wounded, Seaton marched to Bewar, where his convoy was safely collected; on January 3 he was joined by Brigadier Walpole.

Brigadier Walpole left Cawnpore on December 18 with the 2nd and 3rd Battalions Rifle Brigade, the 38th Foot, 2/3 Ben.H.A., and 3/1 Ben.A. His orders were to make a semicircular march by Akbarpore and Etawah to Mainpuri, clearing the district as he went; but he met with little resistance on his way to Bewar.

The Commander-in-Chief's Movements. — After some minor operations had been carried out, Sir Colin Campbell's main body was encamped, January 1, along the Grand Trunk Road a few miles south of the Kali River. Brigadier Grant, in advance, was in command of a force consisting of the *Shannon* Naval Brigade, squadrons of the 9th Lancers and Punjaub

---

[1] Wale's Horse was one of the new levies, raised by Captain F. Wale, Bengal Army.

Cavalry, 7/14 R.A. (Captain C. H. Smith), a company of Engineers, and Hope's Brigade.

Action on the Kali River, January 2.—Early in the morning of January 2 it was discovered that the bridge over the Kali River was damaged, and its repair was at once begun. The ground beyond rose in a gentle slope for about half a mile to the village of Khudagunge, which lay on either side of the Grand Trunk Road. At this village, unknown to the British, a strong force of rebels was assembled, and on the morning of the 2nd, when the repair of the bridge was almost completed, they opened fire with round shot on a picket of the 53rd Foot which had crossed the river to cover the working party. The remainder of the 53rd then crossed the bridge and advanced up the slope in skirmishing order, with 7/14 R.A. on their left and 3 naval guns supporting their right rear. The enemy were held in check, but no further advance was permitted. Sir Colin himself had come up, and he decided to wait for reinforcements. At 11 a.m. Greathed's Brigade arrived and the 8th and 64th Foot crossed the river to support the 53rd. They were, however, taken in flank, and suffered some loss from a heavy gun which the rebels brought into action in front of the village. Captain W. Peel, R.N., then dragged a 24-pounder forward into our skirmishing line, and in a few rounds dismounted the rebels' gun and blew up its ammunition waggon. Still no advance was made, but some horse artillery and cavalry were brought across the bridge and drawn up under cover. Earlier in the day Sir Colin Campbell had ordered the 93rd High-

THE NAVAL BRIGADE AT KHUDAGUNGE.

landers back to camp to get their dinner, after which they were to return and assault Khudagunge. At 4 p.m. they came up to and were about to cross the bridge when the 53rd Foot (composed for the most part of Irishmen) suddenly sprang from cover, assaulted the village and drove the rebels out. Sir

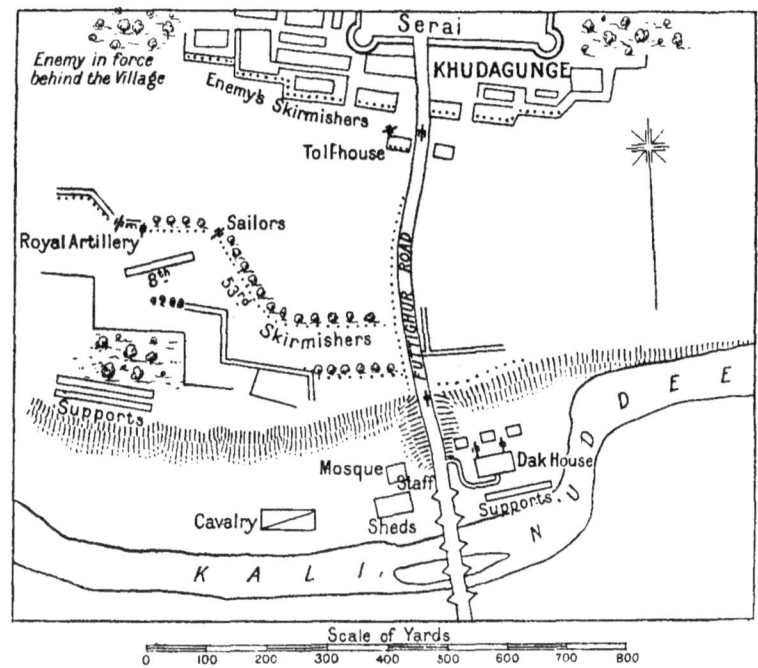

THE KALI RIVER, JANUARY 2.

Colin was much annoyed, but the 93rd were sent forward in support of the 53rd, and the rebels made a hasty retreat. Then Brigadier Grant with the cavalry followed them up on a line parallel to their line of retreat, but hidden from view by intervening crops until suddenly wheeling to his right he burst upon the rebels and pursued them with great slaughter for some miles.¹ Our loss was 10 killed and 30 wounded.

Concentration at Futtehghur.—Next day, January 3, Sir Colin entered Futtehghur, whence all the rebels, including the Nawab of Furruckabad, had fled into Oudh, and so precipitate had been their retreat that neither the bridge of boats across the Ganges nor the valuable gun-carriage factory which stands in the Fort was in any way injured. On January 6 Brigadiers Seaton and Walpole marched into Futtehghur.

Walpole on the Ramgunga.—When the concentration at Futtehghur was completed it was evident that the rebels were perplexed as to Sir Colin's intentions. They assembled in force at Aligunge on the Ramgunga, where they broke the bridge, and on January 12 Brigadier Walpole marched thither with orders to lead them to suppose that an advance on Bareilly was about to be made, without allowing himself to be drawn into an engagement beyond the river. He made ostentatious preparations, therefore, for the repair of the bridge, and the Commander-in-Chief himself carried out a mock reconnaissance. The rebels were completely deceived, and during ten or twelve days they sat watching the bridge while the siege train was being prepared at Agra. At the end of that time they detached two regiments to make an incursion across the Ganges above Walpole's position. Passing the river at a point twelve miles

¹ In this pursuit Lieutenant F. S. Roberts, Ben.A. (afterwards Field-Marshal Earl Roberts), who was on the staff, was engaged in two personal combats. He rescued one of our sowars and also captured a rebel standard. For these acts he received the Victoria Cross.

from Futtehghur, this force marched for Shamshabad, where, being joined by all the disaffected in the neighbourhood, their numbers reached 5,000. Brigadier Hope was at once sent against them from Futtehghur with the following troops :

9th Lancers.	42nd Highlanders.	1/1 Ben. H.A. (Captain and
Hodson's Horse.	53rd Foot.	Brevet-Major F. F. Rem-
	4th Punjaub I.	mington).
		No. 17 F.B. 3/1 Ben.A. (Captain C. H. Blunt).

Action near Shamshabad, January 26.—On January 26 Hope reached the rebel position, half a mile from Shamshabad. The road he was marching on crossed at right angles a nullah which, except at the bridge, was impassable for guns and cavalry. To the right of the road and beyond the nullah the ground rose abruptly, forming an elevated plateau, about a thousand yards square, bounded by the nullah, the road, and the plain, which extended eastwards to the Ganges. On this plateau stood a brick temple surrounded by an old entrenchment which the rebels had converted into a sandbag battery mounting 4 guns, the two principal faces parallel to the road and the nullah. When Hope's advanced guard, composed of Hodson's Horse and Remmington's Troop, came within range the rebel guns opened fire, and Lieutenant T. M. McDowell, second in command of Hodson's Horse, was killed by a round shot. Remmington then led his guns at a gallop across the bridge, and wheeling to his right and mounting the plateau, he came into action at a range of 600 yards. Issuing from the cover afforded by the slopes of the plateau, the rebel sowars suddenly appeared upon Remmington's left; but Hodson's Horse, with the timely support of the 9th Lancers, drove them back into the plain. Meanwhile Blunt brought his guns into action on the brink of the nullah, his Battery and Remmington's Troop being so placed that each enfiladed the face of the rebel work which the other counter-battered. The infantry now deployed behind

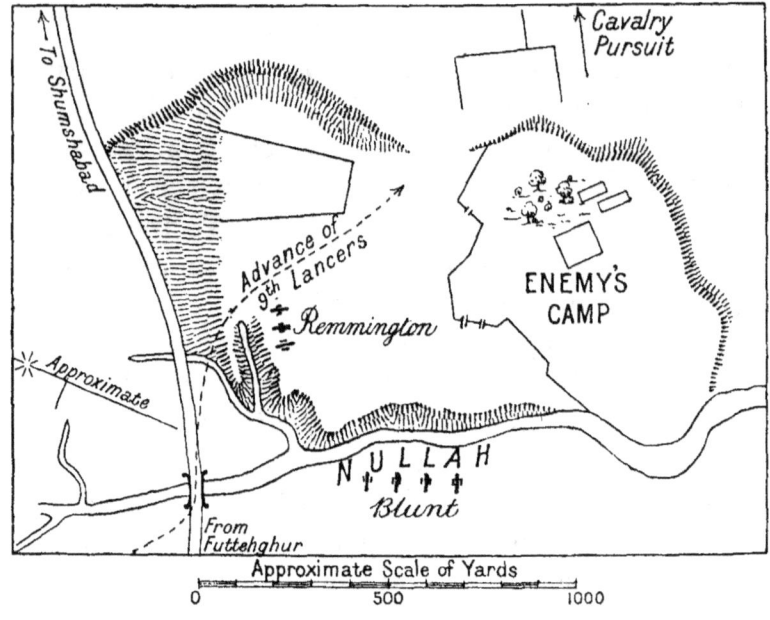

SHAMSHABAD, JANUARY 26.

Blunt's position, the 53rd being in reserve. In half an hour the British guns broke down all resistance, the Punjaub Infantry carried the sandbag battery with a rush, and the 42nd Highlanders and Remmington's Troop, pressing forward to the crest of the plateau, fired on the retreating enemy. Blunt, taking ground to his right, enfiladed the slopes of the plateau, and the cavalry with 2 guns of 1/1 Ben.H.A. (Lieutenant A. H. Lindsay) followed up the rebels, who made for the Ganges. Hope's casualties were 1 officer and 5 men killed and 5 officers and 34 men wounded.

The rebels made no further attempts to cross the Ganges, but on February 4 a considerable force crossed the Jumna at Kalpi and advanced to Bognipore, where they were met by Lieutenant-Colonel G. V. Maxwell with 5 companies of the Connaught Rangers, 50 sowars, and 2 guns of 5/13 R.A. (2nd Captain H. L. Talbot). A running fight took place, and the rebels were driven back across the river.

The Camp at Bantara.—The time was now ripe, and the Commander-in-Chief disclosed his real purpose. Leaving Brigadier Seaton in command at Futtehghur, he returned to Cawnpore and the advance into Oudh began. By the middle of February the main body of his force was across the Ganges and encamped along the Lucknow Road from Unao to Bunnee, while the siege train—fully twelve miles long—made its way to the plain of Bantara, where the force was to concentrate. But the reinforcements from the south-east and east were still some marches distant, and there was time to deliver another blow at the rebels before the great siege began.

Capture of Miangunge, February 23.—The Nana was said to have taken refuge in the fort of Futtehpore Churassie on the Ganges, twenty-five miles north-west of the Cawnpore—Lucknow road, and on February 15

Brigadier Grant was sent to attack him with the following troops :

7th Hussars (2 squadrons).	34th Foot.	E Troop R.H.A.[1] (2nd Captain and Brevet-Major J. E. Michell).
9th Lancers (2 squadrons).	38th Foot. 53rd Foot.	3/3 Ben.H.A. (Lieutenant H. P. Bishop).
		Detachment 5/13 R.A. (1 18-pounder gun and 1 8-inch howitzer) (Lieutenant F. G. E. Warren).

O.C.A., Captain and Brevet-Major J. R. Anderson, R.H.A.

Directing his march across country, Grant reached the fort on the 17th, but the Nana had fled. He then moved eastward, and, clearing the country as he went, he arrived on the 23rd before Miangunge— a strong walled town—where the rebels were concentrated in force. The 18-pounder gun and 8-inch howitzer manned by 5/13 R.A. were at once placed in position by Grant himself, and opened a breaching fire, while Bishop's Troop played upon the town. The 7th Hussars and 4 guns of E Troop R.H.A. were on our left flank watching a detached body of the enemy, the 2 remaining guns of the troop were with the baggage guard. After firing for about an hour Warren made a breach in the wall; Bishop dashed forward to close range and cleared the way with some rounds of case, and the 53rd Foot, entering by the breach, were soon masters of the town. Many rebels were killed by them and more were cut down by the cavalry. Our

---

[1] There was little 6-pounder ammunition available, and if it had not been for the good offices of Major F. Turner, Ben.H.A., E Troop R.H.A. could not have accompanied Grant.

losses were slight. At the end of the month Grant rejoined the Commander-in-Chief and other reinforcements were now close at hand.

II. THE REINFORCEMENTS FROM THE EAST AND SOUTH-EAST

The Gurkhas.—In January, 1858, Jung Bahadur, with 10,000 Gurkhas and 24 guns was at Gorruckpore, and, having driven back a fresh rebel incursion from Oudh, advanced southward to the Gogra to join Brigadier Rowcroft, and after two successful actions on February 17 and February 21 their united forces established themselves on that river. Jung Bahadur then pursued his way to Lucknow, while Rowcroft was ordered to hold Gorruckpore and keep the communication open. On March 10 the Nepaulese column came into touch with Sir Colin Campbell at Lucknow, when the siege was in full progress.

The Jaunpore Field Force.—In January the Jaunpore Field Force was composed of the following troops, which, as already stated, occupied Azimghur, Jaunpore, and Badlapore :

10th Foot	} 2,108 men.	6/13 R.A., 4 9-pounders, 2 24-pounder howitzers (Captain W. A. Middleton).[1]
20th Foot		8/2 R.A., 2 18-pounders, two 9-pounders, 2 5½-inch mortars (Captain J. E. Thring).
97th Foot		4/5 Ben.A., 2 9-pounders (Lieutenant E. Simeon).
Gurkha I., 3,193 men.		4/3 Mad.A., 2 9-pounders, 2 24-pounder howitzers (Captain and Brevet-Major G. S. Cotter).
Volunteer Cavalry		
		Gurkha Artillery, 6 9-pounders.
		O.C.A., Lieutenant-Colonel E. Maberly, R.A.

---

[1] 6/13 R.A. was sent to Jaunpore after the Relief of Lucknow ; the remaining artillery were reinforcements from Calcutta, Allahabad, etc.

The " Nazim " of Sultanpore.—At this period one Mahndi Husain, calling himself the Nazim of Sultanpore, drew together a following of some 15,000 men and established himself near Chanda. His force was composed chiefly of matchlock men, but there was a small proportion of sepoys and he had numerous guns. One of his lieutenants was at Saraon, fourteen miles north of Allahabad, with outposts within four miles of the city. On January 21 the British determined to fall upon this detached force and Franks' left column set out from Badlapore, while Lieutenant-Colonel C. L. D'Aguilar, R.H.A., was ordered to march from Allahabad to Sekundra and join Franks with F Troop R.H.A. and two squadrons of the 2nd Dragoon Guards. The British force at Sekundra on January 22 was thus composed:

2nd D.Gs. (2 squads.).	Detachments: 10th Foot. 20th Foot. 97th Foot.	F Troop R.H.A. (2nd Captain and Brevet-Lieutenant-Colonel H. P. Yates). 8/2 R.A. (Captain J. E. Thring). A/3 Mad.A. (Captain and Brevet-Major G. S. Cotter).

Affair near Sekundra, January 22.—The rebels at Saraon numbered 8,000 men and 14 guns, and on the approach of the British they broke camp and marched to join forces with a neighbouring Talukdar. They then took up a strong position at Nasratpore about two miles from Sekundra, and awaited attack concealed in a dense belt of jungle.

On the morning of the 23rd Franks advanced to the attack and according to his usual custom sent forward a skirmishing line of marksmen, with 2 guns of F Troop on each flank. As ground was gained, the other guns were pushed forward so that when the outskirts of the jungle were carried they all joined in a combined cannonade. D'Aguilar, with 2 guns of F Troop (Lieutenant H. T Arbuthnot) and a squadron of dragoons, was then sent to turn the left flank of the enemy, while Franks, with the main body, attacked the jungle. The rebels fled before our advancing infantry, and D'Aguilar, circling round to their rear, burnt their tents and blew up three cartloads of ammunition. Two guns were captured, the others escaped. Our casualties were 2 men slightly wounded. Next day the cavalry and horse artillery marched on Saraon. The fort, which was found deserted, was destroyed, and the column returned to Allahabad. Franks then set about re-establishing civil authority in the districts bordering that city, and having concentrated his columns at Singramow he waited till the operations of Jung Bahadur and Rowcroft on his right would render a further advance possible. By the middle of February the time had arrived.

Action at Chanda, February 19.—The " Nazim," with 10,000 men, was a few miles in rear of Chanda, where an advanced position consisting of a fort covered by an entrenchment was held by 8,000 men including 2,500 sepoys and 8 guns. On the morning of the 19th Franks attacked this position, and the marksmen of the 10th, 20th, and 97th advanced in a skirmishing line accompanied by 4 of Middleton's and 2 of Cotter's guns supported in rear by the 18-pounders of 8/2 R.A. (2nd Captain W. N. Waller).

The rebels did not stand. Colonel Maberly, R.A., reconnoitring on the left, was able, owing to the friendly cover of a mango grove, to get within 300 yards of their right flank, and he perceived that they were beginning to abandon their works. Cantering forward with his quartermaster, Lieutenant T. B. Strange, R.A., he rode into their flank battery, where two guns stood silent with a dead sepoy lying beside them. With the point of his sword Strange scraped "R.A." on the two guns, and then mounting the parapet waved his handkerchief as a signal to our infantry, whose bullets were falling round the entrenchment. The enemy were still active on their centre and left, but our infantry rushing forward on their right turned their flank and rolled up their line. Middleton and Cotter advanced at a gallop, the gunners clinging to the gun axletree seats or mounted on the off horses,[1] and coming alternately into action fired on the retreating rebels, who broke off into two divisions and followed two lines of retreat. Maberly then placed himself at the head of our handful of cavalry (who at the moment were without a definite leader), and advancing on the right, still accompanied by Middleton and Cotter, pursued the enemy for a mile and a half until the plain merging into mango groves and jungle forbade further progress. On their left the rebels sought to rally on some rising ground, but a few rounds from the 9-pounder guns of 8/2 R.A. and 4/5 Ben.A. put them to flight, and Franks advanced

[1] All the R.A. Companies employed as field batteries had the Indian equipment, comprising seats on the axletrees and saddles on the off horses. It was only the R.H.A. that brought their own guns, horses, and harness to India.

three miles farther. After a halt of two hours he moved to his left and occupied Hamirpore.

Meanwhile, but all too late, the " Nazim " was marching to support his force at Chanda. In the hope, it is supposed, of retrieving the fortunes of the day, he made a détour, and as evening was closing he attacked Franks' left rear, and a shot from an 18-pounder gun fell into camp as our men were pitching their tents. Happily the gun teams were not unhooked, and the infantry, turning out in their shirt sleeves, repelled two successive attacks with ease, but darkness coming on the retreat of the enemy was unmolested. During the day 6 guns were captured, but there is no record of the rebel loss. Our casualties amounted to 11 men wounded.

Action at Sultanpore, February 20.—Next day, February 20, the rebels, forestalled in capturing the strong position and fort of Budhayan, retired to Sultanpore, where a determined stand was to be made. Their numbers here reached 25,000, including 5,000 sepoys and 1,100 sowars with 25 guns. Mirza Guffur Bey, late a general in the service of the King of Oudh, was sent from Lucknow to take the command. Sultanpore, lying in a plain on the right bank of the Gumti, was covered on the south by a winding ravine which ran into this river. Deep at its eastern or river end and bordered by trees and groves for the greater part of its length, the ravine gradually grew shallower until it disappeared in the plain about 4,000 yards from Sultanpore at a point close to the Allahabad Road. Behind this ravine Guffur Bey took up a position which extended 2,500 yards from the Sultanpore bazaar on the left to a serai, Badshahgunge, on the right; his centre was behind some ruined police barracks and his right was buttressed by a line of hillocks. Fourteen of his guns, 5 being of siege calibre, commanded the Chanda—Sultanpore road and were massed in one central battery, with embrasures which admitted of little traverse. The remaining guns were distributed along the front, 6 posted in a sunken battery at Badshahgunge where their muzzles just showed above the terrain. The Bey's arrangements indicated his belief that he would be attacked from the Sultanpore road, for beyond Badshahgunge his right flank lay open and no cavalry, not even a scout, was placed there.

At 9 o'clock on the morning of the 23rd Franks came within sight of the enemy's outposts, and, reconnoitring in person, was not slow in realising the weak point of the position. Occupying their attention with a demonstration against their front, he directed a column of infantry, accompanied by 6/13 R.A. and A/3 Mad.A., to march round the enemy's right flank. A salvo from the Bey's 14 guns thundered down the road, but the column and the batteries had already quitted it, and, almost hidden from the enemy's view by crops and trees, they reached the head of the ravine on the Allahabad Road without any loss. The infantry then deployed and opened fire, the guns in line with them, Middleton engaging the sunken battery at Badshahgunge and Cotter bombarding the serai. After a few rounds our guns limbered up and advanced at a gallop. The infantry, and those gunners who could not find seats on the carriages, etc., were left behind as the batteries quickly traversed the 1,200 yards intervening between them and the flank of the enemy's position. In the sunken battery the guns rested on the bottom of a dried pond, and when the rebels turned them to meet our flank attack the trails and wheels broke through the crust of mud and their fire was very inaccurate. Middleton's last rounds of case shot fired point blank wrought havoc among the rebel gunners;[1] the "cease fire" was sounded, and Franks, who was close beside the guns, went forward with his staff. In his front rode Lieutenant J. J. McLeod Innes, Ben.E., and as he approached the silent battery he saw a sepoy raise a portfire to fire a gun which was pointed direct at the approaching staff. Innes instantly shot the sepoy dead, and covering the gun with his revolver kept other mutineers at bay until assistance arrived, despite the fact that he was exposed to a brisk fire from matchlock men ensconced in cover close by.[2] At length the outpaced infantry arrived and rolled up the rebel line, while Thring's guns and the Gurkha artillery on our right added to their discomfiture. The ravine which had protected their front served only to hem them in, and now with the capture of the central battery—where the sepoys stood by their guns to the last and died beside them—

---

[1] "Middleton always got into action as quickly as he could and also as near as he could get without coming under case fire, and I think it paid against the enemy's artillery. They used to make good practice, but got flurried when we got too close" (General W. Smith [formerly Lieutenant W. Smith 6/13 R.A.] to author).

[2] Lieutenant J. J. McLeod Innes, Ben.E., received the Victoria Cross for this exploit.

all resistance ended. Many rebels were killed in spite of their precipitate flight; indeed, if Franks had possessed cavalry few would have escaped. Twenty guns, including 32-pounders, 24-pounders, and 18-pounders, were captured. The British loss was 2 men killed and 2 wounded.

Next day Franks was joined by a detachment of Punjaub cavalry who a few days later intercepted a roving band of rebels (consisting of 500 infantry, 200 cavalry, and 2 guns), killed some hundred of them, captured the guns, and drove the rest across the Gumti. Then for a time the Nazim disappeared.

Attack on Fort Dhowrara, March 4.—Pressing on to Lucknow, Franks (on March 3) was joined by Captain J. R. J. Coles, with a squadron of the 9th Lancers, some native cavalry and 2 guns of F Troop R.H.A. (Lieutenant H. T. Arbuthnot), which had marched from the Dilkhusha, where Sir Colin Campbell was now encamped. Next day, when about eight miles from Lucknow, the Brigadier learnt that Fort Dhowrara (Munshigunge), two miles to the right of the road, was occupied by the rebels, and he sent Colonel Maberly, R.A., with the horse artillery and cavalry to turn them out. Three companies of the 97th and 2 24-pounder howitzers of A/3 Mad.A. (Lieutenant W. J. Bradford) followed in support.

Arbuthnot first came into action about 900 yards from the fort, but the 6-pounders were ineffective at that distance against the target opposed to them. The fort consisted of a central citadel closed by a strong gate and an exterior entrenchment armed with guns, and after some preliminary manœuvres Arbuthnot got within case range (200 yards) of the entrenchment and was able to take one of its flanks in reverse. When he had been in action for twenty minutes he was joined by Bradford's howitzers, and very soon afterwards the outer works of the fort were in the hands of our infantry, while a number of the fugitive garrison were cut down by our cavalry. The central citadel still held out, and the fire from its loopholes rendered the retirement of the guns (which were powerless against its walls) desirable while preparations were made to destroy the gate. Lieutenant T. B. Strange, R.A., followed by Sergeant Wilkins and Gunners Critchell and Dommett, volunteers from F Troop, R.H.A., went forward and first turned one of the abandoned rebel guns—a 6-pounder—against it, but the fire of this piece was quite useless, and preparations were made to blow in the gate with powder. Lieutenant J. J. McLeod Innes, Ben.E., aided by Strange, then began to fill a bag with powder taken from the ammunition boxes captured in the entrenchment, when he was severely wounded. Captain F. D. Middleton, 29th Foot, rushed to Strange's assistance, and the bag was filled and a piece of quick match inserted. Strange made for the gate with the bag on his back while Middleton accompanied him with a lighted portfire attached to his hog spear.[1] The bag was placed in position with match lighted and the two officers got safely under cover, but the resulting explosion only splintered the gate.[2] A

[1] Captain F. D. Middleton was orderly officer to Brigadier Franks. General T. Bland Strange informs the author that Middleton generally carried a hog spear in action.
[2] The powder was of native manufacture.

peremptory order to retire was then given, and our men quitted the entrenchment followed by triumphant and derisive shouts from the matchlock men in the citadel; but the four guns that we had captured were safely brought away.[1]

About 6 o'clock the same evening Franks' force joined Sir Colin Campbell at the Dilkhusha.

III. OUTRAM'S DEFENCE OF THE ALAM BAGH

When Sir Colin Campbell began his return march to Cawnpore after the Relief of Lucknow, he confided to Major-General Sir James Outram the task of maintaining a military position in Oudh during his absence. The vicinity of the Alam Bagh was selected for this purpose as it covered the communication with Cawnpore and would serve in the fulness of time as a stepping-stone for further operations. His force was thus composed:

Military Train.	5th Foot.	3/8 R.A.
Volunteer C. (European).	75th Foot.	(2nd Captain F. C. Maude).
Detachments.	78th Foot.	4/1 Ben.A. (Detachment)
Newly raised levies.	84th Foot.	(Lieutenant J. Alexander).[2]
	90th Foot.	No. 12 F.B. 2/3 Ben.A.
	1st Mad. E.F.	(Captain W. Olpherts).
	Ferozepore Regt.	No. 3 F.B. 1/5 Ben.A.
	27th Mad.N.I. (Wing)	(Lieutenant E. W. E. Walker).
		No. 20 F. B. 3/5 Ben.A.[3] (Captain
	Mad. Sappers and Miners	G. Moir).
		Details Ben.A. (golandauz) and Oudh Irr.A.
		O.C.A., Major V. Eyre, Ben.A.

---

[1] When the guns retired Arbuthnot and three gunners armed themselves with the four carbines carried on the limbers and exchanged shots with the rebels. One gunner was wounded, but Arbuthnot killed the man opposed to him.

[2] Captain H. Francis assumed command of 4/1 Ben.A. in January, 1858.

[3] This battery arrived in January.

The Alam Bagh enclosure was about 600 yards square, and was occupied by a strong advanced picket; the main body was encamped a mile in rear and the front was protected by swamps, abattis,

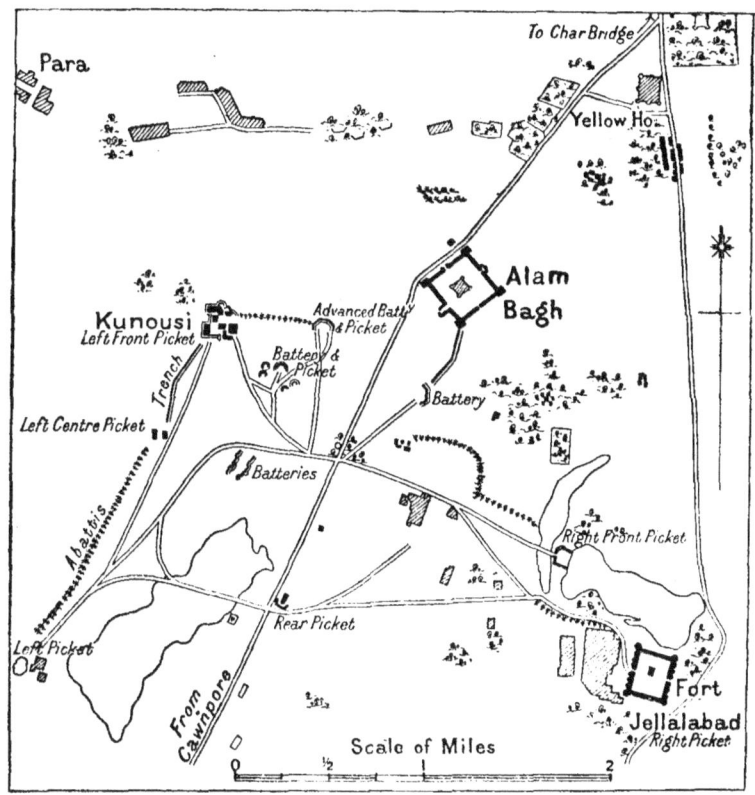

ALAM BAGH DEFENCES, NOV. 28, 1857, TO MARCH 1, 1858

and gun emplacements, with pickets, as shown in the plan. Within the quadrilateral defined by these pickets were supports, lines of abattis, and batteries, the boundary of the position being nearly eleven miles in extent. A detached force held the Bunnee bridge, and convoys every fortnight proceeded to and from Cawnpore.

The enemy occupied a line of earthworks south of the canal, running from Para to the Yellow House, so that Outram's advanced posts were within " gunshot range of the outworks of the vast city, swarming with hosts of mutinous sepoys, with the undisciplined but well-armed soldiers of the rebel Government with many thousand city 'badmashes,' the armed and turbulent scum of a population of 700,000 souls, and with numerous bands of those feudal retainers of the chieftains and great zemindars of Oudh whose normal state for fifty years past had been one of warfare." [1]

The Attacks on his Position.—In the city was the Begum with the young son of the deposed King; at her right hand was the Fyzabad Moulvie who was destined for a time to be a thorn in the side of the British. Appeals, and not in vain, were made to rouse the national spirit, and Hindu jealousy of the ruling race was forgotten in a common patriotism. Depressed for a time by their reverses, the rebels in Oudh remained quiet. But as the days wore on and Sir Colin Campbell's operations appeared to be standing still, while daily accessions were increasing the Lucknow garrison, they determined to surround and annihilate the small force which alone represented British authority in the province. The first attempt was made on December 22, and during January and February their attacks were repeated, but being well served by his spies Outram was always ready and

[1] Outram's Despatch.

always victorious, and while the losses of the enemy were often serious, his were always slight. On February 25 they made their last effort; but by this time Outram was in touch with Sir Colin Campbell and was reinforced by the 7th Hussars, Hodson's Horse, and 1/1 Ben.H.A. (Major F. F. Remmington), and the rebels were utterly defeated and driven back into Lucknow.

During the twelve weeks of Outram's defence of the Alam Bagh position 3/8 R.A. and 4/1 Ben.A. were employed principally on picket duty and in manning guns in the Alam Bagh and in Fort Jelalabad. They had many opportunities of engaging the enemy, but it was the guns under Olpherts, Moir, and Walker that took part in the field operations.

IV. THE DEFENCES OF LUCKNOW

The Inner Line of Defence.—For three months thousands of workmen were ceaselessly employed on the defences of Lucknow. The Kaisar Bagh was not originally fortified, but on the departure of the British garrison it was at once strengthened, and became the rebel citadel. Behind it lay a wide expanse of dense and almost impenetrable city; on its left or western flank were the Chattar Manzil, Farha Baksh, and other strong buildings resting on the river bank; on the right was Hazratgunge and its line of mosques and palaces. In front lay the Chena Bazar, and here a strong entrenchment was made connecting the Chattar Manzil with the King's Brother's House, where a bastion was built to command Hazratgunge. Five other bastions were evenly distributed along the length of this entrenchment, which was armed with cannon and prepared for musketry defence. It constituted the inner or third line of the rebel defence.

Second Line.—Contiguous to the enclosure of the King's Brother's House stood the Little Imambara, and from this point another line of entrenchment— the rebels' second line of defence—ran in a northerly direction to the Gumti, embracing the Khurshid Manzil and the Moti Mahal enclosures. From the Little Imambara this line was continued eastward along Hazratgunge and ended at Banks' House.

First Line.—As the canal approached the Gumti it did not follow the direct way to the river, but bent sharply to the eastward. Its course was now prolonged in its original direction by a deep trench 600 yards long leading straight to the river. From its new point of junction with the Gumti a strong rampart was made along the left bank of the canal as far as the Char Bagh. The banks were scarped and impassable; the bridges were broken down, and bastions were constructed flanking the newly made trench at various points along the line. This was the rebel first line of defence.

Defences in the City.—As a rule all the main streets were commanded by bastions and barricades, and every building of importance, besides being loopholed, had an outer wall protecting its entrance. Houses far in the depths of Lucknow were carefully prepared for defence, some mounting guns so that even should part of the city be lost, resistance could still be continued. In the larger buildings gunpowder was stored in great quantities, and almost every house had its own small supply.

After the siege 127 guns, mostly of small calibre, were captured which had been mounted in the defences.

V. THE ARMY OF OUDH

About the middle of February the organisation of the Army of Oudh was published as follows: [1]

NAVAL BRIGADE (Captain W. Peel, R.N.)

CAVALRY DIVISION (Brigadier J. H. Grant)

1st Brigade (Brigadier A. Little):	2nd Brigade (Brigadier W. Campbell):
9th Lancers.	2nd D.Gs.
Military Train.	7th Hussars.
2nd Punjaub C.	Volunteer C. (Europeans).
5th Punjaub C. (detachment).	1st Punjaub C. (detachment).
Wale's Horse.	Hodson's Horse.

ARTILLERY DIVISION (Major-General Sir A. Wilson, Bt., Ben.A.)

A.A.G., Major E. B. Johnson, R.A.

D.A.Q.M.G., Lieutenant R. Biddulph, R.A.

FIELD ARTILLERY (Brigadier D. E. Wood,[2] R.H.A.)

Captain and Brev.-Lieutenant-Colonel F. Turner, Ben.H.A.

Lieutenant J. S. Frith, Ben.H.A. (*Brigade Major*)

E Troop, R.H.A. (Captain and Brev.-Major J. R. Anderson).	5/12 R.A. (Q Battery) (Captain J. R. Gibbon).
F Troop, R.H.A. (Captain and Brev.-Lieutenant-Colonel C. L. D'Aguilar).	3/14 R.A. (No. 20 Battery) (2nd Captain and Brev.-Major W. G. Le Mesurier).
1/1 Ben.H.A. (Captain and Brev.-Major F. F. Remmington).	6/13 R.A. (Captain W. A. Middleton).
2/1 Ben.H.A. (Captain and Brev.-Lieutenant-Colonel H. Tombs).	No. 12 F.B. 2/3 Ben.A. (Captain and Brev.-Major W. Olpherts).
2/3 Ben.H.A. (Captain W. A. Mackinnon).	A/3 Mad.A. (Captain and Brev.-Major G. S. Cotter).
3/3 Ben.H.A. (Lieutenant H. P. Bishop).	

---

[1] The Artillery of the Jaunpore Field Force is included.

[2] Lieutenant-Colonel and Brevet-Colonel D. E. Wood, R.H.A.

276   THE LUCKNOW CAMPAIGN   [CHAP. VII

SIEGE ARTILLERY (Brigadier G. R. Barker, R.A.)

Lieutenant-Colonel C. J. B. Riddell, R.A.
Lieutenant-Colonel E. Maberly, R.A.
Lieutenant A. Bunny, Ben.H.A. (*Brigade Major*)

8/2 R.A. (Captain J. E. Thring).
3/8 R.A. (2nd Captain F. C. Maude).
6/11 R.A. (2nd Captain W. H. Goodenough).
5/13 R.A. (2nd Captain H. L. Talbot).

4/1 Ben.A. (Captain H. Francis).
1/5 Ben.A. (Lieut. E. W. E. Walker).
3/5 Ben.A. (Captain G. Moir).
4/5 Ben.A. (Lieutenant E. Simeon).
Recruits Ben.A. (Lieut. W. Powell).

ENGINEER BRIGADE (Brigadier R. Napier)[1]

4th and 23rd Companies R.E.   Ben. Sappers and Miners.
Sikh Pioneers.

INFANTRY

1*st Division* (Major-General Sir J. Outram).
  1*st Brigade* (Brigadier D. Russell).
    5th Foot.
    84th Foot.
    1st Mad.E.F.
  2*nd Brigade* (Brigadier C. Franklyn).[2]
    78th Foot.
    90th Foot.
    Ferozepore Regt.

2*nd Division* (Brigadier Sir E. Lugard[3]).
  3*rd Brigade* (Brigadier W. Hamilton).
    34th Foot.
    38th Foot.
    53rd Foot.
  4*th Brigade* (Brigadier the Hon. A. Hope).
    42nd Highlanders.
    93rd Highlanders.
    4th Punjaub I.

3*rd Division* (Brigadier R. Walpole)

5*th Brigade* (Brigadier J. Douglas).[4]
  23rd Foot.
  79th Highlanders.
  1st Ben.E.F.

6*th Brigade* (Brigadier A. H. Horsford).[5]
  2nd Batt. Rifle Brigade.
  3rd Batt. Rifle Brigade.
  2nd Punjaub I.

---

[1] Lieutenant-Colonel R. Napier, Ben.E.
[2] Lieutenant-Colonel and Brevet-Colonel C. Franklyn, 8th Foot.
[3] Major and Brevet-Colonel Sir E. Lugard, 29th Foot.
[4] Lieutenant-Colonel and Brevet-Colonel J. Douglas, 79th Foot.
[5] Lieutenant-Colonel and Brevet-Colonel A. H. Horsford, Rifle Brigade.

VI. THE SIEGE AND CAPTURE OF LUCKNOW

The Advance to the Dilkhusha, March 2.—On the morning of March 2 the Commander-in-Chief set out from Bantara, and passing by the Alam Bagh position directed his march on the Dilkhusha. The advance was led by cavalry and horse artillery, and the Dilkhusha plateau was occupied with little opposition; but as the infantry brigades closed on the advance guard the guns in the canal bastions opened a well-sustained fire and compelled Sir Colin to select a site for his camp as far back as the ravines on the south-east would permit. Two naval guns near the Dilkhusha and two more near the Mohammed Bagh replied to this fire and kept it in check while two advanced infantry pickets were posted in the neighbourhood of the guns.

Occupation of the South of Lucknow, March 3 and 4.—During the next two days the occupation of the position on the south and west of Lucknow was continued. Walpole's Division formed on the left of Lugard's, while Hodson's Horse watched the interval between Walpole's flank and Jelalabad, the right of the Alam Bagh position. The First Division, which was at the Alam Bagh, was broken up. Russell's Brigade, 3/8 R.A., and 3/5 Ben.A. were drawn in to the Dilkhusha; while Franklyn's Brigade with 4/1, 2/3, and 1/5 Ben.A. held the Alam Bagh. On the extreme left Campbell's Cavalry Brigade swept the country to the north-west, while on the south-east connection was made with Brigadier Franks' infantry.

The siege train moved into the Dilkhusha and the engineer park was established at Bibipore on the Gumti.

Sir Colin's Plan of Attack.—Lucknow was strong on the west, east, and south, but no obstacle except the Gumti, 60 yards wide, existed on the north, and all the defences except those of the third line could be turned and taken in enfilade from this side of the city. This glaring defect being at once realised by Sir Colin, he determined to make a strong attack on the south and to send a force across the Gumti to operate on the east and north. The siege would then develop into a left and right attack. To Sir James Outram were confided the operations on the left bank of the Gumti (the Right Attack).

The Siege Batteries begun.—During March 3 the Royal Engineers having made an emplacement [1] (L.a.) in front of the Dilkhusha Palace, Peel maintained a brisk fire on the canal bastions and on three guns which the rebels advanced to the Martinière. Next day all the naval guns were withdrawn to their park while the artillery brought four 18-pounder guns

---

[1] The author has been unable to reconcile the conflicting accounts given of the siege batteries and guns. The " siege batteries " at times were merely positions taken up in the open or behind natural cover, and the guns for which the engineers made certain emplacements were not in every case brought there. The author has therefore designated the batteries, emplacements, etc., by letters (not to conflict with the numbers employed by others) and prefixed an L. or an R. to designate Left or Right attack. In the description of these batteries, etc., he has followed Lieutenant-Colonel W. O. Lennox, R.E., in " Professional Papers, the Corps of Royal Engineers," Vol. X. (new series), p. 80, etc., but with regard to the armament he has depended on the despatches of Colonel G. R. Barker and Colonel C. J. B. Riddell, published in " R.A.I. Occasional Papers," vol. i. 71, 72 ; 81–83.

and two 8-pounder howitzers to the emplacement and continued the fire as before. At sunset the Royal Engineers began two cask-pier bridges, side by side, near Bibipore.

Passage of the Gumti.—Early on March 5 the rebels opened fire from a light field battery upon the men working at the Gumti cask-pier bridges, but the covering party of infantry and two guns of Q Battery R.A. quickly drove them off. The engineers, however, made a *tête du pont* on the left bank, while two naval guns were posted on the right bank, not only to protect the bridges, but to keep down the fire of the Martinière.

About 2 a.m. on March 6 Outram began the passage of the Gumti by the bridges, which were now complete. His force consisted of the 2nd Dragoon Guards, 9th Lancers, and detachments of Punjaub Cavalry; the 3rd Infantry Division and F Troop R.H.A., 1/1 and 2/3 Ben.H.A., 5/12 and 6/13 R.A. The artillery was commanded by Brigadier D. E. Wood, R.A. A siege train was also to accompany him, commanded by Lieutenant-Colonel C. J. B. Riddell, R.A. It consisted of eight 24-pounder guns, four 8-inch howitzers, and ten 8-inch mortars, and its *personnel* was supplied by 8/2, 6/11, and 5/13 R.A. and 4/5 Ben.A.

March 6.—By dawn Outram with his field troops was across the river, and began his march in a northerly direction. Presently the enemy appeared on the left flank, and a smart cavalry skirmish ensued, in which the horse artillery took part. Though the Dragoon Guards in their impetuous advance suffered some loss, the enemy were speedily dispersed, and at nightfall Outram was encamped on the Fyzabad road at Ismailgunge.

During the 6th Brigadier Franks, whose infantry had been in direct communication with the Commander-in-Chief on the previous day, occupied the position that Brigadier Walpole had vacated. The artillery continued in action at the Dilkhusha, and the engineers prepared to move the lower cask-pier bridge to a position farther down the river, where it would be out of sight of the Martinière.

March 7.—On the morning of March 7 the rebels concealed in the groves and ravines in front of Outram's position opened an artillery fire about half-past 8 upon our advanced picket, which was commanded by Major J. F. Pennycuick, who had under him a half battery of 6/13 R.A. Advancing his guns, Pennycuick drove back the enemy in his immediate front, and presently was joined by Middleton with the other half battery. Then the horse artillery under Colonel D'Aguilar came up, F Troop on the left and 1/1 and 2/3 Ben.H.A. on the right. Dashing to the front by alternate half troops, they searched out with case fire the rough ground in which the rebels lay concealed. The Punjaub Cavalry was in support, but the nature of the ground did not permit them to act, and it was by case fire alone that the rebels were driven across the Kokral. They were not followed up, as Outram did not wish to get entangled in the suburbs of Lucknow till his heavy guns arrived.

In the Left Attack, firing from the Dilkhusha plateau was continued during the day.

March 8, Right Attack.—The cask-pier bridge

having been placed in its new position, Lieutenant-Colonel Riddell crossed the Gumti with his siege train and joined Outram on March 8 ; and F Troop, R.H.A., 6/13 R.A., and the 9th Lancers recrossed to the right bank and rejoined the Commander-in-Chief. During the night the Royal Engineers made an emplacement, R.a., for eleven guns at the Kokral Bridge.[1]

March 8, Left Attack.—In the Left Attack the Engineers made two new emplacements, L.b. and L.c., for guns to fire at the Martinière and to keep down fire from Banks' House. L.b. was made on the right of L.a. and L.c. on the right of the Mohammed Bagh. Firing was continued from L.a., but the naval guns near the bridge were brought back to their park.

March 9, Right Attack.—The emplacement R.a. was armed with eight 24-pounder guns and three 8-inch howitzers under Captain Thring, R.A.,[2] and early on the 9th fire was opened on the rebel position which extended along our front, the key being the Race Stand some 600 yards away. Then Brigadier Walpole, with the bulk of the infantry and Q Battery R.A., advanced against the left of the enemy and, sweeping them aside, made a complete wheel to the left and occupied the Fyzabad Road on the north of the Race Stand. Meanwhile Brigadier D. Wood, with a battalion and a half of infantry and 1/1 and 2/3 Ben.H.A., having moved directly to the same spot, the Race Stand was captured, and the

[1] This emplacement (outside the limits of Map V.) is indicated on Map XI.
[2] Captain Thring was assisted by 2nd Captain W. N. Waller, R.A. and Lieutenant E. Simeon, Ben.A., who commanded detached portions of this battery.

left bank of the Gumti up to and including the Badshah Bagh fell into our hands. Towards the afternoon Captain Thring with three 24-pounders and one 8-inch howitzer came into action in the open, near the village of Jagrauli, in a position, R.b.,[1] whence he could enfilade the canal, and subsequently 2nd Captain Waller placed one 24-pounder gun and one 8-inch howitzer in an emplacement,[2] R.c., close to the river and 600 yards west of the Race Stand and opened fire on the Shah Najaf and Moti Mahal. Outram had thus turned the first line of the rebel defences.

March 9, Left Attack.—On the morning of the 9th the artillery placed four 8-inch mortars close to L.a., while the Naval Brigade brought to the new emplacement, L.b., six 8-inch guns, two 24-pounder guns, and four rocket waggons. They also put two 24-pounder guns under natural cover at L.d., some 400 yards to the front of L.a., in order to fire on the Martinière in flank and reverse. The Royal Artillery brought two 18-pounder guns and two 8-inch howitzers to L.c., and early in the morning opened fire on Banks' House. All the other guns, mortars, etc., were directed on the Martinière till about 2 p.m., when the regimental colours of the 1st Bengal European Fusiliers were seen floating from the Race Stand. Sir Colin then knew the progress

[1] 2nd Captain C. F. Young (Quartermaster to Colonel C. J. B. Riddell) and Lieutenant E. Simeon, Ben.A., were at times present in this battery.

[2] The armament at R.c. was subsequently increased to two 24-pounder guns and two 8-inch howitzers. They were commanded successively by Captain A. Pearson, Ben.A., and Lieutenant E. Simeon, Ben.A.

of Outram's attack, and at once gave orders for the assault of the Martinière. The 42nd Highlanders went to the front, supported by the 93rd Highlanders, the 53rd Foot, and the 90th Light Infantry, and the Martinière was promptly captured. Later in the afternoon a British officer was seen to be signalling from the flank bastion of the rebels' line of defence along the canal. He proved to be Lieutenant T. A. Butler, 1st Bengal European Fusiliers. It having been perceived from R.b., where Thring's guns were in action, that the rebels were abandoning their first line, Butler volunteered to swim the Gumti and signal the intelligence to the troops of the Left Attack. The stream was somewhat strong, but Butler landed safely in rear of the bastion, and mounting the parapet, succeeded in attracting the attention of Hope's Brigade.[1] The 4th Punjaub Infantry, followed by the 42nd Highlanders, at once went forward and occupied the abandoned works without encountering any opposition, and then, advancing westward, they became masters of the first line of defence up to the vicinity of Banks' House. The conquered ground was then occupied by the 53rd Foot. During the day the casualties in the Left Attack were 3 killed and 26 wounded. Considering what had been accomplished, this loss might well be considered slight, if it had not included the gallant Captain Peel, the life and leader of the Naval Brigade. He was severely wounded when reconnoitring in advance of his guns by a bullet which struck him in the thigh.[2]

[1] Butler received the Victoria Cross.
[2] Captain Peel (then Sir William Peel, K.C.B.) was recovering from

March 10, *Right Attack.*—On March 10 Outram strengthened his position about the Badshah Bagh, where the suburbs in his front were strongly held by the enemy, and Major-General Grant [1] with the cavalry and horse artillery patrolled the vicinity. Two new siege batteries were armed. An emplacement, R.d.,[2] was made 250 yards to the east of the Badshah Bagh whence five 8-inch mortars kept up an almost constant fire on the Kaisar Bagh until it was captured, and three 24-pounder guns, with a fourth in the open close by, were placed in an existing earthwork, R.e., close to the east gate of the Badshah Bagh, to fire on the buildings along the river in front of the Chena Bazar.[3]

March 10, *Left Attack.*—In the Left Attack a new gun position, L. e., was selected where natural cover existed some 400 yards to the west of the Martinière. Thither the Naval Brigade advanced some guns from L.b., and the artillery their mortars from L.a., and, in conjunction with the guns at L.c., opened fire on Banks' House and its vicinity. The naval guns having made a breach in Banks' House, it was carried

his wound, when by a deplorable misadventure he was placed in a dhooly that had carried a small-pox patient. He took the disease and died, mourned and honoured by all, by none more sincerely than by the Royal Artillery, who witnessed at close quarters his splendid services in the Crimea and the Mutiny.

[1] Colonel James Hope Grant, C.B., was promoted Major-General in February 24, 1858.

[2] The mortar battery was commanded for three days by 2nd Captain W. H. Goodenough, R.A., without relief. Later, 2nd Captain W. N. Waller relieved him. On the 10th two 24-pounder guns were placed close to the mortars.

[3] Captain J. E. Thring commanded this battery, which was originally made by the rebels.

by assault by the 42nd Highlanders and 4th Punjaub Rifles and converted into a military post. The Bungalows and the Karabala were also occupied, and new gun emplacements on the left bank of the canal were prepared. As soon as Banks' House was captured four naval 8-pounder guns were placed under natural cover at L.f. to fire at the Begum's Palace, and two more were installed in an emplacement, L.g., at the end of the Hazratgunge to counter-batter a rebel battery which had been made across the road between the Begum's Palace and a serai opposite to it. At the same time the artillery placed eight $5\frac{1}{2}$-inch mortars at L.h., between L.f. and L.g. During the afternoon the small mortars were taken to the garden of one of the Bungalows at L.i., and were replaced at L.h. by six mortars of 8-inch calibre.[1] Communications were also established between Banks' House and the Bungalows and with the rear.

March 11, Left Attack.—On the morning of March 11 two 8-inch naval guns were advanced to the Bungalow garden and coming into action at L.j., directed a breaching fire against the serai. At the same time the 8-inch mortars at L.h. were reinforced by five of 10-inch calibre, and guns and mortars kept up a hot fire till about 3 o'clock. The breaches were then considered practicable, and the assault was given by two columns at the Begum's Palace and a third at the serai. Engineer parties were at their

[1] Captain G. Moir, Ben.A., was in command of the 8-inch mortars, the fire of which was reported by Brigadier Barker as being very efficient.

head with powder bags and ladders, and the operations were carried out by Brigadier the Hon. A. Hope and Brigadier R. Napier, the Chief Engineer. The 93rd Highlanders, supported by the 4th Punjaub Infantry and 1,000 Gurkhas, captured the Begum's Palace, while the 42nd Highlanders and 53rd Foot carried the serai. This was the sternest struggle which occurred during the siege, and while terrible punishment was inflicted on the rebels, the British loss was serious.[1] As soon as the Palace and serai were captured, the Naval Brigade advanced with two guns along Hazratgunge to the place occupied by the rebel battery—now abandoned—and began firing back through their own embrasures at a confused mass of fugitives which streamed down the road to the inner line of defence. Once within the shelter of the bastion in this line which commanded Hazrutgunge they opened a brisk fire on the sailors, but the latter were well protected by the earthwork which the rebels themselves had made. During the day the Barracks, the Sekundra Bagh, the Kadam Rasul, and Shah Najaf were occupied without opposition. The casualties of the Left Attack amounted to 31 killed and 86 wounded.[2]

[1] The 93rd Highlanders had 52 casualties, including 2 officers and 12 men killed.

[2] Acting Mate H.P. Garvey, R.N., was killed on this day. He was sent with a message from the emplacement L.f. to that at L.g., and riding at speed he passed in front of the mortars at L.h. The 5½-inch mortars were primed with strands of quick match, which were lighted by means of burning tow attached to linstocks. Garvey did not notice that the primings were lighted, and as he passed the mortars went off and a shell struck him on the head. His horse was unhurt. General T. B. Strange has explained to the author that the linstock was used during the siege in preference to a portfire. A portfire when once

March 11, Right Attack.—On the morning of the 11th Outram directed two columns of attack against the suburbs about the Iron and Stone Bridges. The right column formed on the Fyzabad Road under Brigadier Walpole and consisted of the 79th, the Rifles, the 1st Europeans, half Q Battery (Captain Gibbon, R.A.), and two 24-pounders (Captain A. Pearson, Ben.A.). It worked its way, led by its skirmishers, through the suburbs of the town until it reached a mosque [1] on the old cantonment road which commanded the approach to the Iron Bridge. The left column was to advance under Lieutenant-Colonel R. Pratt, 23rd R.W. Fusiliers, by the lower road towards the same point. This column consisted of the 23rd, the 2nd Punjaubis, half Q Battery (2nd Captain A. C. Johnson, R.A.), and two 24-pounder guns (Captain J. E. Thring, R.A.), the two columns being connected by a strong chain of skirmishers which, as well as the left column, met with considerable opposition. Before Pratt's column advanced, Thring opened fire with case shot upon the ground and village in front, and cleared a way for the infantry, who eventually succeeded in occupying the riverside houses and the approach to the Iron Bridge, to the west of which two 24-pounders were placed in a battery, R.g.[2] Leaving the Bengal Fusiliers to

lighted threw off a shower of sparks which at night was sure to draw fire from marksmen posted in minarets and other suitable places.

[1] This mosque was about 700 yards to the north of the Iron Bridge.

[2] These guns till next day were under Captain C. F. Young, R.A. Then Captain A. Pearson, Ben.A., took command. Lieutenant E. C. Cuthbert, R.A., who was with him, on the 13th distinguished himself by putting out a fire in the vicinity of the battery.

hold the mosque in the old cantonment road, Outram led the remainder of the right column along that road to join General Grant, who, with the cavalry and horse artillery, was reconnoitring on the right. The junction being effected, Outram wheeled to his left and made for the Stone Bridge, surprising a rebel cavalry camp and capturing two guns. On reaching the head of the Stone Bridge, without serious opposition, he found it commanded by guns and also by musketry from the opposite side of the river. Thinking the position untenable by the force at his disposal, he withdrew to the mosque occupied by the Bengal Fusiliers; and having made arrangements for holding the Iron Bridge, which included the placing of one 24-pounder gun and one 8-inch howitzer [1] in an emplacement (R.i.) and building an epaulment of sandbags across the bridge, he withdrew the bulk of his force to the camp at Badshah Bagh.

During the previous night the engineers had made a new battery, R.f., in front of the Badshah Bagh, and on the night of the 11th the guns were moved there from R.d. The new battery was made up to a strength of four 24-pounder guns, two 8-inch howitzers, and two 8-inch mortars, three similar mortars being placed outside a few yards away.[2]

March 12, *Right Attack.*—Next day Outram remained stationary, but some additions were made

---

[1] These pieces were removed from the batteries at the Badshah Bagh and were first commanded by Lieutenant E. Simeon, Ben.A., and afterwards by 2nd Captain W. N. Waller, R.A.

[2] These guns, etc., were commanded in succession by Captain J. E. Thring, R.A., 2nd Captain W. N. Waller, R.A., Major F. F. Remmington, Ben.H.A., and Lieutenant A. H. Lindsay, Ben.H.A.

to the heavy guns firing into Lucknow. A battery, R.h., was made on the right front of the Badshah Bagh for four 10-inch mortars,[1] two 10-inch howitzers were added to R.c. (one 8-inch being withdrawn), and an 8-inch howitzer was substituted for one of the 24-pounders in R.h.

March 12, Left Attack.—Early in the morning of the 12th the two naval guns were taken through the rebel battery in Hazratgunge to a new emplacement, L.k., which the Engineers had constructed in front of it. Two openings were made in the wall separating the Begum's Palace from an adjacent mosque and Jaffir Ali's house, which were both occupied without resistance. A small serai on the other side of Hazratgunge, in advance of the one captured on the 11th, was also occupied without resistance, together with a mosque contiguous to and overlooking it. Progress through the walls was usually effected either by charges of powder or by the fire of an 8-inch mortar at close range, the infantry occupying the ground won by these methods. Six 8-inch mortars were moved to the advanced serai, L.l., and five 10-inch mortars were placed at L.m. in that captured on the 11th. The fire of all was directed on the Little Imambara, and the buildings between it and the Kaisar Bagh.

March 13, Left Attack.—During the 13th the work went steadily on, and by noon two naval guns, an 8-inch and 24-pounder, were established at L.n. to breach the Little Imambara, while on their right, at

---

[1] About this time Outram was reinforced by four 18-pounder guns, two 10-inch howitzers, five 10-inch mortars, and four 5½-inch mortars.

L.o., the artillery placed two guns to keep down fire from buildings on the north side of Hazratgunge and five mortars to shell the city in front. Before dark a breach was made in the Little Imambara.

March 13, Right Attack.—On the right attack the batteries maintained their fire, and Outram held his position at the bridge.

March 14, Left Attack.—At daylight on the 14th the breaches in the Little Imambara were considered practicable, and Russell's Brigade of the 1st Division (now commanded by Brigadier Franks)[1] were preparing to assault, when Lieut. F. E. B. Beaumont, R.E., accompanied by some Sikhs of the Ferozepore Regiment, effected an entrance into that building by blowing down the walls on their left front. The enemy fled, and Russell's Brigade passed the breaches without opposition at 9 a.m. and seized the King's Brother's House and adjacent enclosures which overlooked the Kaisar Bagh, as rapidly as openings could be made in them.

The right (or eastern) flank of the second line of defence was thus captured, while the left (or northern) was turned, and in a short time the Khurshid Manzil and the Moti Mahal were in our hands. Here Sir Colin's plans for the day concluded; for he had intended that the Kaisar Bagh should be stormed next day by the 93rd Highlanders. But the impetuous Sikhs followed the retreating enemy into this royal enclosure and the Division Commander was not the man to hold them back. With all his heart Franks

---

[1] When Outram crossed the Gumti the command of the 1st Division was conferred upon Brigadier Franks.

threw supports forward, and by 11 o'clock the Kaisar Bagh was ours. Lieutenant F. G. E. Warren and a party of 5/13 R.A. entered with the Sikhs and turned two of their own guns upon the enemy with great effect; a tumultuous scene of slaughter and plunder followed; the Palaces were sacked, and, with a loss of 20 killed and 100 wounded, the rebel citadel was gained. And now Fortune approached Sir Colin with both hands full. The city was almost in his grasp, the garrison were wholly at his mercy; yet the opportunity was allowed to pass by.

March 14, *The Right Attack.*—A heavy fire was kept up on the Kaisar Bagh and neighbouring buildings until they were entered by our troops; then the guns from R.c. and R.d. were moved forward into the open about R.j., and were turned on the Machi Bhawan, the Residency, and the buildings on their right.[1] At the Iron Bridge there was great excitement, and orders for an immediate advance were momentarily expected, an advance that would hem in the Lucknow garrison and cut off fugitives from the Stone Bridge. With great bravery and with great good fortune—for he did not lose a man— Lieutenant C. E. Wynne and a party of Royal Engineers removed the sandbag epaulment made across the bridge, the infantry were told off, and the guns of Q Battery, R.A., were hooked in, when Sir James Outram approached a group of officers

---

[1] They were manned by 50 men of 1/1 Ben.H.A., who performed excellent service. Brigadier Walpole also detached 50 men of the Rifle Brigade to reinforce the Siege Train. They were a most useful addition.

and said, "I am afraid, gentlemen, you will be disappointed when I tell you that I am not going to attack to-day."[1] He had in fact been forbidden to cross the river by Sir Colin if there was a chance of losing a single man.[2] Surely the most astounding order ever sent to a British general. The Stone Bridge was left open to the Begum and her followers, and in consequence another year of desultory fighting was quite needlessly imposed upon the British Army in Oudh.

On the 14th Jung Bahadur, with 9,000 Gurkhas and 24 field guns, crossed the canal and operated against the southern suburbs of the city, some distance to the west of Banks' House.

March 15, Left Attack.—The 15th was spent in securing the captured buildings, in destroying mines and disposing of the great mass of gunpowder[3] that was found stored in various places. A battery for three $5\frac{1}{2}$-inch mortars, L.p., was made in the Kaisar Bagh to play on the city. In the Right Attack four $5\frac{1}{2}$-inch mortars were sent to the Iron Bridge to keep down fire from houses on the opposite bank, and four 8-inch mortars, two 10-inch howitzers, two 8-inch howitzers, and two 24-pounder guns under

---

[1] "Up among the Pandies," p. 212. The author, Lieutenant V. D. Majendie, R.A., was present on the occasion in Q Battery, R.A.

[2] "In the camp the order was attributed to the counsels of Mansfield; but the responsibility rested and still rests with Sir Colin" (Mallison, ii. p. 398).

[3] On March 16 Shoeing Smith Lever, of 3/14 R.A., was mainly instrumental in emptying into a well a quantity of captured powder in a room of the Kaisar Bagh, while the building was on fire at no great distance from him. A second room was then reported to contain a large quantity of ammunition; to this he was proceeding, when an explosion took place. Lever was mortally wounded ("Occasional Papers, R.A.I.," vol. i. p. 51).

2nd Captains W. H. Goodenough and H. L. Talbot, R.A., were taken to R.j., to reinforce the guns already there. Outram continued to hold his position, and the guns at R.f. and R.h., under 2nd Captain W. N. Waller, R.A., and Captain A. Pearson, Ben.A., were busily employed. General Grant with the bulk of the cavalry was sent out towards Seetapore to intercept fugitives, and Brigadier Campbell from the Alam Bagh moved towards Sundeela on a similar duty.

March 16.—On the morning of March 16 Outram, with the 5th Infantry Brigade, crossed to the right bank of the Gumti by a newly constructed bridge of boats, and, supported by the Ferozepore Regiment and the 20th Foot, advanced through the Chattar Manzil to attack the Residency. He was accompanied by 6/13 R.A. and A/3 Mad.A. The Residency was gained with but little opposition, and Cotter, leading his guns to the top of the high ground within the enclosure, opened fire on the Machi Bhawan. The Lucknow end of the Stone Bridge was then captured, and later in the day the Machi Bhawan and the Great Imambara were occupied and the approaches to the Stone Bridge at last fell into our hands. But it was too late. As Outram re-entered Lucknow by the bridge of boats, vast numbers of the enemy were seen hurriedly crossing the Stone Bridge; our cavalry had been sent in the wrong direction; the 5th Brigade left a gap in the line of investment when it crossed the river, and through this gap the fugitives poured unmolested into the open country by way of the Fyzabad Road. During the afternoon the guns of A/3 Mad.A. were replaced by two naval 8-inch guns. The Royal Artillery brought five 10-inch mortars to the Residency and five 8-inch mortars to the Great Imambara, and the bombardment of the western portion of the city was carried on during the night.

During the day the batteries near the Iron Bridge and the guns in the open at R.j. kept up a brisk fire on the western portion of the city, and the 6th Brigade easily repulsed an attack made on its pickets by fugitives from Lucknow.

The absence of Brigadier Campbell from the Alam Bagh appeared to offer an opportunity for the rebels to attack that position; but their cavalry, which were in advance, were quickly routed by 2/3 Ben.A., while the guns of 1/5 Ben.A. dealt very effectually with the main body of their infantry, and their whole force was soon in retreat. Their position along the canal west of the Char Bagh, from which they had so long threatened the Alam Bagh, was now about to be taken in reverse by Jung Bahadur.

March 17–18.—For this purpose the Gurkha force occupied the Char Bagh and Ganeshgunge, but they were at once attacked by the rebels. The Jung, however, turned their flank and completely defeated them, taking ten guns. He then established himself in his position. At the same time Outram continued his westerly advance with a portion of the 5th Brigade, 6/13 R.A. and two 8-inch howitzers, under Major H. A. Carleton, with a detachment of 2/3 Ben.A. He reached the vicinity of the Jumma Musjid without serious opposition, but most unfortunately 17 officers and men of the Royal

Engineers were killed by an accidental gunpowder explosion. This day the cavalry columns under Grant and Campbell returned to their former positions. The next day, March 18, was spent in further clearing the rebels out of their retreats.

March 19.—On the 19th Outram, reinforced by the 9th Lancers, pushed on along the Gumti to attack the Musa Bagh, where some thousands of rebels were assembled. But on the advance of skirmishers of the 23rd and 79th, supported by 6/13 R.A., their whole force took to flight, leaving 2 guns behind them. The 9th Lancers pursued them for four miles, and captured 6 guns, and the infantry and field artillery following up captured four more. But the bulk of the fugitives escaped. Brigadier Campbell, with a force of cavalry and horse artillery on our extreme right, should have cut off their retreat, as well as that of other fugitives from the city, but that officer's movements on this and the following day failed to achieve this purpose.

March 21.—On March 21 the 93rd Highlanders, the 4th Punjaub Infantry, and 2 guns of 6/13 R.A. (Lieutenant W. D. Milman) were sent into the western suburbs to Sadatgunge to attack the Moulvie, who the previous day re-entered the city from the west and barricaded himself and his followers in a mosque. After a sharp contest they were driven out, and some of the fugitives were cut up by Campbell's cavalry, but the Moulvie escaped.

Action at Kursi, March 23.—At midnight on March 22 General Grant marched for Kursi, twenty-five miles to the north of Lucknow, where

4,000 rebels were said to be collected. His troops consisted of:

2nd D.Gs.	53rd Foot.	1/1 Ben.H.A. (Captain and Brevet-Major F. F. Remmington).
1st Punjaub C.	2nd Batt. Rifle Brig.	
2nd Punjaub C.	3rd Batt. Rifle Brig.	
		2/3 Ben.H.A. (Captain W. A. Mackinnon).
	Ben. Sappers and Miners.	No. 20 H.F.B. 3/5 Ben.A. (Captain G. Moir).

O.C.A., Lieutenant-Colonel E. Maberly, R.A.
S.O., Lieutenant T. B. Strange, R.A.

Grant reached Kursi on the afternoon of March 23 [1] and found the enemy in full retreat. Halting his infantry and 3/5 Ben. A. in front of the village, he pushed on to the left of it with his cavalry and horse artillery [2] and, forming line to his right, charged the rebels. The Punjaub Cavalry, supported by the Dragoon Guards and 2 guns of 1/1 Ben.H.A. (Lieutenant G. R. Manderson), rode five or six times through the rebel lines, killed over 200 of them, and captured 13 guns. Grant's casualties were 4 killed and 10 wounded. On the 25th he marched back to Lucknow.

The Siege of Lucknow was now at an end. At the comparatively small cost of 127 killed and 595 wounded the Commander-in-Chief had dealt a heavy blow to the prestige of the rebels, but the fruits of his victory were in great part lost; for when it lay in his power to crush all hope of further resistance, he deliberately allowed the main body to escape, only to reassemble elsewhere in their thousands.

---

[1] The march was delayed by the difficulty experienced in getting the 18-pounder guns of 3/5 Ben.A., which had a heavy field battery equipment, along the bad roads.

[2] During this advance Lieutenant C. Hunter, detached to the left front with two guns of 2/3 Ben.H.A., did good service.

CHAPTER VIII

CENTRAL INDIA

I. THE CENTRAL INDIA FIELD FORCE

Sir Hugh Rose at Sehore.—The Central India Field Force was organised in two brigades, the first being at Mhow, the second at Sehore, where it was joined by Major-General Sir Hugh Rose on January 8. Having learnt that a strong body of mutineers were approaching Saugor, Rose determined to march to its relief with the 2nd Brigade, for Brigadier Whitlock, to whom this duty had been entrusted, was still at Kampti with the Saugor Field Force. The 2nd Brigade Central India Field Force was thus composed:

Brigadier C. Steuart [1]

Head-quarters.	3rd Bom.E.R.	1st Troop Bom.H.A. (4 guns)
14th L.Ds.	24th Bom.N.I.	(Captain and Brev.-Lieutenant-Colonel S. Turnbull).
3rd Bom.L.C.	Mad. Sappers and miners.	
3rd Cavalry.		No. 18 F.B. 2/Res.[2] Bom.A. (3 guns) (Captain J. G. Lightfoot).
Hyderabad Contingent.		
		Hyderabad Cont. Art. (3 guns) (Capt. C. Douglas, Ben.A.).
		Siege Train. Detachment 4/2 Bom.A. (Lieutenant C. H. Strutt).

[1] Lieutenant-Colonel and Brevet-Colonel C. Steuart, 14th Light Dragoons.

[2] Three European companies of Bombay Artillery were raised in November 1857 and called the Reserve Artillery.

Siege of Rahatghur.—Sir Hugh set out on January 16 and next day entered Bhopal, where the Begum reinforced him with 700 men and three 9-pounder guns which were attached to 2/Res. Bom.A.[1] On January 24 he arrived before the rock fortress of Rahatghur on the River Bina, which for months past was held by the rebels. The rock, a mile and a half in length, was covered and surrounded by thick jungle, and sloped from the west, where its precipitous side was washed by the river, to the east, where it was accessible. On the north side lay the town. On January 26 Rose established himself on the east side of the fort, and opened fire with light guns and small mortars to cover the construction of his breaching batteries. At the same time he drove the enemy out of the town into the fort, the north side of which was attacked by 8-inch mortars and howitzers and the 9-pounders of No. 2/Res. Bom.A. On the morning of the 28th the batteries on the east were armed with two 18-pounder guns and an 8-inch howitzer[2] and a practicable breach was almost made, when a momentary diversion was caused by the Rajah of Banpore. The Rajah with about 2,000 men came from the north and menaced the right rear of the British camp, but he was at once repulsed by the cavalry and field guns, whilst the heavy pieces continued their fire on Rahatghur. A second but half-hearted attempt made on the left flank of the camp

---

[1] The 1st Troop Bom.H.A. was completed to 6 guns at Saugor by two 6-pounder guns supplied by Lieutenant T. Nicholls, Ben.A., Commissary of Ordnance.

[2] These pieces were worked by 26 men of 4/2 Bom.A.

after nightfall was easily dealt with by the British, and when day broke the Rajah was in full retreat to Barodia. It was also discovered that during the darkness inhabitants and garrison had alike deserted Rahatghur and had followed the Rajah.[1]

Barodia was situated on the left bank of the Bina, about twelve miles to the north-west of Saugor, and as a rebel concentration there would menace the left

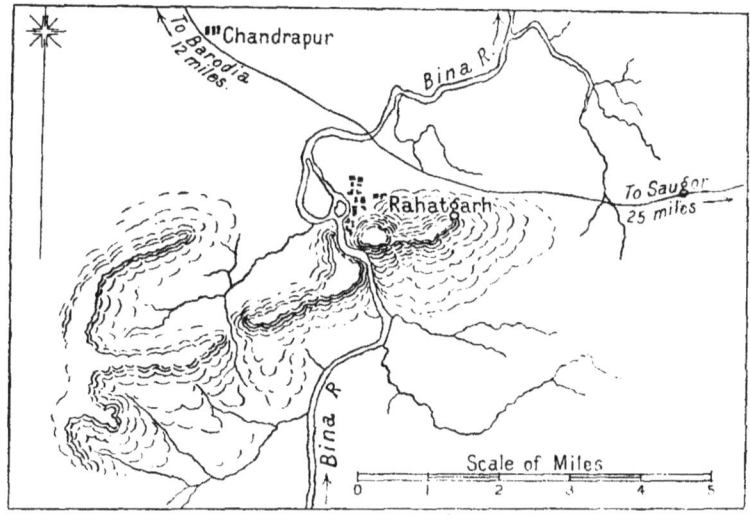

RAHATGHUR, JANUARY 28

flank of Sir Hugh's forward march, it was necessary to attack it. Leaving Brigadier Steuart to protect Rahatghur with the 24th Bom.N.I., some guns and cavalry, Sir Hugh set out for Barodia on January 31.

[1] Rose had not sufficient troops for investment, and the garrison and inhabitants escaped by a precipitous and dangerous path to a ridge above Chandrapur, whence they fled to Barodia. Letters found in Rahatghur showed that Feroz Shah (see p. 245) had been there before the siege.

Action at Barodia, January 31.—It was about midday when he began his march with the following force :

14th L.Ds. (3 troops).	3rd Bom.E.R. Mad. Sappers and Miners.	1st Troop Bom.H.A. (Captain and Brev. - Lieutenant - Colonel S. Turnbull).
3rd Bom.L.C. (2 troops).		No. 18 F.B. 2/Res. Bom.A. (Captain J. G. Lightfoot).
		Detachment 4/2 Bom.A. with two 5½-inch mortars (Lieutenant C. H. Strutt).

Detachment (all arms) Hyderabad Contingent.

The advanced guard, which included two guns of the 1st Troop (Lieutenant R. Pittman), was under the command of Lieutenant-Colonel S. Turnbull, Bom.H.A., and as Barodia was approached the enemy were discovered in thick jungle covering the ford of the Bina. They were so well hidden, however, that our infantry and cavalry advancing in skirmishing order were close upon them before they were perceived, and in the rapid fusillade which they opened on our troops Turnbull's horse was shot. Then Pittman's guns were rapidly brought into action, and though presently reinforced by two more under Lieutenant T. C. Crowe, who galloped up from the rear guard, the rebels held their ground until the 3rd Europeans advancing with the bayonet drove them across the river.

Covered by the skirmishers, Captain J. Forbes then led the 3rd Bombay Cavalry and the cavalry of the Hyderabad Contingent across the ford, and finding the enemy in a second position where, though their flanks were covered, their centre lay open, a spirited charge upon their weak spot drove them back towards Barodia—a village surrounded on three sides by thick jungle, with a wet ditch along its front, and further protected by a small fort containing two guns.

Although the ford was a bad one, the 1st Troop at once followed the cavalry and opened fire, to which the rebels replied with round shot and rockets while their infantry occupied the jungle and lined the wet ditch with skirmishers. It was here that Captain G. Neville, R.E., who was acting as A.D.C. to Sir Hugh Rose, was killed at his side by a round shot, and Pittman was severely wounded by another. The troop had marched without waggons, and its ammunition was all expended[1] when 2/Res. Bom.A. came up and took its place, and about the same time Lieutenant C. H. Strutt opened fire on the fort with his mortars.

The rebels were soon driven from the jungle and the ditch, and were forced to seek shelter in the village, but when our infantry advanced and surrounded it most of the defenders had already made their escape. They lost about 500 men. The British casualties were 2 killed and 21 wounded. It was now getting dark, and Rose marched back to Rahatghur, where Steuart's small force was in a somewhat exposed position. But their defeat at Barodia for a time disheartened the rebels; they fled behind the passes of Bundelkhund and quitted a district which for eight months was at their mercy.

[1] A similar incident occurred at the Battle of Balaclava, where I Troop R.H.A. was replaced by W Battery R.A.

Relief of Saugor.—On February 3 Saugor was relieved, and Rose, continuing his easterly march, on February 12 drove the Rajah of Sharghur's troops out of the strong fort of Garakota. He then returned to Saugor, and as the roads to the west and north were now clear he was able to open communications with Major Orr,[1] who with the bulk of the Hyderabad Contingent was at Goonah on the Grand Trunk Road, along which the 1st Brigade was advancing. Orr was acting as advanced guard to Brigadier Stuart[2] who had left Mhow on February 6, and was due to arrive at Goonah on the 26th. But before Rose was in a position to resume his march much had to be done. The urgent repairs required by the carriages of the siege train were completed in four or five days, but the organisation of some kind of transport for the supplies he was obliged to take with him was a matter of difficulty and it was only on February 27 that he was able to move. This interval, however, was utilised by Orr in reconnoitring the passes into Bundelkhund with his cavalry.

The delay at Saugor raised the spirits of the rebels, and the Rajahs of Banpore and Sharghur proposed to block Rose's further progress at the passes leading into Bundelkhund. The direct road to Jhansi led over the pass at Narhat, where the main body of the rebels entrenched themselves; but Rose, making a feint attack which held the enemy in this position,

---

[1] Captain and Brevet-Major W. A. Orr, Mad.A.

[2] It is necessary to avoid confusion between the two Brigadiers. Stuart was the officer who previously commanded the Malwa Field Force (see p. 244).

threw his main force against the pass of Madanpore and, easily brushing aside the Rajah of Sharghur who held it, turned the line of the rebels' defences. By the first week in March he was in possession of the forts of Surahi and Maraura, and had annexed the Sharghur territories [1] to the British Government. On March 10 the palace of the Rajah at Banpore was captured, and on March 14 the 2nd Brigade reached Tal Bahat, where it halted to await the 1st Brigade.

Movements of the 1st Brigade.—The 1st Brigade Central India Field Force was composed as follows:

Brigadier C. Stuart

14 L.Ds. (1 squad.).	86th Foot.	5/14 R.A. (Captain F. M. M. Ommaney).
3rd Bom.L.C. (1 troop).	25th Bom.N.I. 21st Company R.E. Bom. Sappers and Miners.	4/2 Bom.A. (Captain J. D. Woollcombe). Siege Train (Lieutenant T. T. Haggard, Bom.A.).

Brigadier Stuart duly reached Goonah on February 26, where he awaited further instructions. It was Rose's intention to combine his two brigades for an attack on Jhansi; he therefore ordered Stuart to leave the Trunk Road and turn off in an easterly direction towards Tal Bahat, and to capture Chanderi on his way. On March 5 Stuart arrived within striking distance of that place.

Siege of Chanderi, March 5 to 17.—The fort of Chanderi was a strong place, and rebels dispersed by the advance of the 2nd Brigade flocked there in

[1] Surahi was the fortified palace of the Rajah of Sharghur. Maraura was an old fort with a double line of defence. It was important from its position commanding the high road between Saugor and Jhansi.

great numbers. Some twelve pieces of ordnance were mounted on its walls. Two miles to the westward lay the small village of Futtehabad, and here on March 5 Stuart had his first successful skirmish. He then moved to his right front and occupied Ramnagar, another small village lying in a valley two miles S.W. of Chanderi, a view of which was obtained when a rocky ridge to the northwards was cleared of the enemy and ascended by our troops. The fort was built on an isolated hill of the same altitude as the ridge, and was separated from it by a jungle-clad ravine some 800 yards wide. The ridge and the hill were connected by a rocky col, the farther end of which was enclosed by the fortifications and defended by a solid curtain of masonry flanked by two towers. Stuart determined to breach this curtain. The ascent of the ridge was impeded by precipitous rocks and thick jungle, and great difficulty was experienced in getting the heavy guns into position. On the evening of the 7th two 8-inch mortars were brought up from Ramnagar and on the 10th two more mortars and an 8-inch howitzer were added, after the sappers had cut a way through the jungle. These pieces came into action on the right and left of the col, along which working parties were employed in making a road for the advance of the 18-pounder guns. The vertical fire directed on the interior of the fort was so successful that on the 11th two 18-pounder guns were placed in battery 400 yards from the curtain, and an 8-inch mortar and an 8-inch howitzer shifted to positions on the right and left, so as to bring the concentrated fire of four pieces to bear upon the wall.

On the evening of the 16th a practicable breach was reported, and next morning the assault was successfully made by the 86th Foot (Royal County Down) and the 25th Bom.N.I. The enemy fled from Chanderi leaving some 100 killed behind them, and in their flight to the jungle suffered further loss at the hands of the cavalry. The British casualties were 2 killed and 23 wounded. On March 11 the

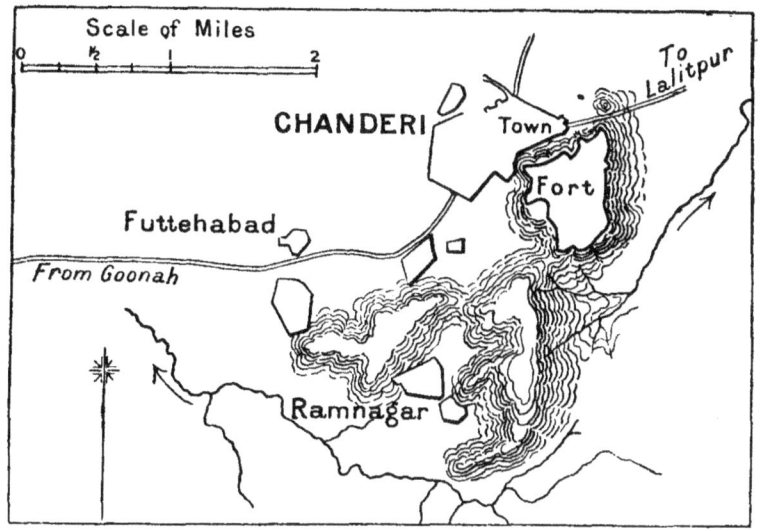

CHANDERI, MARCH 17

Royal Artillery lost Lieutenant P. Moresby, who was struck on the head by a round shot whilst doing duty with the 18-pounder guns. On the 19th Stuart marched towards Jhansi to join Sir Hugh Rose, leaving Sindiah's Agent in charge of Chanderi.

Jhansi.—The city of Jhansi was protected and commanded by a fort standing on a rock, which rises from the plain on the west of the town. The fort was of massive masonry with granite walls 10 to 12 feet thick, and was surrounded by elaborate outworks and turrets bearing guns. Except in the west, where the fort rock was an insurmountable barrier, a continuous wall ran round the city. It was $4\frac{1}{2}$ miles in circumference, 6 to 12 feet thick, 18 to 30 feet high, and provided with flanking bastions. The most important bastion was situated due south of the fort, where the wall passed over a huge mound. It was semicircular in shape and protected by a formidable ditch strengthened with masonry. The bastion carried 5 guns. The garrison of Jhansi consisted of some 10,000 native levies and 1,500 Bengal sepoys, of whom 400 were cavalry.[1] On the south front 27 guns were mounted worked by golandauz under an expert commander.[2] Two rocky hillocks rose from the plain on the south of the town; that on the east 1,800 yards, that on the west 600 yards distant from the fort, the summit of which was on the same level as their highest points. Between the hillocks lay a line of lesser mamelons. This strip of elevated ground was about to be occupied by the British.[3]

[1] The sepoys consisted principally of the 14th Bengal Irregular Cavalry and the 12th Ben.N.I., who had mutinied at Now Gong and Jhansi.

[2] These were the men of 4/9 Ben.A., reinforced possibly by others on leave.

[3] It was recognised by the Governor-General and the Commander-in-Chief that the capture of Jhansi presented difficulties of a most formidable character, and discretion was allowed to Rose to pass it by merely leaving a force to observe it. Furthermore, just when the siege was about to begin pressure was put upon him to divert his march eastwards to Charkhari, of which the loyal Rajah was in danger from the rebels at Kalpi. But Sir Hugh with unfailing instinct realised that to turn aside now would be fatal to British prestige, while on the other hand the capture of the famous city would be a crushing blow to the rebel hopes.

The Siege of Jhansi, March 23 to April 3.—On March 23 Jhansi was completely invested by the cavalry of the 2nd Brigade, and a squadron of the 14th Light Dragoons pushed on in advance of the 1st Brigade. The cavalry occupied seven flying camps, forming a ring of mutually supporting outposts. Next day the western hillock, the Left Attack, was occupied by a strong picket of the

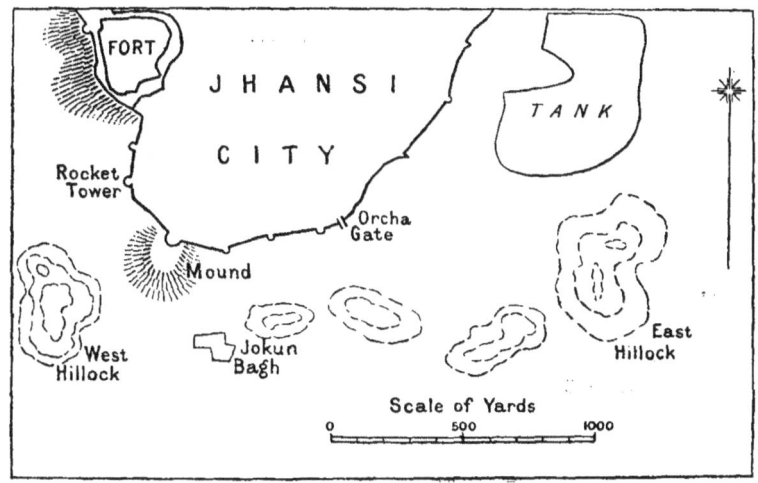

JHANSI

Hyderabad Contingent, and two of their 5½-inch mortars were brought to bear upon the Mound Bastion and the adjoining buildings. At the same time the 2nd Brigade encamped in rear of the east hillock, the Right Attack, where an 8-inch howitzer and two 8-inch mortars were placed in battery, and afterwards reinforced by two 24-pounder howitzers. On the 25th the 1st Brigade encamped behind the west hillock, about two miles from the camp of the

2nd Brigade. Lieutenant J. C. J. Lowry with two guns of 5/14 R.A. was sent to the cavalry camp commanded by Major R. H. Gall, 14th L.Ds., on the north of the city, and two guns of No. 1 Troop Bom.H.A. (Lieutenant T. C. Crowe) were attached to Major A. Scudamore, 14th L.Ds., who commanded a similar camp on the east. Four siege batteries were begun on the western hillock, the Left Attack, and were armed with three 18-pounder guns, two 10-inch mortars, and two 8-inch mortars. On the 28th a bombardment was opened from both Attacks, and by sunset the towers were in ruins. The mortars under Lieutenant R. Pittman, Bom.H.A.,[1] had exploded a magazine in the fort, and the town wall, where the breach was to be made, began to fall away, disclosing a pallisade revetment which was at once attacked with red-hot shot; while picked marksmen kept up a constant rifle fire upon the defenders. But the rebels fought with great determination, and their guns were well served until most of their trained gunners were killed or wounded.[2] In order to bring an enfilade fire to bear upon the breach, a new battery, No. 5 Left Attack, was begun near the Jokun Bagh, where two 5½-inch mortars were in action. It was determined to replace them by two 10-inch

---

[1] The 1st Troop was employed on picket and patrol duties as a rule, but all artillery officers took their turn in the siege batteries.

[2] Women carried ammunition to the batteries, one of which displayed the black flag of the Fakirs. In the cool of the evening the Begum and her ladies, brilliantly dressed, could be seen on the battlements amongst the men serving the guns. The defence was conducted with great spirit and, as the siege progressed, with the courage of despair. Seventy-six foully murdered Christians lay in the Jokun Bagh, and the avengers of blood were at hand.

THE FORT AND CITY OF JHANSI

mortars for vertical and a 9-pounder and 24-pounder howitzer supplied by the Hyderabad Artillery for direct fire. While the new battery was being made the two field guns were worked in the open under the command of Bombardier P. Brennan, of 5/14 R.A. The bearing of the young non-commissioned officer attracted the notice of Sir Hugh Rose, who there and then promoted him, and he was afterwards awarded the Victoria Cross. On March 30 the breach at the Mound appeared practicable, and arrangements for the assault were in progress, when Sir Hugh was informed that a large force was advancing to the relief of Jhansi.

Tantia Topi's Attempt to relieve Jhansi.—After his defeat at Cawnpore, December 6, Tantia Topi hid himself in Bundelkhund, but early in March he repaired to Kalpi on the summons of the Rao Sahib. The Rajah of Charkhari had never wavered in allegiance to the Government, and the representative of the Peshwa sent Tantia Topi to punish him. After a siege of eleven days Charkhari was captured, and the rebel general carried off 24 guns and 3 lakhs of rupees. Still acting under the orders of the Rao Sahib, Tantia Topi set out for Jhansi, his force being increased to 22,000 men and 25 guns, by Gwalior mutineers and native levies. On March 30 he reached Barwa Sagar on the right bank of the Betwa.

Battle of the Betwa, April 1.—Sir Hugh's mind was at once made up. He would neither reduce the bombardment nor relax the investment, but would march to meet the enemy even though his available force would not exceed 1,500 men. Towards evening he reached Basopa with detachments from the 2nd Brigade, and took up a position which commanded Rajpur and Kolwan, the two adjacent fords of the Betwa. But learning from his outposts next morning, March 31, that the enemy showed no inclination to attempt the passage, and desiring to force them to fight with their backs to the river, he withdrew his troops in order to lure them on. With this intention

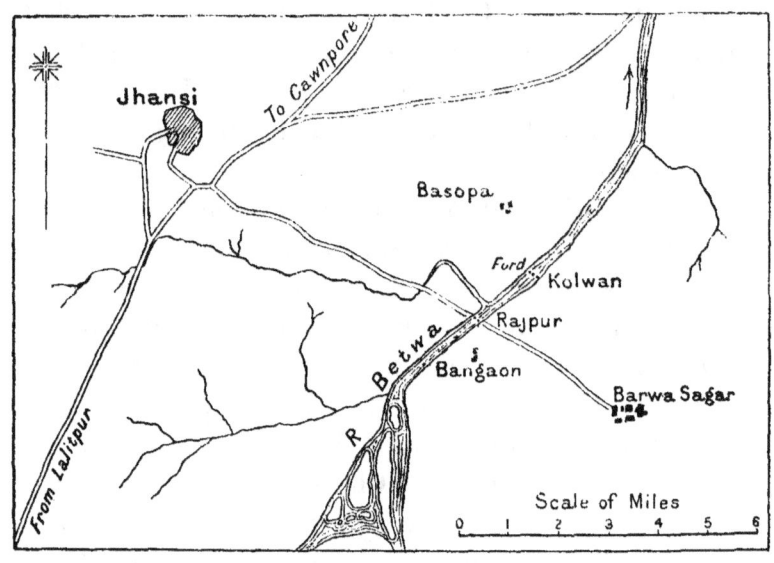

THE BETWA, APRIL 1.

he retired to a position a mile to the south of the camp of the 2nd Brigade, and ordered Brigadier Stuart with detachments of the 1st Brigade to take post behind him. Tantia Topi then crossed the Betwa and by sunset was in position opposite to the British. Bonfires were lighted on the hills in his rear, and all night sounds of rejoicing were heard in Jhansi. About midnight Sir Hugh was informed that numbers of the enemy were crossing at Kolwan, with the evident intention of outflanking the British, and Brigadier Stuart was ordered to march at once and check this movement.

The rebels in the immediate front of Sir Hugh Rose were formed in two lines, the second for some unaccountable reason three miles in rear of the first. But Tantia Topi was under the impression that he had only a small force to deal with, the larger part of Rose's army being employed in the siege, and when morning broke on April 1 he determined to attack. Rose had a like intention, and when the rebels came on the British came on too. In their front line were less than 400 bayonets of the 3rd Europeans and the 24th Bom.N.I. with 3 siege guns; on the right were 5 troops of cavalry, 14th Light Dragoons and Hyderabad Cavalry, and the 1st Troop Bom.H.A.; on the left No. 2/Res. Bom.A. and 2 troops of the 14th Light Dragoons. In second line were 200 of the 24th Bom.N.I., the total force not exceeding 900 men. Rose met the rebel onset with a counterattack on both flanks with his field guns and cavalry, and presently throwing his centre forward broke the first line of the enemy, and following them up drove them back on their second line.

Meanwhile Brigadier Stuart had been marching all night. He had 150 of the 14th Light Dragoons and Hyderabad Cavalry, 400 of the 86th Foot and 25th Bom.N.I., and his artillery under Captain F. M. M. Ommaney, R.A., consisted of 2 guns of 5/14 R.A. (Lieutenant L. C. A. A. de Cetto) and 4/2 Bom.A. (Captain J. D. Woollcombe). Finding no enemy near Kolwan, he counter-marched towards the sounds of battle that reached him, and after a march of more than an hour he encountered some 2,000 rebels holding a village in his left front with guns in position. Our artillery came into action at 600 yards range and bombarded the village; the infantry then advanced in skirmishing order, the cavalry hovering on their flanks, and soon a bayonet charge drove out the rebels, who fled, leaving 6 guns behind them. They rallied, however, at another village a little in rear, but the 86th Foot ejected them and captured some camels and elephants. The rebels then formed themselves in a compact body covered by a strong rear guard and retired towards Rajpur. Stuart's force was completely exhausted. It had been under arms for over thirty hours and could do no more.

Meanwhile Tantia Topi, who personally commanded his second line, saw his first line routed and his right flank driven back by Stuart. He had with him 2 heavy pieces drawn by elephants, an 18-pounder and a 9-inch mortar, which opened fire on our advancing troops. But on the approach of Rose's field guns they at once relinquished the contest, and the rebel general, firing the jungle behind him, retired towards the Betwa. Undeterred by the flames, our cavalry and artillery pressed forward, and the guns, coming within range of the Rajpur ford, over which the enemy were in full retreat, opened a cross fire upon them. Then some of our cavalry got across the river and the rebel retreat became a helpless rout, in which guns and stores were abandoned. The pursuit continued for nine miles, 18 guns being captured and 1,500 rebels slain. The British loss was 19 killed and 62 wounded, the 14th Light Dragoons being the chief sufferers.

Storming of Jhansi, April 3.—The battle of the Betwa gave Jhansi a respite of forty-eight hours. On April 3 all was ready for the assault. Two columns of stormers under Major W. K. Stuart and Lieutenant-Colonel R. H. Lowth, 86th Foot, were to issue from the Left Attack and make for the Rocket Tower (the S.W. bastion of the city) and the breach at the Mound. Simultaneously two other columns from the Right Attack under Captain H. G. Robison and Lieutenant-Colonel J. Liddell, 3rd Europeans, were to escalade the wall on the right and on the left of the Orcha gate. During the night a 24-pounder howitzer of No. 5/14 R.A., run out in front of the Jokun Bagh, enfiladed the wall between the breach and the Rocket Tower. At daybreak three rounds from the 18-pounders on the left gave the signal, and while all the mortars and howitzers poured shells into the town, the stormers went forward. The attacks led by Lowth and Stuart were at once successful; they penetrated into the town, and then the 86th, swinging round to their right, took in rear the rebels who were opposing the advance of the stormers from the Right Attack. Liddell and Robison encountered serious difficulties. The scaling ladders were found to be too short, and in some instances broke under the heavy weight put upon them; and though the engineer officers who accompanied the assaults displayed the most devoted courage, the first attacks were repelled.[1] Eventually Robison effected an entry on the left of the Orchha gate, and Liddell following him, the four storming parties united and drove the rebels out of the town with great slaughter. On April 4 all the city, the fort excepted, was in our hands. That night, through the connivance of some of our native allies, the Begum escaped. Carrying her adopted son in her arms, she was let down by a rope into the ditch of the fort where a horse awaited her, and, accompanied by a few adherents, she made her way to Kalpi in spite of all efforts to intercept her. On April 5 it was discovered that the fort was evacuated. Nine guns were found inside and 26 in the town. The rebels lost about 5,000 killed. Exclusive of the casualties at the Betwa, the British had 50 killed and about 200 wounded, principally in the 86th Foot and 3rd Europeans. Lieutenant G. B. C. Simpson, 23rd Ben.N.I., who was attached to the artillery, was severely wounded. Lieutenant-Colonel S. Turnbull was mortally wounded, and was succeeded by Captain Ommaney in command of the artillery of the field force. Captain J. G. Lightfoot was transferred to the 1st Troop Bom.H.A.

Though Jhansi was captured, it was some time before Sir Hugh Rose was in a position to continue

---

[1] The walls were 25 and 30 feet high where Colonel Liddell's stormers first attempted the escalade. Three ladders only were used. Lieutenant H. R. Micklejohn, Bom.E., led the way up one, but when he reached the top he was dragged over the wall and cut to pieces. Lieutenant W. G. D. Dick, Bom.E., was shot through the head at the top of another. Lieutenant F. R. Fox, Mad.E., was also seriously wounded. The engineer loss was 5 killed and 16 wounded.

his advance; his left flank was not safe, and he was obliged to await the advance of the Rajputana Field Force.

II. THE RAJPUTANA FIELD FORCE

Capture of Awah.—In January, 1858, reinforcements from Bombay began to enter Rajputana, and the rebels were speedily driven from Awah. The place was invested, and after five days' siege a practicable breach was made in the walls, but the garrison escaped under cover of darkness.

General Roberts' Force.—March had come, however, before a sufficient force assembled to deal with the Kotah rebels. In that month Major-General H. G. Roberts arrived in Nusseerabad and relieved Brigadier G. St. P. Lawrence in the military command. The following troops were under his orders:

8th Hussars.	72nd Highlanders.	2nd Troop Bom.H.A. (Captain J. G. Petrie).
1st Bom.L.C.	83rd Foot.	
2nd Bom.L.C.	95th Foot.	3rd Troop Bom.H.A. (Captain and Brev.-Lieutenant-Colonel E. S. Blake).
1st Sind Horse.	10th Bom.N.I.	
2nd Beluchi Hse.	12th Bom.N.I.	
	13th Bom.N.I.	No. 8 F.B. 2/6 Ben.A. (3 guns) (2nd Captain C. S. Lemarchand).
	11th Company R.E.	
		1/2 Bom.A. (Captain W. D. Aitken).
		4/3 Bom.A. (2 guns) (Lieut. G. G. Brown).
Siege Train:		
Six 18-pounder guns.	Four 8-inch howitzers.	
Four 12-pounder guns.	Four 8-inch mortars.	

O.C.A., Lieutenant-Colonel E. Price, R.A.[1]

---

[1] Lieutenant-Colonel Price, R.A., followed the Field Force to Kotah and assumed command of the artillery the day before the assault.

Siege of Kotah, March 24 to 30.—The city of Kotah on the right bank of the Chumbal had a river front of two miles, and was defended by bastioned walls. The palace of the Maharao was in the south of the city, and dominated that quarter, being separated from the city proper by a walled enclosure. When the revolt took place the Maharao temporised with his rebellious followers, but secretly communicated with the Rajah of Kerowlee, his feudatory, who promptly came to his aid and drove the mutineers out of that quarter of the city in which the palace stood. The rebels, however, numbered 6,000 and were well provided with guns.

After passing through difficult country Roberts arrived before Kotah, and on the morning of March 24 his siege batteries opened fire from the left bank of the Chumbal.[1] On the 26th detachments of the 83rd Foot and 13th Bom.N.I. were sent across the river to reinforce the Maharao, whose position had been twice assaulted since the arrival of Roberts' forces.

On the morning of March 30 the bulk of the British infantry were transferred to the right bank of the river while a general bombardment of the town was carried on. At half-past 11 the firing ceased by signal, the assault was delivered by three columns which were drawn up in the Maharao's quarter, and by half-past 2 o'clock Kotah was in Roberts' possession. The rebels escaped by the east gate and were pursued to the Parbati River sixty

---

[1] The men of the field artillery assisted to work the siege guns. General Roberts in his despatch described the work of the artillery as "having been almost beyond belief."

miles away. Some 400 of them were killed and numbers of prisoners were taken. The British casualties were 14 killed and 46 wounded. Roberts now returned to Nusseerabad and the Rajputana field force was distributed, a brigade being sent to Neemuch and another to Sipri. The capture of Kotah not only

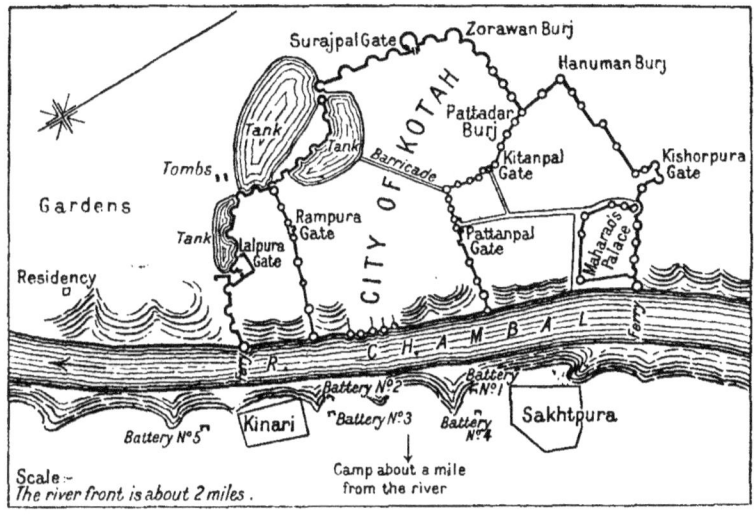

SIEGE OF KOTAH

restored British prestige in Rajputana, it also cleared the way for a further advance by Sir Hugh Rose.

III. THE ADVANCE ON KUNCH

Occupation of Poonch.—While these events were in progress Sir Hugh Rose laid in supplies of all kinds. He also despatched two columns [1] under Major W. A.

---

[1] Major Orr took with him the bulk of the Hyderabad Contingent. Major R. H. Gall had 1 squadron 14th Light Dragoons, 3rd Hyderabad Cavalry, and 3 guns Bom.A. (It is not stated whether these belonged to No. 4/2 or to No. 2/Res. Bom.A.)

Orr and Major Gall to clear the line of the Betwa towards Kalpi, and to prevent the Rajahs of Banpore and Sharghar from annoying his right flank, or from breaking back to Bundelkhund where they might again cause disorder. Leaving a garrison in Jhansi,[1] the General, with the 1st Brigade, on the night of April 25 began his march towards Poonch, whither the 2nd Brigade was to follow two days later. Gall had already reached Poonch and ascertained that the rebels were assembled at Kunch, a town fourteen miles distant. Orr was at Ait, ten miles N.E. of Poonch. He had engaged the Rajahs of Banpore and Sharghar and taken a gun from them, but in spite of his efforts they effected their escape to the south, where later on they were again heard of. On May 1 Rose with the 1st Brigade marched into Poonch, his 2nd Brigade, which had been reinforced by the 71st Highland Light Infantry, being a few marches behind him.

Affair of Lohari, May 2.—Before proceeding farther, however, he determined to punish an act of treachery. A few days previously Gall had stationed a picket of Hyderabad cavalry in Lohari, a village 8 miles N.W. of Poonch, where a fort was garrisoned by men in the service of the Rajah of Santhar, one of the lesser chiefs who professed loyalty to the Government. On Gall's departure this garrison betrayed the picket to the rebels at Kunch, and though the troopers successfully cut their way out, it was

[1] The Jhansi garrison under Lieutenant-Colonel J. Liddell consisted of a wing of the 3rd Bombay Cavalry, 100 Hyderabad Cavalry, a wing of the 3rd Europeans, a wing 24th Bom.N.I., ½ Company Bom. Sappers and Miners, and the 3 Bhopal guns which have been attached to 2/Res. Bom.A.

determined to make an example of the traitors. On May 2 Gall invested Lohari with a small mixed force including 4 guns of 5/14 R.A. (Captain T. S. P. Field).[1] The guns shelled the enemy from their walls, and Lieutenant J. Bonus, Bombay Engineers, blew in the fort gate with a blacksmith's bellows full of powder taken from the limbers. The infantry then rushed the fort and all the garrison were put to the sword.

Reappearance of the Begum of Jhansi.—It was owing to the efforts of the indefatigable Begum of Jhansi that the rebels were assembled at Kunch. When she reached Kalpi after her escape she besought the Rao Sahib to give her an army that she might go and fight, and her request was granted. At Kalpi were assembled mutineers from the Bengal Army and Gwalior Contingent; cavalry from Kotah and native levies, a heterogeneous mass formidable at least in number. Tantia Topi, after flying from the Betwa, reached Kalpi the same day as the Begum, and being at once placed in command was ordered to take up a position at Kunch to block the road to Kalpi. Kunch, an open town, was difficult to attack, as it was surrounded by woods and walled enclosures; but (as was their custom) the rebels only prepared to receive an attack by the direct road, namely, that from the south, and on this side of the town only were entrenchments constructed.

Battle of Kunch, May 5.—On May 5, his 2nd

---

[1] Gall's force consisted of 1 squadron 14th Light Dragoons, 100 sabres Hyderabad Cavalry, wing 3rd Europeans, wing 25th Bom.N.I., 4 guns No. 5/14 R.A., 20 Bombay Sappers and Miners.

Brigade having come up, Rose resumed his march, but instead of following the direct road he inclined to the westward, thereby not only turning Tantia Topi's entrenchments, but threatening his line of retreat. On the morning of May 7 the 1st Brigade was at Nagepuri, the 2nd Brigade at Chamra, and Orr with the Hyderabad Contingent at Umri, while Major A. Scudamore, with a squadron of the 14th Light Dragoons and 3 guns of the 1st Troop Bom.H.A. (Lieutenant T. C. Crowe) was on the extreme right. Rose determined to make a combined attack on Kunch from these three directions. The turning movement was perceived by the rebels, who conformed to it, and consequently when the 1st Brigade advanced from Nagepuri they found in their immediate front a long wall with a wood on the left of it both occupied by infantry. With the 1st Brigade were 3 guns of the 1st Troop Bom.H.A. (Captain J. G. Lightfoot), two 18-pounder guns and an 8-inch howitzer from the siege train, together with 5/14 R.A. (Captain T. S. P. Field) and 4/2 Bom.A. (Lieutenant C. H. Strutt) in reserve. The heavy guns opened against the rebel position, and the horse artillery galloped to the left front in order to bring an oblique fire to bear on them. But the effect of the fire could not be estimated, as the trees completely concealed the rebels, and Major R. H. Gall, accompanied by Lieutenant G. Arbuthnot, R.A., rode into the wood to reconnoitre.[1] They found the

---

[1] This was a very dangerous duty, and both these officers were specially mentioned in despatches, and later Sir Hugh Rose recommended them for the Victoria Cross. This recommendation was not acted on.

rebels had left their first position, but were holding the northern outskirts of the town. Rose then deployed the 86th Foot and the 25th Bom.N.I., and, supporting them with all his field guns, sent them forward in two lines. The rebel line fell back except the left, which occupied a patch of cultivated ground, and 5/14 R.A. was sent to shell them out of it; but the rebels stood firm until the advance of the 2nd

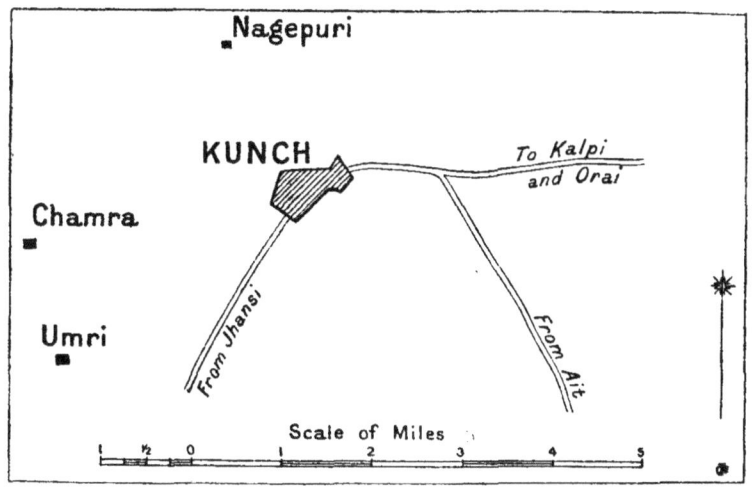

KUNCH, MAY 7

Brigade from Chamra threatened their rear. They then broke and fled, pursued by the cavalry of the 1st Brigade, whose advance was much impeded by the trees. The 86th Foot and 25th N.I. then entered the town, and throwing their right shoulders forward, forced the enemy away from their comrades holding the southern portion. The 2nd Brigade for some unexplained reason did not advance into the town; they marched round it and took up a position on the south, but the artillery, 2/Res. Bom.A. (Lieutenant P. H. Harcourt),[1] and cavalry attached to it joined in the subsequent pursuit. Meanwhile Orr attacked straight to his front; his guns,[2] under Captain C. Douglas, Ben.A., prepared the way for him, and his cavalry in conjunction with Scudamore's force protected his right flank from the attacks of the rebel horse. His infantry had some initial success, but the rebels held on to the south of the town till the advance of the 1st Brigade forced them to evacuate it. Then they all retired towards Orai and Kalpi.

The retreat was at first conducted in an orderly manner, in strict accordance with the existing drill regulations, and Rose's infantry, exhausted by the great heat, were unable to pursue; but the cavalry and guns advanced, and some sharp encounters ensued. Gradually the skirmishers of the enemy were killed or scattered, and one by one their guns were abandoned, when the retreat became a rout. Overborne by heat and fatigue, our artillery were unable to gallop up to case-shot range, and our cavalry were reduced to a walk, but five or six hundred of the rebels were killed. On reaching some woods eight miles from Kunch, the bulk of them broke up and scattered. They lost 9 guns. The British casualties were 9 killed and 47 wounded. There were also 45 fatal cases of sunstroke.

[1] 2/Res. Bom.A. had only three guns owing to the Bhopal guns having been left at Jhansi. It was commanded by Harcourt on Lightfoot's transfer to the 1st Troop.

[2] These consisted of five 6-pounder guns, two 12-pounder howitzers, and two small mortars.

The Nawab of Banda.—Mutual distrust now arose among the heterogeneous components of the rebel army. Tantia Topi, as at the Betwa, early sought safety in flight, and the Rao and the Begum were almost deserted in their stronghold of Kalpi, when help reached them from an unexpected quarter. The Nawab of Banda with a considerable force suddenly appeared upon the scene, and despair was replaced by confidence.

IV. THE SAUGOR FIELD FORCE

Whitlock's Advance on Banda.—It was intended that the Saugor Field Force should be organised in two brigades, but their formation was a slow process. On January 23 Brigadier G. C. Whitlock set out from Kampti to Jubbulpore with his 1st Brigade and joined the Nagpore Movable Column. On February 3 the following troops were under his orders:

12th Lancers.	3rd Mad.E.R.	A Troop Mad.H.A. (Captain and Brev.-Major J. D. Mien).
4th Mad.L.C.	1st Mad.N.I.	
6th Mad.L.C.	Nagpore Rifles.	E Troop Mad.H.A. (Captain and Brev.-Major E. Brice).
7th Mad.L.C.		
2nd Hyderabad Cavalry.	Mad. Sappers and Miners	6/14 R.A. (Captain E. Palmer).
		No. 1 F.B. A/4 Mad.A. (Lieutenant R. Pope).
		No. 8 F.B. B/4 Mad.A. (Lieut. F. H. Thompson).

O.C.A., Brigadier W. H. Miller, Mad.A.

The regiments of Madras Cavalry were sent to Allahabad, and Whitlock, leaving a small garrison in Jubbulpore, began his northward march on February 17 at the head of 1,900 men. His movements were slow and circumspect. At one time it was intended that he should co-operate in the siege of Jhansi, but when that place fell he was at Panna, more than a hundred miles away. On his march thither he had certainly brushed aside such rebels as he chanced to meet, but he never inflicted a serious blow upon them, and the districts traversed by his force fell again into lawlessness as soon as he had passed by.

Battle of Banda, April 18.—After a skirmish at Kubrai on April 17, Whitlock next day found the Nawab of Banda in position eight miles west of the city, and five miles from the River Ken, with 6,000 men in first line and 3,000 in reserve. His guns commanded the main road to the city and his front and flanks were covered by ravines and broken ground. Whitlock's advanced guard, coming under a sharp artillery fire, deployed into line, 600 yards from the position of the enemy, when it was discovered that two of his guns would enfilade our further advance. Captain A. W. Macintyre, Mad.A., commanding a squadron of Hyderabad Cavalry, charged them at once, capturing one and driving off the other. The 3rd Madras Europeans then went forward, but were hard pressed until Whitlock's main body came up on their left and opened a flanking fire, while the cavalry and artillery turned the enemy's left flank. The rebels then retreated towards the River Ken. In the subsequent advance Brigadier Miller was wounded and the command of the artillery devolved on Major T. Lavie,[1] Mad.A. The Nawab, however, fell back slowly and occupied

[1] Lavie was in command of the Artillery Division made up by A/4 and B/4 Mad.A.

all ground favourable for defence. The fight endured for four hours, and then a last stand was made on a line of hills close to the Ken.

Captain E. Palmer with 6/14 R.A. was in charge of the park on the night of the 17th, and orders for the advance of his own guns did not reach him till 4 o'clock next morning. He at once started with his four 18-pounder guns and two 8-inch howitzers drawn by elephants, while his centre division, broken up into an advanced and rear guard, was armed with $5\tfrac{1}{2}$-inch mortars carried by coolies. His road to the front was much intersected by nullahs and one of the elephants gave considerable trouble; but he covered eight miles in three hours, and was able to join with his 18-pounder guns in the last conflict of the day. He opened fire at 2,000 yards range, when a shot falling near the rebel guns caused their instant withdrawal. Our whole line then advanced, and the enemy, pursued by the cavalry and light guns, made haste to cross the river, while the 18-pounders kept up a fire on them till they got out of range. Once the Ken was behind them the retreat of the rebels became orderly, and they marched off northwards, having lost some 1,000 killed and wounded and 17 guns. The British had 5 killed and 29 wounded. Whitlock then entered Banda, where he awaited the arrival of his 2nd Brigade (1,000 strong with 4 guns), which, leaving Jubbulpore on March 18, entered the Banda District on May 27.

Flight of the Nawab to Kalpi.—The Nawab, abandoning his palace and all its treasures, fled to Jalalpur on the lower Betwa, where he busied himself in reorganising his followers, and so successful were his efforts that when he marched into Kalpi, about May 10, his arrival was greeted with enthusiasm, and numbers of the disaffected flocked to the standard of the Peshwa.

V. Capture of Kalpi

Sir Hugh Rose's Movements.—Kalpi, standing on a precipitous rock that rises from the Jumna, was the only arsenal of warlike stores on the right bank of the river. It was surrounded at a distance of eight miles by a ring of solidly built temples, eighty-four in number, each being in a walled enclosure, and both in front and in rear of this ring were networks of ravines, the roads through which the rebels broke up and barricaded. The direct attack (from the south-west) would be extremely difficult, besides which absence of water forbade a concentration of troops in this region. The most favourable line of attack was from the south-east, where the approaches to the town were more open; but it was not this consideration only which caused Sir Hugh Rose to turn aside to the right of the main road and march on Gulauli, a village four and a half miles from Kalpi on the banks of the Jumna, over which river there was a ford close by. Lieutenant-Colonel G. V. Maxwell, of the Connaught Rangers, with a small column detached from the Commander-in-Chief's army was marching on the left bank of the river towards this ford, and when Rose could get in communication with him the junction of the Bombay and Bengal armies would be effected.

Skirmishes outside Kalpi, May 15 *to* 17.—On May 15 the 1st Brigade encamped east of the village of Gulauli. Major Orr with the Hyderabad contingent occupied Tehri, and the 2nd Brigade, under Lieutenant-Colonel R. D. Campbell, was marching to Deopura.[1] After the battle of Kunch 5/14 R.A. was attached to this Brigade, and two guns under Lieutenant J. C. J. Lowry were with the rear guard commanded by Major J. Forbes, 3rd Bombay Light Cavalry. On the 16th the rear guard halted about a mile from Itaura, and whilst the baggage was being dragged over a deep and wide ravine, only passable by carts in single file, a force of the enemy about 5,000 strong, including 1,000 cavalry and some heavy guns drawn by elephants, suddenly appeared on the left and left rear and occupied Itaura. When Orr at Tehri became aware of this menace, he at once reinforced Forbes with some Hyderabad cavalry, and so firm a front was shown that the rebels made no forward movement during the two hours occupied in getting the baggage across the nullah. The guns and infantry followed, and Lowry unlimbered on the far side, the cavalry, first at a walk and then (when concealed from view) at a gallop, bringing up the rear. The enemy at once occupied the vacated position, but were checked by the rifles of the 3rd Europeans and the fire of Lowry's guns, until the baggage train was out of danger. During the remainder of the march

[1] A few days previously the 2nd Brigade lost its way, and in consequence had to make a double march. Numbers were prostrated by the extreme heat, including Brigadier Steuart and his whole staff. Lieutenant-Colonel Campbell, 71st Highland Light Infantry, then took command of the Brigade.

Lowry was several times obliged to halt and fire, but as Deopura was approached the pursuit died away and the enemy turned their attention to the village of Matra, which was occupied by Campbell as a protection to his left flank. Being reinforced from Kalpi, the rebels now attacked Matra with great determination, but were driven back by the 2nd Brigade with the help of a detachment from the 1st Brigade. Next day, the 17th, the attack was renewed. Matra was held by the 71st and 2/Res. Bom.A. (Lieutenant P. H. Harcourt) while the remainder of the 2nd Brigade defended Deopura. A squadron of the 14th Light Dragoons and two guns of 5/14 R.A. under Lieutenant J. C. J. Lowry were sent to the left front, while Captain T. S. P. Field with the remaining four guns came into action on the right, where he remained till dusk, having successfully driven the enemy from a small village in his immediate front. Meanwhile the cavalry on the left, with whom the Brigadier was present, encountered a force of rebel horse some 2,000 strong with three guns, advancing on a wide front and threatening our left rear. Campbell ordered Lowry to open fire at once, at a range of 2,000 yards, but after a few rounds the guns limbered up and, coming into action at 800 yards, effectually prevented any further advance on the part of the enemy until darkness closed the operations.

Concentration of the Field Force.—But in their position on the exposed flank the 2nd Brigade were liable to constant attack, and moreover the wells at Deopura and Tehri were showing signs of exhaustion.

Sir Hugh therefore determined to concentrate his forces, and encamped the 2nd Brigade and the Hyderabad Contingent on the left of his 1st Brigade. Two mortars protected by a strong picket were placed in battery on a commanding hillock on the west of Gulauli, to bombard a fortified ridge halfway to Kalpi and the adjacent ravines.

Maxwell reached the left bank of the Jumna on May 18, with the 88th Foot, some Sikh infantry and cavalry, No. 17 F.B. 3/1 Ben.A. (Lieutenant-Colonel C. H. Blunt [1]) and 682 men of the Rifle Brigade and 8th Foot mounted on camels; in all 2,000 men with four field guns and eight 10-inch and 8-inch mortars. He at once sent two companies of the 88th, some Sikh Infantry, and the Camel Corps to reinforce Sir Hugh Rose, and on the 21st placed mortars in battery to bombard the fort and town of Kalpi and the village of Rayar, where the enemy had guns posted to protect their left flank. These mortar batteries, three in number, were commanded one by Blunt himself, one by Lieutenant R. T. Hare of the same company, and the third by Captain A. D. Turnbull, Ben.E.

Plans of the Rebels.—On May 20 the rebels made another attack and were driven back by our pickets, but such skirmishes were only a preliminary to the great effort they intended to make on the

[1] Blunt at the Siege of Delhi was a 1st Lieutenant, and in September was promoted Captain, and commanded 2/3 Ben.H.A. after Captain E. K. Money was wounded. When that officer was replaced in command of the troop by Major W. A. Mackinnon, Blunt got the command of 3/1 Ben.A. on the transfer of Major G. Bourchier to 3/2 Ben.H.A. at Lahore. In March, 1858, he received two brevets.

22nd, when, as they swore on the sacred waters of the Jumna, they would drive the British into the river or perish in the attempt. Sir Hugh was playing a waiting game; he had repeatedly driven back the minor attacks made on his position, but had never followed up his success, for he had no intention of exposing his troops to the terrific heat [1] until he was in a position to strike a crushing blow that would give him the possession of Kalpi. His plans were almost ripe when the news of the impending sortie reached him on the 21st, and he at once prepared to receive the rebels.

Battle of Gulauli, May 22.—The line Deopura—Tehri—Gulauli separated the net of ravines surrounding Kalpi from the open plain which, except for a narrow fringe of high and broken ground along the Jumna, extended southwards from that river to the Kalpi—Banda Road. On the morning of May 22 Sir Hugh disposed his forces in two wings, the right, under Brigadier Stuart, on a line perpendicular to the river and facing the ravines; the left under his own immediate command thrown back *en potence* across the plain. Stuart's front extended from the river at Gulauli to a point nearly opposite to Tehri, and comprised three detached posts. Lieutenant-Colonel Lowth with a detachment of the 86th Foot and Major G. H. Robertson with a wing of the 25th Bom.N.I.

---

[1] At Kunch the thermometer was at 115° F. Before Kalpi 116° F. On the subsequent march to Gwalior it burst in an officer's tent at 130° F. On May 19 out of 36 of the 14th Light Dragoons forming an escort for forage 17 were brought back to camp in dhoolies after only two hours' exposure. On the march to Gulauli 200 out of 400 men in the 25th Bom.N.I. fell out of the ranks.

were on the right and left flanks, and the central body, 86th Foot and 3rd Europeans, were about the hillock the mortar battery was placed on. This hillock was not only the key of the right wing, it was the key of the whole position. In reserve were the 21st Company R.E. (Lieutenant J. B. Edwards), the other wing of the 25th Bom.N.I., two troops of cavalry (14th Light Dragoons and 3rd Bombay Cavalry), and 4/2 Bom.A. (Lieutenant C. H. Strutt). Sir Hugh's left wing faced the open country south-east of Tehri. On its right flank were the 1st Troop Bom.H.A. (Captain J. G. Lightfoot) and a squadron of the 14th Light Dragoons. On their left, with a detachment of the 3rd Europeans on each flank, were five siege guns (Lieutenant T. Haggard), and on their left again were the Camel Corps (Major J. Ross, Rifle Brigade), the Sikh Infantry, and 2/Res. Bom.A. (Lieutenant P. H. Harcourt). In support were three troops of cavalry (14th Light Dragoons and 3rd Bombay Cavalry), a wing of the 71st Highland Light Infantry, and 5/14 R.A. (Captain T. S. P. Field). The Hyderabad Contingent watched Sir Hugh's left flank and two companies of the 88th Foot and a detachment of the 25th Bom.N.I. supplied the camp guard. Captain F. M. M. Ommaney, R.A., was in command of the artillery.[1]

[1] Though Ommaney was the officer commanding the artillery of the 1st Brigade from the beginning of the campaign and of the Central India Field Force since the siege of Jhansi, and though he was mentioned in Despatches, he was not given the Companionship of the Bath which was his due; he received a barren brevet-majority. The authorities certainly did not err on the side of generosity in rewarding those who had taken part in the Central India Campaign.

In order that the battle should be fought during the intense heat of an Indian May day the rebels made no sign till after 8 a.m., when their presence in the ravines was reported by Stuart's pickets. Three guns of 4/2 Bom.A. (Sergeant-Major Graham [1]) were then unlimbered on a small piece of level ground just in front of the mortar battery. The three others were sent under Strutt to the left to support Robertson; and lest the hillock should be outflanked, the Brigadier advanced the wing of the 25th Bom.N.I. which was in reserve to a position between that held by Robertson and the main body. Stuart then reported to Sir Hugh that he felt his position was secure.

To all appearance indeed the rebels were about to attack our left. Shortly after 8 o'clock a large mixed force issued from Kalpi and marched down the Banda Road, headed by musicians and the entourage of the rebel leaders. When they reached Deopura the troops extended across the plain, and their numerous cavalry appeared to be making an outflanking movement. Sir Hugh, having massed his cavalry in front of his left, ordered them to retire as the enemy advanced, so as to lure them under the fire of the siege guns. This plan, however, succeeded only partially; for though Haggard's fire caused casualties and disorder in the ranks of the sowars, they refused to be drawn within striking distance of the British cavalry. But Sir Hugh had

[1] Captain Woollcombe was invalided after Jhansi and, as Lieutenant B. Christie (severely wounded at Dhar) did not rejoin till after Gulauli, Strutt was the only officer present with No. 4/2 Bom.A. at that battle.

no anxiety about his left flank; **he knew** that in the open he could more than hold **his own** without

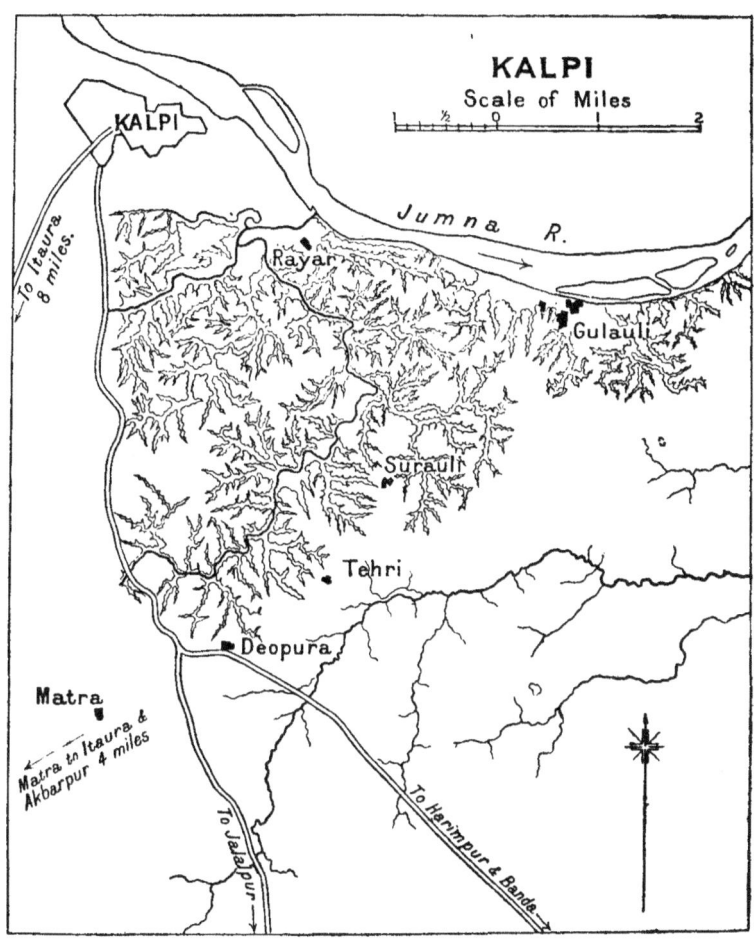

moving a man from his right; but he was anxious about the ravines and with good reason.

While the cavalry demonstration was in progress the enemy secretly introduced large reinforcements of infantry into the network of ravines behind the line Tehri—Gulauli; and there they lay concealed until Stuart disturbed them in their lair by sending forward a company of the 3rd Europeans to reconnoitre. The sepoys then rose from their hiding-places in thick chains of skirmishers, followed by large supports and columns in mass at a distance. In an instant a general engagement began along the whole line from the Jumna to Tehri and the ravines became enveloped in smoke and fire. All our guns came into action and the supports closed up.

Sir Hugh was watching the attack made on the centre of his position, when, noticing a slackening of fire on his right, he sent to Brigadier Stuart to inquire if he desired the support of half the Camel Corps. The Brigadier replied that he did; but even as he received the answer Sir Hugh realised that the sound of the enemy's fire was now coming from within the line our pickets had held, and that his right, the vital point, was in danger. He at once determined to bring the whole Camel Corps to the threatened quarter, and placing himself at their head he made for the mortar battery hillock. On his way he met an orderly riding at full speed with a message from Stuart saying there was not a moment to lose. The Camel Corps pressed on and presently dismounted at the foot of the hillock. They arrived just in time.

When the general engagement began, the enemy on our right directed their main efforts against Lowth, who was reinforced by another company of the 86th.

Indeed for a time Stuart saw no reason to change his opinion as to the security of his position. But as the sun mounted higher, the terrific heat began dangerously to affect his men, who were struck down or incapacitated in ever-increasing numbers. Ere long it was discovered that many of the rifles were leaded and could not be fired, and presently Stuart could only muster a handful of soldiers to defend the hillock and work the guns. About noon came the crisis. The ravines and gullies surrounding the hillock were crowded with sepoys who, seeing the small force opposed to them, raised a triumphant shout and rapidly neared the guns. The Brigadier called upon the gunners to draw their swords and defend them, and dismounting from his horse he placed himself beside them. It was at this supreme moment that Sir Hugh Rose and the Camel Corps appeared upon the scene.

Major Ross had dismounted his men at the foot of the hillock and led them up the slope in line at the double, accompanied by Sir Hugh and his staff, whose horses (with one exception) were all killed or wounded as they topped the rise. With a rousing cheer the General, the Staff, and the Camel Corps charged the dense masses of the mutineers, ten times their numbers, driving them back to the ravines; and Graham's guns, being speedily brought forward to a convenient knoll, pursued them first with case and then with round shot. Those rebels who had been pressing upon Lowth were now caught in a trap; for their right flank was laid open by the retreat of the attackers of the hillock, and

Lowth, seizing his opportunity, drove them into the Jumna.

When these events were in progress the wing of the 25th Bom.N.I., which had been moved up from the reserve, was attacked by overwhelming numbers, and was being pressed back, when Edwards brought up the 21st Company R.E., and by a brilliant charge defeated the assailants and pursued them to the ravines. The other wing of the 21st, under Robertson, was also attacked, and offered a determined resistance. The rebel sepoys pressed close up to the men, halted, and flung reproaches at them, couched in the most revolting language, for their unshaken fidelity to the British. But the 25th answered the maledictions of the mutineers in a manner worthy of their reputation and English discipline, with a volley, a cheer, and a charge with the bayonet.

The whole British line now advanced, a convergent attack being made on the right of the enemy, who, thoroughly beaten and disheartened, broke up and scattered. Their loss must have been severe, but there is no record of it. Our casualties were 31 killed and 57 wounded.

Capture of Kalpi, May 23.—Kalpi was shelled so vigorously by the mortars [1] on the left bank of the Jumna that it presented no attraction as a place of retreat, and next day was occupied without resistance after 5/14 R.A. had cleared the ravines of the

---

[1] Gunner Farrell, of 5/14 R.A., was specially mentioned in despatches for good service with these mortars, as was also Sergeant Judgson, Hyderabad Contingent Artillery.

enemy, who fled by the Jalaon Road pursued by a small column [1] under Major Robertson, 25th Bom.I.

Affair at Sher Ghat, May 16.—While the operations round Kalpi were in progress the Jumna was successfully bridged at Sher Ghat, whither Colonel W. Riddell [2] with the 3rd Bengal European Regiment, some Sikh horse and foot, and two guns of 7/14 R.A. (2nd Captain and Brevet-Major C. E. Walcott) had been sent from Futtehghur.

Dispersal of the Troops.—After the occupation of Kalpi, which was found to contain war stores worth from twenty to thirty thousand pounds, the Hyderabad Contingent was ordered back to the Deccan, and Sir Hugh Rose, about to proceed on sick leave, issued a farewell order. But the Begum of Jhansi was still unsubdued.

VI. THE GWALIOR CAMPAIGN

The Seizure of Gwalior.—After their defeat at Kalpi the rebel chiefs fled westward to Gopalpur, forty-six miles south-west of Gwalior, and Tantia Topi was summoned from the secret retreat he had occupied since the battle of Kunch. Then an audacious plan was conceived which is usually attributed to the energetic Begum. Not only was it audacious, it was simple—the rebels were to seize Gwalior and hold it in the name of the Peshwa! As if by magic, the beaten rebels reassembled, and on May 30 the Begum and Tantia Topi entered Morar

[1] This column was composed of : Detachments 14th Light Dragoons, 3rd Bombay Light Cavalry, Hyderabad Cavalry, No. 2/Res. Bom.A. (Lieutenant P. H. Harcourt), and the 25th Bom.N.I.

[2] Lieutenant-Colonel and Brevet-Colonel.

at the head of 7,000 infantry, 4,000 cavalry, and 12 guns. Next day 8,000 of Sindiah's troops joined them after a mock resistance, and the Maharajah escaped to Agra. The Begum then occupied Gwalior, becoming mistress of its treasures and its arsenals, and thus dealing a heavy blow to British prestige. But her triumph was brief.

Rose's Plans.—On June 3 the news reached Sir Hugh Rose, who, ill as he was, at once resumed his command : at the same time the Government recalled the Hyderabad Contingent and placed troops of the Rajputana Field Force and Bengal Army at his disposal. Leaving a small garrison[1] in Kalpi, he began his advance on Sindiah's capital after two days' preparation. Major W. A. Orr was already south of Jhansi when the orders of recall reached him, and Rose directed him to march to Punniar and cut off the retreat to the south of fugitives from Gwalior. Brigadier M. W. Smith[2] was ordered to march from Sipri to Kotah-ki-Serai, seven miles south-east of Gwalior, and halt there. Colonel W. Riddell was to move with his column and a siege train from Agra to a point seven miles north of Gwalior, and was then to invest the city on the west. Rose himself proposed to march on Morar. It was hoped that by June 19 the troops would be in position and the investment completed; then the city would be attacked from the hills lying

---

[1] This consisted at first of Detachments 3rd Bombay Cavalry, 3rd Bombay Europeans, 24th Bom.N.I., the 21st Comp.R.E., and 5/14 R.A.; but later on the 3rd Bombay Cavalry and the 24th Bom.N.I. followed Sir Hugh Rose and reached Morar on June 18.

[2] Lieutenant-Colonel and Brevet-Colonel M. W. Smith, 3rd Dragoon Guards, commanded the brigade of the Rajputana Field Force at Sipri.

on the south and south-east. Riddell, however, was delayed at Dholpore by the rising of the Chumbal, and was unable to carry out his portion of the plan.

Battle of Morar, June 16.—After an arduous march Rose [1] reached Behahdurpore five miles east of Morar on June 16. The troops had been long on the road, and the sun was already high when Captain H. D. Abbot, who was reconnoitring in front with the Hyderabad cavalry, reported that the rebels were in the Morar cantonments and Rose determined to attack them at once.

The Morar Cantonments lay on the right bank of the river from which they got their name, and were traversed from east to west by the road to Gwalior. Between them and Behahdurpore the country was open and level, except that a belt of nullahs ran along their east front. On the south-east the country was broken and hilly. The enemy were holding the cantonments with cavalry while they occupied the diversified ground on their right and right front with infantry, strong bodies being hidden in the nullahs. A battery of guns was in their centre in a concealed position, and another was on their right.

It was Rose's intention while masking the dangerous ground on his left front to outflank and roll up the cavalry which formed the rebel left wing. He deployed the 86th Foot and the 25th Bom.N.I. in his front line, some siege guns [2] under Lieutenant C. H. Strutt,

[1] Sir Hugh Rose in his Despatch says that on one day the heat rose to 130° in the shade.

[2] In Major-General T. C. Crowe's opinion Strutt had only one 8-inch howitzer on this occasion. He further states that both horse and field batteries were accustomed to take their turn with the siege guns.

4/2 Bom.A. being in the centre, and detachments of the 14th Light Dragoons on the flanks. Brigadier R. Napier was in command of the second line, his orders being to give special heed to the dangerous country on his left. His infantry, consisting of the 71st Highland Light Infantry and some Madras Sappers and Miners, were ordered to advance in échelon from the right. A wing of the 14th Light Dragoons protected the left flank of the second line, which was accompanied by the 1st Troop Bom.H.A. (Captain J. G. Lightfoot) and 2/Res. Bom.A. (Lieutenant P. H. Harcourt). The 3rd Bombay Cavalry formed the rear guard.

As he approached Morar Rose moved his front line diagonally to his right, but speedily found himself involved in broken ground [1] and subject to the direct fire of the rebel central battery and to enfilade fire from that on their right. But Strutt brought his heavy guns into action with great promptitude, and being speedily reinforced by Harcourt, the fire of the central battery of the rebels, after causing some casualties amongst the horses, was effectively checked, while Lightfoot galloping up on the left engaged the rebel right flank battery. Under cover of this fire Rose, while holding firmly on his left with the 25th Bom.N.I., sent forward the 86th Foot in skirmishing order, who, swinging round their right flank, cleared the rebel cavalry out of the cantonments. Meanwhile Abbot with the Hyderabad cavalry, after a somewhat lengthy détour to the

---

[1] Rose while marching to outflank the rebels was guided by Sindiah's Agent, who lost his way.

right, found a way across the nullahs and broken ground in front of the cantonments and came down upon the rebel left; but as soon as the advance of the 86th was perceived, the enemy withdrew their crippled guns across the bridge over the Morar, and, before Abbot got within striking distance, the bulk of the rebel cavalry followed the guns. Napier following up with the second line now found himself close to the nullahs, and by a timely reconnaissance discovered that they were strongly held by a concealed foe. The 71st Highland Light Infantry went forward and were received with a fusillade, but the 25th Bom.N.I. brought an enfilade fire to bear upon the nullahs, the 71st charged with the bayonet, and, suffering some loss themselves, drove the rebels out, killing many of them. Soon after a wing of the 14th Light Dragoons, wheeling round the British left flank, caught a number of fugitives in the open and inflicted heavy loss upon them.

Thus after two hours' fighting Morar was captured, and the weary troops found welcome shelter in the cantonment buildings which had not been destroyed. The victory gave Sir Hugh Rose command of the line of the Morar River and of the road to Agra, and enabled him to communicate with Dholpore on the right, and Kotah-ki-Serai on the left.

Brigadier Smith at Kotah-ki-Serai.—On the morning of June 17 Brigadier Smith reached Kotah-ki-Serai, but found himself in a cramped and insecure position. North of that place the road to Gwalior after crossing the Morar by a ford ran alongside of a canal through a deep defile in the hills for about two miles, and then crossed the open plain about the Phul Bagh. A preliminary reconnaissance revealed the fact that the rebels were in position on the hills east of the canal, and that their guns occupied a ridge about 1,500 yards north of Kotah-ki-Serai. They thus commanded Smith's proposed halting-place, and his only course was to attack them. The following troops were under his command :

8th Hussars.	95th Foot.	No. 3 Troop Bom.H.A.
1st Bom. Lancers.	10th Bom.N.I.	(Captain and Brevet-Lieutenant-Colonel E. S. Blake).

O.C.A., Lieutenant-Colonel T. W. Hicks, Bom.A.[1]

Action of Kotah-ki-Serai, June 17.—Brigadier Smith sent forward the horse artillery escorted by the cavalry (less a troop of Hussars and a squadron of Bombay Lancers left as camp guard), and after a few rounds the rebel guns were seen to retire. Then Lieutenant-Colonel J. A. R. Raines, 95th Foot, was placed in command of the infantry and ordered to clear the hills, while the Brigadier himself with the cavalry and guns advanced through the defile. Raines skirmishing over the hills reached the ridge the rebels had occupied, when he perceived they had taken up a new position a thousand yards in its rear, that their guns were posted in his front and on his left, and that at the same time a force of their cavalry was moving southwards along the Sirol—Tahia road on the right flank of the British advance. Brigadier Smith at once sent a troop of hussars and two guns to reinforce

[1] This officer had joined Smith on the march. He came from Jhansi in command of details.

his camp guard, and ordered two other guns to the hills to support Raines' further advance, while two companies of the 95th checked the rebels upon the Sirol road. Presently Raines' skirmishers won their way to the margin of the Phul Bagh plain, and Smith at the same time reached the northern end of the defile. At this moment the rebel horse were retreating from Sirol to their cantonments in the Phul Bagh, and the 8th Hussars were launched against them. In their impetuous charge the Hussars cut down many of the flying sowars[1] captured two 9-pounder guns that accompanied them, and for a time were masters of the rebel camp. Then they returned, but so exhausted that they could hardly sit their horses. But now the enemy, issuing from the Lashkar,[2] appeared both in front and on the flanks, and Smith withdrew his cavalry under cover of his guns. Then occupying both ends of the defile with infantry, he bivouacked on the hills whence he had driven the enemy and, reporting his action to Sir Hugh Rose, he asked for reinforcements.

Meanwhile the rebels occupied an entrenched position on the hills lying on the west of the defile; where, with the deep canal running along their front, they would be on the flank of a British advance from the south, while their artillery commanded the British guns and baggage parked in

[1] In this charge the Begum of Jhansi was killed. She was riding in the ranks dressed as a man. The circumstances of her death are given by Forrest, iii. p. 281.

[2] As the army was quartered in this locality, it naturally was called لشکر (lashkar, *Pers.*), the camp or encampment; and this name was colloquially extended to the " new city," which was close to it.

the defile. Strong reinforcements of all arms were also about Gurhi.

Battle of Gwalior, June 19.—When Sir Hugh Rose received Smith's message he at once sent off a small mixed force to his assistance, and leaving Brigadier Napier with the bulk of the cavalry, a small force of infantry, and the 1st Troop Bom.H.A. at Morar, he himself followed, and on the evening of the 18th bivouacked with the remainder of his force opposite to Kotah-ki-Serai on the left bank of the Morar.

When Sir Hugh Rose realised the position of the enemy his first intention was to preserve a defensive attitude during the 19th, to cross the canal when it became dark, and by a night march round the enemy's right flank, to cut them off from Gwalior. Accordingly, during the night of the 18th two 18-pounder guns and an 8-inch howitzer[1] were sent up a steep hill above the bivouac to keep down the fire of the rebel guns; but next morning it was discovered that the carriage of the 8-inch howitzer was so shaken by hard usage that the piece had to be fired with a reduced charge. The 18-pounders also were unprovided with sights, and, as a result, the heavy guns were ineffective. At the same time the attitude of the enemy was threatening. With his usual grasp of the situation Sir Hugh at once took the bold and the safe course. He would not await the possibility of attack, he would assume the offensive himself. Brigadier Stuart with the 86th Foot and 25th Bom.N.I., and Brigadier

---

[1] These guns were commanded on this occasion by Lieutenant T. T. Haggard, Bom.A., the Commissary of Ordnance. The author is not aware how they were manned.

Smith with the 95th Foot and the 10th Bom.N.I., were ordered to attack the rebel position on the hills; the 3rd Troop Bom.H.A. and a squadron of the 8th Hussars were to occupy the northern end of the defile, and 4/2 Bom.A. was to support the infantry. The remainder of the force was to be in reserve on the left rear.

The infantry attack was at once successful; the 86th, turning the right flank of the rebels, captured their guns, and, attacking the entrenchments, drove out the enemy, who fled partly towards Gwalior and partly to the hills on their right rear. The 95th came up on the right of the 86th, and the captured ordnance — three excellent English 9-pounders — were turned upon the enemy and, being manned by the infantry, made good practice.[1] The 25th Bom.N.I. advanced in support of the 86th and 95th, and the 1st Bombay Lancers, getting over the difficult ground with rapidity and dexterity, ascended the heights as soon as the guns were taken, and covered the left flank. Farther to the right the 10th Bom.N.I. ascended the hills and came under a mixed musketry and gun fire from the enemy's left, but, quickly clearing the rebels from their immediate front, they captured two field guns and three mortars. Strutt, however, with 4/2 Bom.A. was for a time unable to find a way over the canal. But the success of the infantry attack placed Rose in possession of the highest range of heights on the

---

[1] Lieutenant J. D. Brockman of the 86th and Lieutenants J. Budgen and J. Sexton of the 95th, as well as some of their men, had received instruction in gunnery.

south of Gwalior, and being convinced that he could take the place before sunset he resolved on a further advance.

The rebels now occupied a line extending from the Phul Bagh on their left to the Lashkar parade ground on their right, which was supported by guns including two 18-pounders. Rose's intention was to attack the enemy's left flank with the bulk of his cavalry supported by the horse artillery, and at the same time to launch his infantry down the slopes and, after capturing the Lashkar parade ground, to force an entry into the New City. The 18-pounder guns, however, were formidable opponents; their fire was answered by the captured ordnance worked by the infantry, and occasionally the Enfield rifles drove their gunners away; but it was not until after the arrival of one of Strutt's guns, which had at last been got over the canal, that a lucky shrapnel shell was burst over them and caused their final disappearance. Then our infantry poured down the slopes, and the Bombay Lancers, who had advanced by the Chirwai—Lashkar road and had taken up a position at its issue from the hills, galloped forward and after a brilliant charge cleared the rebels from the parade ground and pursued them into the streets of the Lashkar. The infantry followed as rapidly as possible, but no resistance was met with in the town. On the right, however, the cavalry and the horse artillery had a tough encounter as they advanced in accordance with their orders. But by sunset the town of Gwalior was in Rose's hands; the fort, however, was still occupied.

Capture of the Fort, June 20.—Next morning Lieutenants W. Rose and W. F. F. Waller with a few men of the 25th Bom.N.I. on their own initiative attacked the celebrated stronghold. They burst open the main gateway, and surprising the other gates before the garrison, a party of fanatical Mussulmen artillerymen, could shut them, they took the fort by storm. In this brilliant feat of arms Rose lost his life and Waller won the Victoria Cross.

Tantia Topi, while his troops were still fighting, headed the retreat on the 19th; but owing to Riddell being detained on the Chumbal, the bulk of the rebels made good their escape westward, and Rose, having sent orders to Napier to pursue them from Morar, reinstated Sindiah in possession of his capital.

Action at Jaora-Alipore, June 22.—Napier's orders reached him early on June 20, and, leaving his infantry to protect Morar, he marched at once with his cavalry and horse artillery. Skirting Gwalior, he came under fire from the fort, where the Moslem gunners still held sway; but taking no notice of them he pursued his course westward, and on the morning of the 22nd he came up with the enemy—reported to be 12,000 strong, and to possess 22 guns—at Jaora-Alipore.

Rising ground hid Napier's approach, so that he was enabled to reconnoitre the enemy from a distance of 1,200 yards. He saw that they were drawn up with guns and infantry in the centre and cavalry on both flanks, the left flank being open. He at once ordered Captain Lightfoot to come into action with the 1st Troop Bom.H.A. against this flank and then act as circumstances might dictate. Lightfoot dashed forward and opened fire at 250 yards, while the cavalry formed line to the left. The rebel guns made a feeble reply, but soon desisted. Then, seeing that the enemy had already begun to waver, Lightfoot without pausing to limber up ordered " Detachments Front," and advanced at the gallop while the cavalry instantly charged.[1] The rebels at once took to flight; some 400 were killed and all their guns (26 in number) fell into the hands of the British. Napier had only 3 men killed and 8 wounded. One gunner died of sunstroke. In the Gwalior operations the British casualties between June 16 and 20 amounted to 29 killed and 65 wounded.

On June 29 Sir Hugh Rose again relinquished the command of the Central India Field Force and was succeeded by Brigadier-General R. Napier.

VII. THE SAUGOR FIELD FORCE

Movements of the Saugor Field Force.—On May 27 Major-General **Whitlock** was joined at Banda by Brigadier **MacDuff**[2] with the 2nd Brigade, consisting of the 43rd Light Infantry, some Irregular Cavalry, and Foot and Mounted Police. A few days later Whitlock marched on Kirwi, where the rulers of the state, the Kirwi Raos, preserved a hostile attitude. On the approach of the British, however,

---

[1] " We always maintained that we reached the enemy before the cavalry " (Letter from Major-General T. C. Crowe to author).

[2] Lieutenant-Colonel and Brevet-Colonel J. MacDuff, 74th Foot.

they at once surrendered, Kirwi was annexed by the Government, and Whitlock brought back to Banda treasure and jewels of fabulous amount. The Saugor Field Force was then distributed about the districts of Kalpi, Banda, and Kirwi. But in the southern portions of Central India and Bundelkhund no regular British troops, the weak garrison of Jhansi excepted, were to be found north of the line Mhow—Saugor, where the Malwa Division was in process of formation. During Sir Hugh Rose's advance all his force had been required to fight his way forward, and as soon as his back was turned disorder again broke out. After the fall of Jhansi the Rajahs of Banpore and Sharghur, having evaded Major Orr's pursuit, escaped to their own territories, and Chanderi was occupied by the rebels, who, however, were ejected by Brigadier Smith on his way from Kotah to Sipri. But when, early in June, that officer set out for Gwalior, though the city was held for the British Government, the district of Chanderi fell into open revolt. Colonel Liddell in command at Jhansi had no men to spare for duty outside the walls, and, in that neighbourhood and also at Kunch, robber chieftains pillaged at their pleasure. Orr indeed on his homeward march with the Hyderabad Contingent had scattered some of these bands, but on his recall to Gwalior they again drew to a head. The short-lived revolution in that city had a disastrous effect on peace and good government from the Jumna to the Nurbudda, and until it was quelled no active measures could be taken to suppress rebellion.

VIII. BEYT ISLAND

Attack on Beyt Island.[1]—Beyt Island, situated in the Gulf of Cutch, was at this time a nest of insurgents, fanatical Hindus called Waghis, the descendants of West Coast pirates. Early in March it was determined by the Bombay Government to reduce the fort in this island, and a British force consisting of 2/11 R.A., two companies of the 4th Foot, and the 10th and 11th Bom.N.I. was sent there for the purpose. The Royal Artillery were on their way to Karachi from England and were commanded by Captain N. S. K. Bayly, who as senior officer also commanded the expedition. The ordnance available consisted of two 6-pounder ship's guns, but there was no round shot and only a scanty supply of shrapnel and case.

On April 2 a landing was effected, and when within 300 yards of the fort Bayly placed the infantry under cover of a convenient wall and then went forward with Captain J. Paton, 4th Foot, to reconnoitre. But he was seriously wounded. Paton then took command and ordered 2nd Captain A. W. Johnson, R.A., to bring the guns into action. Johnson opened fire at a range of 250 yards upon a loopholed tower in the fort; but the ammunition was quite unsuitable, and even when one gun was run forward to a sheltered position 100 yards from the tower no useful result was obtained. An attempt to blow in the gate of the fort with bags of gunpowder also failed, and Paton reluctantly withdrew the troops.

[1] See "Occ. Papers, R.A.I.," i. p. 65.

THE CENTRAL INDIA CAMPAIGN

Capture of Beyt Island.—Beyt Island remained unsubdued until the autumn of 1859, when it was captured on October 9 by Lieutenant-Colonel Donovan [1] with the Kattywar [2] Field Force, which consisted of the 28th Foot, the 6th Bom.N.I., a detachment of Madras Sappers and Miners, and a detachment of 2/3 R.A.[3] (Lieutenant A. J. Raitt).

The operations narrated in this chapter are hardly even now regarded in their proper proportion by general readers of the history of the Great Mutiny, though they appeal with irresistible force to every student of tactics and strategy. Countless books have dealt with Lucknow and Delhi, pictures and photographs exist in profusion ; but with the exception of some slight memoirs it is in the despatches alone—not always very clear—that one can trace the story of the campaign which saved Southern India, included the destruction of three rebel armies, and rendered possible the pacification of India north of the Jumna. Sir Hugh Rose was undoubtedly the greatest commander of his day.

[1] Major and Brevet-Lieutenant Colonel E. W. Donovan, 33rd Foot.
[2] Kattywar is the western region of Gujerat.
[3] As the Brigade system was introduced into the Royal Artillery in April 1859, the designation of this company had become No. 7 Battery, 13th Brigade.

CHAPTER IX

THE EXTINCTION OF THE MUTINY

I. THE SITUATION IN UPPER INDIA, MARCH, 1858

The Forces of the Enemy.—While the unhindered escape of the rebels after the capture of Lucknow was alone sufficient to put the whole province in a blaze, an ill-judged proclamation added fuel to the fire. Lord Canning's decree, promulgated when the city was captured, confiscated the property of all who had risen against the Government and bore hardly and even unjustly on the Oudh Talukdars, who, so long as they dared, were friendly to the British. Even during the time when the existence of a strong native government left them no option between rebellion and ruin, their participation in the war was more nominal than real; but now, regarding the Governor-General's edict as intolerable, they rose, and for the first time, in full revolt.[1] Now their jungle-girt strongholds would afford ample shelter to the wandering bands of mutineers, while their armed followers would swell the number of our enemies, of whom there was no lack. The Nana Sahib still possessed some relics of prestige, and though he never ventured his person within reach of British troops, his brother Bala Rao was active in the field, and many supported

[1] See "Sepoy Revolt," by General M'Leod Innes, p. 244.

his cause; the Moulvie of Fyzabad had his own peculiar following of Mohammedan fanatics; and the Begum in the name of the King attempted to resuscitate the disbanded army of Oudh. Furthermore, rebellion in Rohilkhund was still triumphant; even as the last defences of Lucknow were beaten down serious trouble arose in Behar, and Jhansi and Kalpi were still in rebel hands.[1]

The Plans of the Commander-in-Chief.—On the other hand, the Commander-in-Chief was at the head of a powerful army, and the line of the Ganges and Jumna was safeguarded by Allahabad, Futtehpore, Cawnpore, and Futtehghur. In the opinion of Sir Colin Campbell the subjugation of Oudh was the first thing to accomplish, but Lord Canning strongly adhered to the original plan[2] of marching on Bareilly as soon as Lucknow was captured. The Commander-in-Chief bowed at once to this decision and took the following steps to deal with the existing situation :

1. He sent a force under Brigadier Sir E. Lugard to the Benares District and Behar.

[1] Early in the year General Mansfield, Chief of the Staff, wrote to Sir Hugh Rose as follows : " Sir Colin would be glad to learn if Jhansi is to be fairly tackled during the present campaign. To us it is all-important. Until it takes place Sir Colin's rear will always be inconvenienced, and he will be constantly obliged to look back over his shoulder as when he relieved Lucknow. The stiff neck this gives to the Commander-in-Chief and the increased difficulty of his operations you will understand " (Clyde and Strathnairne, p. 105).

[2] The Governor-General was influenced by the fact that in Rohilkhund there were two parties, the Mohammedans, who were mainly hostile, and the Hindus, who were mainly friendly to the British, and he did not wish to see the loyalty of the former put to too great a strain, or permit them to be crushed by their powerful neighbours (Forrest, iii. p. 336, etc.).

2. He ordered the defences of Lucknow to be improved.

3. He put Major-General J. H. Grant in command of a force strong enough to garrison Lucknow and provide mobile columns for operations in the districts.

4. He himself prepared for the reconquest of Rohilkhund and issued the following orders :

(*a*) A force under Brigadier R. Walpole was to advance from Lucknow and sweep the left bank of the Ganges up to the frontier of Rohilkhund.

(*b*) A force under Major-General N. Penny was to march from Meerut into Rohilkhund.

(*c*) A force under Brigadier J. Jones [1] was to march from Rurkhi into Rohilkhund.

(*d*) A siege train was to march from Cawnpore by Futtehghur to Aligunge.

The Commander-in-Chief, with such troops as could be spared from Lucknow and Cawnpore, proposed to join Walpole near Aligunge when that officer had carried out his orders, and in company with the siege train and Penny's force to advance on Bareilly from the south while Jones attacked it from the north.

---

[1] Lieutenant-Colonel and Brevet-Colonel J. Jones, 60th King's Royal Rifle Corps.

II. ROHILKHUND, MARCH TO JUNE, 1858

Action at Kankar, April 7.—The garrison of Futtehghur under Brigadier T. Seaton consisted of the following troops :

| Sikh C. | 82nd Foot. | 7/14 R.A. (4 guns)[1] (Captain and |
| | 7th Punjaub I. | Brevet-Major C. H. Smith). |

At the beginning of April the rebels in Rohilkhund, excited by the activity of their comrades in Oudh, assumed a threatening attitude, and crossing the Ramgunga pushed an advanced guard to within twenty-two miles of Futtehghur. Seaton determined to attack them, and marching out with the greater part of his garrison reached Kankar on April 6, where the enemy occupied some villages with infantry and guns, while their cavalry was thrown out in front. At daylight on the 7th the rebel horse were brushed aside by the Sikh cavalry, and Smith opened fire to prepare the way for an infantry attack on the villages. As our infantry advanced rebel cavalry threatened their right flank, but three guns detached[2] under Lieutenant C. J. Auchinleck, R.A., kept them at bay, and the 82nd and the Punjaubis drove the enemy from Kankar. In clearing the villages Lieutenant F. Whinyates, R.A., captured a 10-pounder brass gun, and two wall pieces were also taken. The rebels lost 250 men ; the British had 5 killed and 17 wounded.

[1] Two guns of 7/14 R.A. under 2nd Captain and Brevet-Major C. E. Walcott and Lieutenant H. Edmeades, were on detachment at Mynpori.

[2] One of these pieces was a 24-pounder howitzer worked by a detachment of the 82nd Foot under Lieutenant J. S. Hand ("Occ. Papers, R.A.I.," i. p. 66).

356 THE EXTINCTION OF THE MUTINY [CHAP. IX

This spirited action of Seaton's seemed to dishearten the rebels, for on the defeat of their advanced guard they all retired into Rohilkhund.

Walpole at Ruiya.—On April 7 Brigadier R. Walpole marched out of Lucknow in a north-westerly direction with the following force :

9th Lancers.	42nd Highlanders.	2/1 Ben.H.A. (Capt. and Brev.-Lieut.-Colonel H. Tombs).
2nd Punjaub C.	79th Highlanders.	
	93rd Highlanders.	3/3 Ben.H.A. (Capt. and Brev.-Lieut.-Col. F.F. Remmington).
	4th Punjaub I.	
	23rd Compy. R.E.	Detachment 1/5 Ben.A. (Lieutenant E. W. E. Walker).
	Ben. Sappers and Miners.	4/1 Ben.A. (two 18-pounders, two 8-inch howitzers, two 8-inch mortars) (Captain H. Francis).

O.C.A., Major J. Brind, Ben.A.
S.O., Lieutenant E. W. E. Walker, Ben.A.

After a week's march the Brigadier was before the fort of Ruiya, which was held by a rebel Talukdar. It was a typical native fort with thick mud walls surrounded by a ditch and jungle ; its faces were of varying strength, but reconnaissance was impossible and Walpole appears to have got misleading information. On April 15 an infantry attack without artillery preparation was made on one of the strong faces of Ruiya and met with a disastrous repulse. Whether Walpole was entirely responsible for the method of the infantry advance is open to question, but blame must ever rest upon him for not making use of his heavy guns. His loss amounted to 94 killed and wounded, Brigadier the Hon. A. Hope being among the slain.[1] This heavy loss was incurred in

---

[1] Brigadier-General the Hon. Adrian Hope had distinguished himself on many occasions and was idolised by the Highland Brigade. His

vain, for the enemy evacuated Ruiya when darkness came on.

Affair at Sirsa.—Resuming his march, Walpole on April 22 again encountered the enemy at Sirsa, a village on the right bank of the Ramgunga, a few miles from Aligunge. The cavalry and horse artillery were at once sent forward and captured four guns, and on the advance of the infantry the rebels retired. Next day, the 23rd, they were driven across the bridge of boats at Aligunge, and Walpole, following them up, rapidly secured the bridge and encamped on the left bank of the river. Here he was joined by other troops, and on April 27 the Commander-in-Chief arrived and took command of the whole force, the additional troops [1] being :

1st Punjaub C. (1 squad).	78th Highlanders.	*Siege Train :*
	82nd Foot.	*Ordnance :* eight 24-pounders,
5th Punjaub C. (1 squad).	2nd Punjaub I.	twelve 8-inch mortars, eight 5½ mortars.
17th Ben.Irr.C. (1 squad.).	Punjaub Pioneers.	*Personnel :* 3/14 R.A. (2nd Captain and Brevet-
Lahore Light Horse.		Major W. G. Le Mesurier). Detachment 4/6 Ben.A. (Captain C. Cookworthy).

Sir Colin's Advance on Bareilly.—On April 29 Sir Colin reached Shahjehanpore, where a halt was

death provoked bitter resentment against Walpole, who, if the popular General had not fallen, would probably have escaped censure. There has been much controversy regarding Ruiya. (Malleson, ii. 509, etc. ; Forrest, iii. 339, etc.)

[1] A wing of the 80th Foot formed the infantry escort of the Commander-in-Chief from Lucknow to Futtehghur. On reaching that place the 80th Foot were left in garrison, and the 82nd Foot and some Sikh cavalry under Lieutenant C. A. de Kantzow accompanied Sir Colin in his march to Aligunge.

made, and Lieutenant-Colonel E. B. Hale, with the 82nd Foot, some Sikh cavalry, and two 9-pounder and two siege guns under Lieutenant H. O. Hitchens and a detachment of 4/6 Ben.A., was left in command. On May 3 the Commander-in-Chief passed through Miranpore Kutra, where he was joined by the Meerut Column, and next day he was before Bareilly.

The Meerut Column.—Major-General Penny was ordered to march from Meerut through Badaon to Miranpore Kutra with the following force :

6th D.Gs. (Carabineers) (2 squads.).	64th Foot.	No. 14 F.B. 3/3 Ben.A. (Captain and Brev.-Major H. Hammond).
Mooltani Horse.	22nd Punjaub I. Beluchi Batt.	2/4 Ben.A. (two 18-pounders, two 8-inch howitzers) (Lieutenant H. M. Cadell).

Penny marched to Badaon, and pushing on to the south-east encountered the rebels at Kakraula, where he was killed in a skirmish.[1] He was succeeded in the command by Colonel H. R. Jones,[2] who led the column to the appointed meeting-place without further incident.

The Rurkhi Column.—The Rurkhi Column was composed of troops stationed at Rurkhi under Brigadier J. Jones and of a detachment at Manglaur under Lieutenant-Colonel J. H. Smyth, Ben. H.A.

---

[1] General Penny "was riding in front of the advanced guard, unsuspicious of evil, when, on nearing the village of Kakraula, a discharge of grape was fired from a gun in front. Lieutenant Curtis with his Carabineers . . . took the gun. The village was shelled, but the General was not again seen alive" (Stubbs, iii. 416).

[2] Lieutenant-Colonel and Brevet-Colonel H. R. Jones, 6th D.Gs. (Carabineers).

At Rurkhi

6th D.Gs. (Carabineers) (2 squads).	60th K.R.R. 1st Sikh I.	No. 7 F.B. 1/1 Ben.A. (Captain A. G. Austen).
Mooltani Horse.	1st Punjaub I. 17th Punjaub 1.	4/4 Ben.A. ⎫ two 18-pounder guns. Detachmts.: ⎨ two 8-inch howitzers. 1/4 Ben.A. ⎧ two 8-inch mortars. 3/4 Ben.A. ⎩ two 5½-inch mortars.

(Lieutenant F. W. Stubbs).

At Manglaur

Detachments:
6th D.Gs.   1st Sikh I.   5/1 Ben.H.A. (Lieutenant G.
1st Punjaub C.   A. Renny).

On April 13 Jones crossed the Ganges, and encountering the rebels in the neighbourhood of Najibabad, captured four guns. On April 21, being joined by the troops at Manglaur, he defeated the rebels at Nagina, where, with few casualties, he inflicted upon them a loss of some 700 men, and captured 15 guns.[1] Leaving Smyth with about 2,000 of the native troops and 5/1 Ben.H.A. to protect his rear, Jones marched to Moradabad, where he halted a few days in order to synchronise his advance on Bareilly with that of the Commander-in-Chief. On May 5 he was close to that city, having had a slight skirmish at Mirgunge in which his cavalry and 1/1 Ben.A. were engaged, and some 60 rebels killed.

The Battle of Bareilly.—At an early hour on May 5 Sir Colin's troops were in motion along the road leading from the south to Bareilly. About two and

---

[1] The night before the combat at Nagina, Smyth brought Brigadier J. Jones a plan drawn by one of his native gunners showing the position of the enemy, particularly of their guns. This plan was very useful to the Brigadier. ("Records of the R.H.A.," p. 38.)

a half miles from the city this road crosses at right angles a small stream, the Nerkuttea, along the south bank of which rebel infantry supported by guns were in position. The British advance was covered by cavalry skirmishers, supported by 2/1 and 3/3 Ben.H.A. and 3/3 Ben.A., and the rebels, after firing a few rounds, withdrew to the north bank of the Nerkuttea, leaving their guns behind them. Numerous topes of trees obscured the view of the north bank, which gradually rose from the river's side to the site of the Bareilly cantonments, to the westward of which was a small fort. The position of the enemy was uncertain, but strong masses of cavalry were visible on the British left. A halt was therefore made to allow the siege train and baggage to close up, during which a reconnaissance pushed round the south-west of the ruined cantonment discovered that while the fort was abandoned, the cantonments were occupied. The 4th Punjab Infantry then crossed the stream and were moving on the old cavalry lines, when they were suddenly charged by 300 ghazis and driven back in confusion. The 42nd Highlanders, who were in reserve, formed line rapidly and met the ghazis with a volley which laid many of them low, but the desperate fanatics, springing over the bodies of their fallen comrades, hurled themselves against the levelled bayonets, where they perished to a man. About the same time the rebel horse threatened the siege train on the south bank of the Nerkuttea, but our cavalry, supported by the horse artillery, easily checked their efforts. Then the rebels began to fall back slowly to the enclosed gardens that surrounded the city, and Sir Colin, having occupied the fort with a detachment of the 42nd Highlanders, encamped on the north bank of the Nerkuttea.

During the day the Rurkhi Column was engaged on

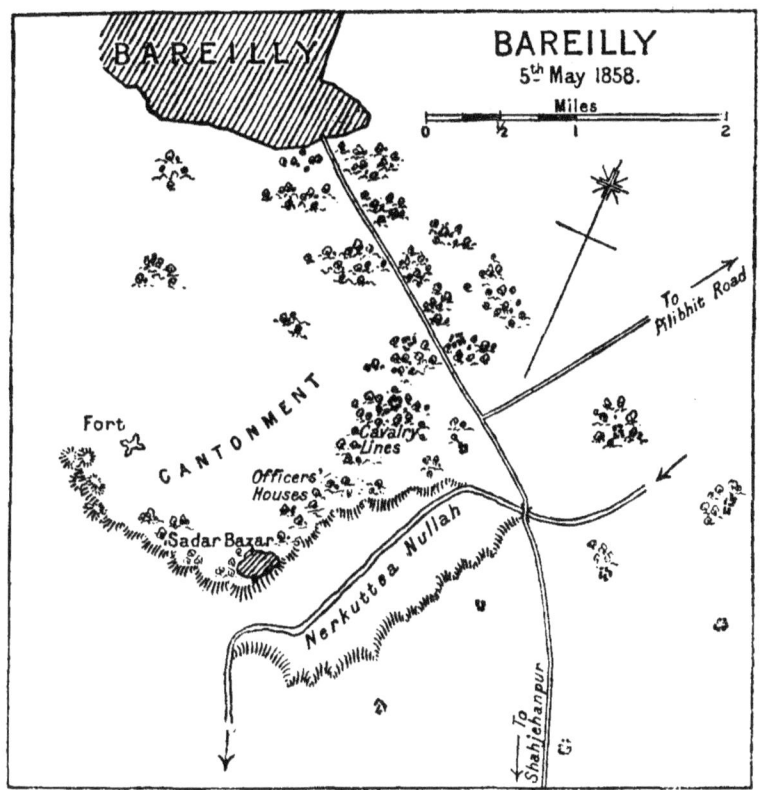

the north side of the city. After the heavy and light batteries had both been in action, the infantry captured two guns and got a foothold within the city, communication with Sir Colin's force being established during the afternoon.

Next day (May 6) the Commander-in-Chief sent a cavalry patrol along the Pilibhit road, and on the 7th he occupied Bareilly. The rebels retreated to Pilibhit and Bassowli.

Combats round Shahjehanpore.—Meanwhile Colonel Hale was attacked at Shahjehanpore. The Fyzabad Moulvie had been driven out of Oudh in April,[1] and in the beginning of May was in the neighbourhood of Shahjehanpore with 8,000 men and 12 guns. Hale, warned by his spies, vacated the town and took up a position in the jail and in an entrenchment made beside it, where he remained in a state of siege for eight days.

On May 8 Sir Colin despatched Brigadier J. Jones to relieve Shahjehanpore with the following force:

6th D.Gs.	60th K.R.R.	1/1 Ben.A. (Captain A. Austen).
Mooltani Horse.	79th Highlanders.	
	82nd Foot (wing).	H.F.B. 4/4 Ben.A., etc. (Lieutenant F. W. Stubbs).
	22nd Punjaub I.	

On May 11 after a slight skirmish the relief was carried out, but Jones soon discovered that he had only engaged a portion of the enemy; for the Moulvie was now reinforced by Prince Feroz Shah, the Begum of Oudh and some followers of the Nana Sahib.[2] On May 15 after a day's desultory fighting Jones, realising that he could do little more than maintain his position, asked for reinforcements.

Meanwhile the Commander-in-Chief at Bareilly was engaged in the pacification of Rohilkhund. On

---

[1] See p. 373.
[2] Malleson, ii. p. 536, etc. See also pp. 245, 299 n.

May 11 Lieutenant-Colonel J. Coke, 1st Punjaub Infantry, being appointed Brigadier, was despatched towards Pilibhit with a mixed force, while Walpole was directed to garrison Bareilly with the following troops :

2nd Punjaub C.	42nd Highlanders.	3/3 Ben.H.A. (Capt. and Brev.-
	78th Highlanders.	Lieut.-Colonel F. F. Rem-
	93rd Highlanders.	mington).
	17th Punjaub I.	4/1 Ben.A. (18-pounders, etc.)
		(Captain H. Francis).

NOTE.—At a later date 5/1 Ben.H.A. (Lieutenant G. A. Renny) joined the garrison.

On May 15 Sir Colin set out for Shahjehanpore with the remainder of the army, and while on the march received Jones' request for reinforcements. He at once sent a message of recall to Coke, and pressing onwards, reached Shahjehanpore on May 18, where, being joined by that officer on the 22nd, his whole force was concentrated with the exception of the Bareilly garrison. On May 23 Sir Colin left Brigadier J. Jones in command, with orders for the further prosecution of the operations, and set out for Futtehghur, where he would be in touch with the operations which were being carried out in other parts of India.[1]

Pacification of Rohilkhund.—On May 24 Brigadier J. Jones marched towards Mohamdee, a town in Oudh, close to the eastern frontier of Rohilkhund, and encountered and drove back the rebels at Burnai.

[1] The operations which were being carried out in the other parts of India rendered it necessary that the Commander-in-Chief should be near the telegraph wire. The advance on Gwalior was in progress, and our columns were traversing Oudh and Behar. These reasons induced Sir Colin to hasten to Futtehghur. ("State Papers," Appendix D, p. xxxi.)

Next day Mohamdee was found abandoned, the fort was blown up and Jones returned to Shahjehanpore.¹ On the 31st a column which included Tombs's Troop and Stubbs's Heavy Battery marched to Shahabad and scattered the last assembly of the Rohilkhund rebels, who were all driven into Oudh ; on June 4 the Field Force was broken up and the troops were distributed in quarters.²

III. BENARES DISTRICT AND BEHAR

Mahndi Husain and Kunwar Sing.—Towards the end of February Mahndi Husain reappeared and, encouraged by the departure of British troops to Lucknow, got together 15,000 men including 4,000 sepoys (Cawnpore and Futtehghur mutineers) and 9 guns. Having entrenched himself at Belwa, he was about to attack Azimghur, but on March 5 was met at Amorha by Brigadier Rowcroft, who drove him back to Belwa and captured 7 guns. Rowcroft, though lately reinforced by a useful body of European Cavalry (the Bengal Yeomanry) and some Sikhs, was not strong enough to attack the works at Belwa, but he kept them under observation. Meanwhile Kunwar Sing, who since his defeat at Jugdispore was lurking in western Behar, set out with 1,200 sepoys (Dinapore mutineers) and a

---

¹ After Mohamdee the Moulvie made for Powain, a town on the frontier of Oudh and Rohilkhund, where he was shot dead from the walls by the brother of the rajah of Powain. (Malleson, ii. p. 543, etc.)

² Le Mesurier went to Allahabad, where 3/14 R.A. became a heavy field battery.

motley following of irregulars to join Mahndi Husain. The defeat at Amorha interfered with the rebel plans, but the "Nazim," while holding on to Belwa, was able to detach a considerable force to the southeast which joined Kunwar Sing at Atraulia about the middle of March.

Combats round Azimghur, March 22 to April 6.— At this time Lieutenant-Colonel Milman was holding Azimghur with part of the 37th Foot, the 4th Madras Cavalry, and two guns of 4/5 Ben.A. (Lieutenant D. J. Welsh); on March 22 he attacked Kunwar Sing at Atraulia, but was driven back to Azimghur.

Here he was reinforced by the rest of the 37th Foot under Lieutenant-Colonel W. L. Dames, who assumed command.[1] On the 27th Dames made an unsuccessful sortie, and Azimghur was in extreme danger till its relief by Colonel Lord Mark Kerr on April 6. This officer made a rapid march from Allahabad with the 13th Foot (369 men), a troop of the 2nd Dragoon Guards, and two 6-pounder guns and two 5½-inch mortars under Lieutenant J. Robertson and a detachment of 8/2 R.A. Before a junction with Dames could be effected, Lord Mark Kerr had a sharp encounter with the rebels in which he lost 8 men killed and 34 were wounded.

Lugard's Column.—On March 29 Brigadier Sir E. Lugard left Lucknow with the following force :

---

[1] Major and Brevet-Colonel W. L. Dames, though regimentally junior to Lieutenant-Colonel and Brevet-Colonel E. C. W. M. Milman, was his senior by brevet, and therefore took command of the mixed force at Azimghur.

Military Train.	10th Foot.	½ E Troop R.H.A. (2nd Captain
3rd Sikh C.	34th Foot.	and Brev.-Major J. E. Michell).
12th Ben.Irr.C.	84th Foot.	A/3 Mad. A. (Captain G. S.
(1 squad).		Cotter).
	Detachment R.E.	8/2 R.A. (Captain J.⎫
	Punjaubi Pioneers.	E. Thring) ⎪ Heavy
		Det. 5/13 R.A. (Lieu- ⎬ Battery
		tenant W. H. Fitz ⎪
		Maurice). ⎭

O.C.A., Lieutenant-Colonel C. J. B. Riddell, R.A.
S.O., Brev.-Major N.O.S. Turner, R.A.

As the bridge at Sultanpore had been burnt by the rebels, Lugard marched to Jaunpore, and learning (April 11) that the enemy were in force at Tigra, a village a few miles to the north, he attacked and drove them off, inflicting a loss of 80 men and 3 guns. On April 14 he was close to Azimghur, where in his immediate front flowed the Tons River with its bridge of boats intact and 13,000 rebels on the farther bank. Kunwar Sing, who was in command of them, defended the bridge with skill and determination, and when at length the British passed over it, his main body was far away. Major J. E. Michell, R.H.A., with the cavalry and horse artillery went in pursuit, and after a twelve-miles chase overtook their rear guard. It was composed of sepoys of the Dinapore Brigade, who retired in good order and, disputing their ground with great obstinacy, were only driven from point to point by case-shot fire at close range. The rebels lost 80 men and 3 guns, while Michell had 25 casualties. He halted at Nathupore and asked for reinforcements. Next day the 34th Foot and a detachment of 8/2 R.A. with two 5½-inch mortars under Lieutenant S. Kyle were sent to him, and on the 16th

Lugard despatched Brigadier J. Douglas[1] to take the command, accompanied by the 84th Foot, the Madras rifles,[2] some Madras Sappers and Miners, and four guns of A/3 Mad.A. Douglas continued the pursuit, and on the 17th came up with the rebels at Azimutghur (Naghar), where they held a strong position covered by thick topes of trees surrounded by high banks. Michell was sent to the front with the cavalry and field guns; but finding himself under a heavy fire and unable to produce any apparent effect on the sheltered enemy, while at the same time his own left flank was threatened, he fell back slowly until the infantry came up. Then before a general advance of the British line the enemy gave way and broke up into separate bodies, who, flying in different directions, reunited when they had left pursuit behind.

Kunwar Sing's Retreat on Jugdispore.—Douglas on the 20th again came in touch with the rebels who were in position at Mannahar (Munnear). After falling back for a mile before a general advance of the British the rebels again broke up, and in their retreat abandoned two 9-pounder guns with waggons complete.[3]

Kunwar Sing now determined to cross the Ganges and make for Jugdispore, artfully spreading a rumour that he meant to effect the crossing by means

---

[1] This officer (see p. 276) commanded the infantry of Lugard's column.

[2] The Rifles and Sappers and Miners came from Carthew's Brigade at Futtehpore.

[3] These guns belonged to 4/1 Ben.H.A., the native troop that mutinied at Neemuch in June, '57.

of elephants at Bulliah. Colonel Cumberlege [1] was accordingly despatched to intercept him with some Madras Cavalry and a detachment of 8/2 R.A. under Lieutenant J. Robertson with two 6-pounder guns drawn by bullocks.[2] But Kunwar Sing, having secretly collected a number of boats at Sheopur Ghat, ten miles below Bulliah, evaded Cumberlege, and the bulk of his force was on the right bank of the river when Douglas' advance guard appeared upon the scene. After the skirmish at Mannahar Douglas had continued the pursuit, and when he arrived on the left bank of the Ganges E Troop R.H.A. caused considerable loss to the rebel rear guard, and Kunwar Sing himself, who was crossing on an elephant, was wounded in the hand by a fragment of shell.

Defeat of Le Grand.—The rebel Talukdar, though deserted by some of his followers when he retreated across the Ganges, found many new adherents on his arrival at Jugdispore. On April 23 Captain A. J. Le Grand marched from Arrah with 150 men of the 35th Foot, 50 men of H.M.S. *Pearl*, Naval Brigade, and two 12-pounder howitzers under Corporal J. Hoolahan, 4/5 Ben.A., and three men of 1/6 Ben.A. He attacked Kunwar Sing in the jungle, but met with a disastrous defeat; he himself was killed, and two-thirds of his force were slain by the enemy or struck down by the sun. The guns were spiked and

---

[1] Lieutenant-Colonel and Brevet-Colonel B. W. Cumberlege, 4th Madras Cavalry. See p. 323.

[2] The 6-pounder guns were taken from Ghazipore, where they were station or post guns. When Kunwar Sing crossed the Ganges these guns were sent back and the two captured 9-pounder guns with their equipments were handed over to Robertson, who rejoined 8/2 R.A.

abandoned, and Corporal Hoolahan and his three gunners were killed beside them.[1]

Almost immediately after this exploit Kunwar Sing died from the effects of his wound, but Le Grand's defeat caused consternation in the district, and Lugard's force was hurried to Arrah, where it assembled on May 5, with the exception of the 34th Regiment, which remained under Lieutenant-Colonel R. D. Kelly at Azimghur.

Clearing the Jugdispore Jungle.—The jungle of Jugdispore ran north and south for about twenty miles with a breadth from three to five miles, and on May 7 Lugard moved the bulk of his force across its north end and halted until Colonel F. B. Corfield[2] (who at this date was a couple of marches from Jugdispore) had time to take up a position at the south end with a small mixed force consisting of some irregular cavalry, a detachment of the 6th Foot, some Sikhs, and the Indian Naval Brigade and two guns of 5/3 R.A. (J Battery) under Lieutenant C. R. Franklen.

On May 9 Lugard attacked Kunwar Sing's house and village, which lay within the jungle. Captain W. N. Waller, R.A. bombarded the buildings with 8-inch howitzers, while Cotter's guns and the $5\frac{1}{2}$-inch mortars under Kyle cleared the topes on either side. The house and village were captured and the enemy retired southwards. Then the horse

---

[1] The three gunners were G. Bygrave, S. Watson, and A. Houghey.
[2] Lieutenant-Colonel and Colonel F. B. Corfield, 6th Ben.N.I. His force marched from Sasseram and had some skirmishes with the rebels on its way to Jugdispore.

artillery and cavalry were sent to reconnoitre round the jungle, while the infantry, accompanied by Kyle's mortars and A/3 Mad.A., drove the rebels through it, and Corfield, who had reached Peroo, attacked it on the south-east. During the remainder of the month numerous small affairs took place in which the infantry, supported by the mortars, operated inside the jungle and the cavalry and field guns outside; and after some severe handling the rebels quitted the jungle early in June, and Lugard began to clear it away by coolie labour. Then the Field Force was broken up and the artillery went to Benares; but the district remained in a disturbed state till the end of November, when the rebels were finally subdued by mounted infantry.

Mahndi Husain driven across the Rapti.—After his defeat on March 5 Mahndi Husain held his ground at Belwa, but all his attempts to move eastward were repulsed by Rowcroft. When the pressure of our advance in Oudh [1] rendered his position dangerous the "Nazim" moved northwards, while Rowcroft advanced westward and in September reached Basti. Meanwhile Colonel Fischer arrived at Gorruckpore with the 27th Mad.N.I., the Ferozepore regiment, and A/3 Mad.A. (2nd Captain R. Cadell)[2], whither a wing of the 13th Foot and 2 guns of 4/5 Ben.A. (Lieutenant D. J. Welsh) had preceded him. Thus reinforced, Rowcroft pressed the "Nazim" back, and on November 26 drove him across the Rapti at Dumurreagunge.

[1] See pp. 376, etc.
[2] Major G. S. Cotter having gone on sick leave, Cadell was transferred from C/5 to A/3 Mad.A.

IV. The Summer Campaign in Oudh

Sir Colin's Plans.—While the events narrated in the previous sections were in progress on the eastern and western borders of Oudh, that distracted province was the scene of constant encounters. After the capture of Lucknow the Commander-in-Chief's first object was to safeguard that city, the next to strike at the rebels whenever they drew to a head until the time should come when he would be free to sweep the country from the Ganges to the Nepaul frontier. The first object was accomplished by engineer works on an extensive scale and by the provision of an adequate garrison, the second by including in that garrison mobile troops fitted for rapid operations in the districts.

The Defences of Lucknow.—On March 23 Brigadier R. Napier, the Chief Engineer, submitted the results of an extensive survey he had made of Lucknow, and in accordance with his advice the following works were begun and prosecuted during the year.

1. A line of military posts extending from the Dilkhusha along the right bank of the Gumti to the Musa Bagh.

2. A large fort near the Stone Bridge enclosing the Machi Bhawan, and the Imambara, and a smaller fort near the Iron Bridge containing a new building for barracks.

3. Extensive demolitions leaving open spaces along both banks of the Gumti.

372 THE EXTINCTION OF THE MUTINY [CHAP. IX

4. The construction of wide roads round and through the city.

The garrison, which was placed under the command of Major-General J. H. Grant, C.B., consisted of the following troops:

Lucknow Garrison, April, 1858

2nd D.Gs.	23rd Foot.	F Troop R.H.A. (2nd Captain and Brevet-Major H.P.Yates).
7th Hussars.	38th Foot.	1/1 Ben.H.A. (Captain and Brevet-Major H. LeG. Bruce).
Hodson's Horse.	53rd Foot.	
Wale's Horse.	90th Foot.	
	2nd Batt. Rifle Brig.	2/3 Ben.H.A. (Captain W. A. Mackinnon).
	3rd Batt. Rifle Brig.	
	1st Ben.E.F.	5/12 R.A. (Q Battery) (Captain J. R. Gibbon).
	1st Mad.E.F.	
	Ferozepore Regt.	6/13 R.A. (Captain and Brevet-Major W. A. Middleton).
	5th Punjaub I.	
		No. 12 F.B. 2/3 Ben.A. (Captain and Brevet-Major H. A. Carleton).
		6/11 R.A. (Heavy Battery) (2nd Captain W. H. Goodenough).
		5/13 R.A. (Heavy Battery) (2nd Captain H. L. Talbot).
		3/8 R.A. (Garrison Battery) (2nd Captain and Brev.-Major F. C. Maude).
		3/5 Ben.A. (Garrison Battery) (Captain G. Moir).

O.C.A., Brigadier G. R. Barker, R.A.

Action at Bari, April 12.—The Commander-in-Chief set out for Rohilkhund on April 9, and on the eve of his departure ordered General Grant to march with a column to Bari (twenty-nine miles from Lucknow on the Seetapore road), where the Moulvie of Fyzabad was reported to be with 7,000 men. Leaving Brigadier G. R. Barker, R.A., in command of the garrison, Grant left Lucknow on April 11 with the following troops:

[APRIL 12, 1858] BARI

2nd D.Gs. (1 squadron).	38th Foot.	2/3 Ben.H.A. (Captain W. A. Mackinnon).
7th Hussars.	2nd Batt. Rifle Brig.	6/13 R.A. (Captain and Brev.-Major W. A. Middleton).
Hodson's Horse.	1st Ben.E.F.	5/13 R.A. (2nd Captain H. L. Talbot).
Wale's Horse.	5th Punjaub I.	

O.C.A., Lieutenant-Colonel E. Maberly, R.A.
S.O., Lieutenant T. B. Strange, R.A.

Early on the 12th Grant's advance guard, led by Wale's Horse and two guns of 2/3 Ben.H.A. (Lieutenant C. Hunter), was approaching a village on the Lucknow side of Bari when a force of rebel cavalry made a sudden swoop upon the guns. The Moulvie was holding this village with his infantry and, as it afterwards transpired, had ordered his cavalry to circle round and fall on the British rear when they saw they were committed to an attack. But the chance of capturing the guns tempted them to neglect these orders and their sudden attack might have succeeded if it had not been met by a timely charge of the 7th Hussars, who afterwards frustrated a second attempt made on the rear of the British column; and a third and last attack was defeated by steady volleys from the Bengal Fusiliers, delivered at thirty paces. The Moulvie's cavalry, active as they were in attack, by their premature action spoilt his plans, and after a feeble resistance he withdrew his infantry from the village and retreated towards the east. Grant followed him to Ramnuggur, but there were no rebels to be found; both the Begum and the Moulvie had escaped to Rohilkhund.

Affair of Simri, May 10.—At the end of April Grant's column was again on the move, for there was trouble in the Baswarra district. The same troops accompanied him, except that he substituted the 90th Light Infantry for the 1st Bengal Fusiliers, and Q Battery R.A. for 6/13 R.A. On May 10 he arrived before a strong fort at Dundiakhera only to find that the rebels had left it and assembled (about 3,000 strong) at Simri, twelve miles to the eastward, where they made a poor resistance and lost two guns.[1] Grant then marched back to Lucknow, his force having suffered considerably from sunstroke.

Action at Newabgunge-Barabanki, June 12.—The rebels made their next appearance at Newabgunge-Barabanki, and on June 12 Grant was within striking distance of them with the following force :

2nd D.Gs.	2nd Batt. Rifle Brig.	2/3 Ben.H.A. (Captain W. A. Mackinnon).
7th Hussars.	3rd Batt. Rifle Brig.	
Hodson's Horse.	5th Punjaub I.	5/12 R.A. (Q Battery) (2nd Captain A. C. Johnson).
Wale's Horse.		
Oudh Mounted Police.		No. 12 F.B. 2/3 Ben.A. (Major H. A. Carleton).

O.C.A., Major H. A. Carleton, Ben.A.
S.O., Lieutenant T. B. Strange, R.A.

A stream crossed by a bridge ran in front of the rebels' position on an extensive plateau, and Grant, advancing at dawn on the 13th, took them by surprise. Some of their guns opened an ineffective fire on our troops as they were crossing the bridge, but they were speedily silenced by **Q Battery**, which dismounted

[1] During the night after the action at Simri an extraordinary scene occurred in the British camp owing to a false alarm, and considerable firing took place, but only one man, a bullock driver, was killed. Captain Gibbon, commanding Q Battery R.A., was twice knocked down, and finally wounded himself badly with his own revolver. The command of the battery devolved upon 2nd Captain A. C. Johnson.

one of them. Then four companies of the Rifle Brigade went forward in support of the horse artillery and cavalry, who led the advance, and the main body were following up when some mounted zemindaris and two guns made a sudden attack on the British rear guard, composed of Hodson's Horse and two guns 2/3 Ben.A. (Lieutenant J. Perceval). Hodson's Horse were unsteady, and Perceval's guns were in extreme danger when Grant sent the 7th Hussars to the rescue, and Major Carleton joined Perceval with the other four guns of 2/3 Ben.A. The rebels now planted two green flags beside their guns and rallied round them, while Carleton shelled them at a range of 500 yards, and the 7th Hussars rode through them twice. But with conspicuous bravery the zemindaris held their ground until overwhelmed by the advance of the Rifle Brigade, 125 of their dead being found lying beside the green flags. There was no further resistance, and Grant entered Newabgunge-Barabanki, having dispersed 15,000 rebels, who lost 600 men killed and 6 guns. The British casualties were 67 killed and wounded; 33 men died of sunstroke, and 250 were taken into hospital. This victory for a time broke the spirit of the rebels, who fled in all directions, but in the fierce heat and with a force of only 3,500 men Grant was unable to pursue them. He, however, had but a short respite.

Relief of Maun Sing.—A certain Talukdar named Maun Sing,[1] who was now an avowed adherent of the Government, was at this time besieged by rebels in

---

[1] Maun Sing, an Oudh Talukdar, must not be confused with another of the same name who was a feudatory of Sindiah's (see p. 412, etc.).

his stronghold at Shahgunge close to Fyzabad, and on July 29 Grant entered that city at the head of the following troops:

7th Hussars.	2nd Batt. Rifle Brig.	F Troop R.H.A. (2nd Capt. and Brev.-Major H. P. Yates).
Hodson's Horse.	1st Mad.E.F.	5/12 R.A. (Q Battery) (2nd Capt. A. C. Johnson).
		5/13 R.A. (Heavy Battery) (2nd Capt. and Brev.-Major H. L. Talbot).

As the British approached Fyzabad the enemy fled from Shahgunge; but Grant sent the horse artillery and cavalry in pursuit to Ajudiah Ghat, a few miles lower down the Gogra, and F Troop R.H.A. put a few rounds into the rear guard of the rebels as they were escaping in boats to join the Begum, who, having returned to Oudh, was in the trans-Gogra district.

Operations at Sultanpore.—But another body of rebels, 8,000 strong, had gone southwards to Sultanpore, and Grant ordered Brigadier A. H. Horsford to follow them up. Owing to heavy rain Horsford's departure was delayed till August 7, when with F Troop R.H.A., the cavalry, the Madras Fusiliers, and the 5th Punjaubis, he marched out of Fyzabad, and on August 12 arrived close to Sultanpore, which is situated on the right bank of the Gumti. On August 20 Grant reinforced him with the 2nd Battalion Rifle Brigade and two 9-pounder guns of Q Battery (Lieutenant T. B. Strange, R.A.), and on the 23rd he himself

---

The former had played very successfully a double game. He first professed loyalty, then joined the rebels and then deserted them (Knollys, p. 294, etc.).

followed with the heavy battery (5/13 R.A.) and the 53rd Foot which he had summoned from the rear.

On August 25 Colonel Galwey [1] with the Madras Fusiliers and Strange's guns crossed the Gumti, while Lieutenant F. G. E. Warren with two 18-pounder guns of 5/13 R.A. covered the passage.[2] The operation was safely concluded in two hours, and the Fusiliers advancing under the protection of the field guns drove the rebel pickets out of two villages they were occupying. Galwey's force then bivouacked. The main position of the rebels was a few miles south of Sultanpore, but though their numbers were now swelled to some 14,000 men with 15 guns, they did not molest the small British force except by a desultory musketry fire at night. On the 29th Grant's complete force, less the 53rd Foot and 5/13 R.A., was on the right bank of the Gumti, but the enemy having melted away, they went into quarters at Sultanpore and Fyzabad, which were linked to Lucknow by posts at Newabgunge-Barabanki and Derriabad.

Affairs at Salimpore, September 22, 23.—Towards the end of September the rebels collected on both banks of the Gumti at Salimpore, a large village near Amithi, fourteen miles south-east of Lucknow. On the night of September 22 Major E. G. Bulwer, with

[1] Major and Brevet-Lieutenant-Colonel M. Galwey, 1st Madras European Fusiliers.

[2] The infantry crossed in large flat boats, and the artillery horses were swum over by volunteers from the 5th Punjaubis. Only two small canoes with planks across were available for the transport of the guns, which, when dismounted, were safely taken over piecemeal. Previously, through the boundless zeal of a staff officer, a gun complete was run on to one of these canoes, which promptly sank gun and all; but luckily the place was not deep and the gun was rescued from the mud by means of the team and limber.

a detachment of the 23rd Royal Welsh Fusiliers, two 9-pounder guns (Lieutenant E. C. Cuthbert, 6/11 R.A.), and two small mortars (Lieutenant E. Maitland, 3/8 R.A.), marched from the Dilkhusha, while Lieutenant-Colonel R. Pratt, with another detachment of the 23rd and four guns of 2/3 Ben.H.A. (Captain W. A. Mackinnon), set out from Newabgunge-Barabanki. At daybreak on the 23rd Bulwer attacked the rebels at Salimpore on the north bank, and shortly afterwards those on the south bank (who were strongly entrenched) were assailed by Pratt. In both cases the rebels were broken up and scattered.

Skirmishes about Derriabad.—About the same time some small skirmishes took place in the neighbourhood of Derriabad, where Captain A. Hume, 1st Bengal European Fusiliers, was in command. Two guns of 2/3 Ben.A. (Lieutenant A. Swinton) and three 6-pounder guns of the Karpurthala Contingent (Ensign P. Roddy) were engaged in them.[1]

Brigadier Berkeley's Operations.—While the above operations were in progress under the orders of Major-General J. H. Grant, a separate force based on Allahabad and Benares was formed under Brigadier C. A. F. Berkeley.[2] It was composed as follows:

6th Mad. C.	Detachments:	E Troop R.H.A. (3 guns) (2nd Captain
Lahore Light	32nd Foot.	and Brev.-Major J. E. Michell).
Horse.	54th Foot.	4/14 R.A. (4 guns) (Lieutenant
	7th Punjaub I.	E. R. Cottingham).

---

[1] Ensign P. Roddy was promoted from Acting Staff Sergeant 2/3 Ben.H.A. and received the V.C. while serving with the Karpurthala Contingent under Major-General J. H. Grant, in September, '58. This Contingent was placed at the disposal of the British in May, '58, and was ordered at first to Jamson Poorwah on the Sye.

[2] Lieutenant-Colonel and Brevet-Colonel C. A. F. Berkeley, 32nd Foot.

The rebel Talukdars were inclined to deal hardly with any of their number who refused to join in the general revolt, and for this reason an attack was made on the Rajah of Soraon, his town was burnt, and he and his family made prisoners. Berkeley being ordered to deal with the situation, he on July 14 attacked and drove the rebels from Dhainawan (Dehain), seven miles north-west of Saraon in Oudh territory, and on the 16th arrived before Tiroul, a miniature mud fortress with a central citadel holding three guns. The place was hidden in jungle, but on the 16th it was assailed with vertical fire by 4/14 R.A. and during the night was abandoned by its garrison. Berkeley then broke up several rebel strongholds accessible from Allahabad, and throwing out pickets linked himself to Grant at Sultanpore.

V. OPERATIONS IN OUDH DURING OCTOBER

Sir Colin's Plans.—When Sir Colin Campbell quitted Rohilkhund on June 5, the situation was far from clear. Sir Hugh Rose's advance on Gwalior was on the point of beginning, and operations in Oudh and Behar were in progress, but when he arrived at Allahabad and met the Governor-General in consultation on June 15, a great change had taken place in the situation, and the time had come for the discussion of the final steps in the campaign. Sir Colin's plan was simple. Oudh forms a rough parallelogram bounded on the east and west by the Benares District and Rohilkhund, on the south by the Ganges, on the north by Nepaul, and is traversed from west to east by the Gogra. The majority of the rebels in Rohilkhund had been driven into Oudh; those in the Benares District had been scattered: the Commander-in-Chief therefore intended to guard the eastern and western frontiers of Oudh while he swept that province, first from the Ganges to the Gogra and then from the Gogra to the Nepaul hills. His intention was to begin operations in November after the season of the rains, but October proved a busy time for the troops under Major-General Grant.

Skirmishes at Miangunge and Kantur, October 5 and 7.—Since June Brigadier F. C. Evelegh[1] was stationed on the Cawnpore road at Newabgunge with the 1st Sikh Cavalry and Oudh Mounted Police, the 20th Foot, and 1/1 Ben.H.A. (Major H. Le G. Bruce). On October 5 he had a successful skirmish at Miangunge, and scattered the rebels there assembled. On October 7 Captain Hume from Derriabad hearing that the rebels had collected at Kantur, seven miles off, marched out and attacked them, Ensign Roddy distinguishing himself by the capture of a 9-pounder gun, and the rebels again scattered.

Affair at Doadpore, October 20.—On October 20 Brigadier Horsford marched from Sultanpore towards the Kandu River with a mixed force[2] including two guns of F Troop R.H.A. (Lieutenant F. Lyon) and two guns of Q Battery (Lieutenant T. B. Strange). Rebels estimated at 4,200 men with six guns were

[1] Lieutenant-Colonel F. C. Evelegh, 20th Foot.

[2] Horsford's force consisted of detachments of the 7th Hussars, Hodson's Horse, Oudh Mounted Police, the 32nd Foot, 1st Madras European Fusiliers, 5th Punjaub Infantry, and Oudh Police. In all some 1,400 men, and the four guns mentioned in the text.

reported at the village of Doadpore; but when Horsford reached that place he found that they had retired towards the Kandu River, and he determined to pursue them with the four guns and a small body of horse, for the bulk of his cavalry were already carrying out a wide turning movement. The track he followed led through a jungle unsuited for the movements of mounted troops, but fresh wheel ruts of the rebel guns were plainly to be seen, and the advance was as rapid as possible. Presently small bodies of the rebels came in view, and Lyon got in a round at a range of 100 yards. Brigadier Horsford then ordered Strange to follow up the cavalry,[1] and the advance was continued at racing speed. On the left of the road a body of the enemy opened fire and emptied a few cavalry saddles, but they were scattered by Strange with a couple of rounds of case, and the cavalry went on. Soon afterwards two abandoned ammunition waggons on the roadside showed that the guns were not far off, and a little farther on it was discovered that the rebels had made an emplacement[2] right across the road. It was unoccupied, but the handful of cavalry turned aside to find an easier way through the jungle and scrub. The leading gun of Q Battery, however,

---

[1] The guns of Q Battery were well horsed, with teams of Gulf Arabs and Cape horses. "The Remount Committee at Calcutta in the latter part of 1857, of which Captains Middleton and Sarsfield Greene were members, had done their best to make the batteries efficient in this respect" (Stubbs, iii. p. 447 n.).

[2] In his despatch Brigadier Horsford refers to the batteries made by the rebels. Evidently the ditch and parapet on the road had been made for guns to check the British advance; but the rebels' hearts failed them at the last moment.

went straight at the obstacle and the leaders fell. Nothing daunted, a second attempt was made, the gun got across, the second followed with two ammunition waggons, and the ditch and parapet were soon trampled down. The course of events had thus placed two field battery guns at the head of a cavalry chase, and blind Fortune gave them the prize. At a turn of the road Strange suddenly came upon the rebel guns. One was limbered up and continued its flight; the other was in action on the road. Some sepoys stood round it, while others were trying to unhook the traces of the horses in order to escape on their backs. The gun was loaded [1] and might have been fired, but Strange and a few mounted N.C. officers were upon it in an instant. The naik in charge of the escort raised his musket as Strange approached, but that officer passed his sword through his body, and a little later the quartermaster-sergeant of F Troop overtook and slew the native commandant, who wore the full-dress belts of an artillery officer. The Staff, the handful of cavalry, and Lyon's guns now came up, and at some distance away was found the second piece, a howitzer, abandoned on the road. Then, the main body of the cavalry having now carried out their flanking movement and rejoined the column with some captured elephants, Horsford returned to Sultanpore, his cavalry and artillery having marched thirty-five miles. About 250 rebels were killed; yet the British casualties were trifling, and the captured guns received a great ovation. They belonged to

[1] "The gun was certainly loaded, for it had to be upended to get out the case shot" (Major-General T. B. Strange to author).

2/7 Ben.A. which had mutinied at Nusseerabad in June, and they bore the mural crown which commemorated the good service of the battery at Jellalabad in 1841–2. They were brought in by their own gun teams, one driven by men from F Troop and the others by men of Q Battery.

Skirmish at the Kandu River.—A few days later, October 27, the rebels were again encountered on the Kandu River by the cavalry and F. Troop R.H.A., who took two guns and killed some fifty rebels.

Operations on the Western Frontier of Oudh.—On October 4 the rebels attacked a police post at Sundeela, but were driven back to Jamo by Major E. G. Maynard, 88th Connaught Rangers, commanding the advanced guard of a column under Brigadier J. R. Barker, who marched from Lucknow and arrived at Sundeela on October 7. His force was composed as follows :

2nd D.Gs.	Detachment	No. 17 F.B. 3/1 Ben.A. (Captain and Brev.-Major A. Pearson).	
Oudh Mounted Police.	88th Foot. 3rd Batt. Rifle Brig. Oudh Police. Ben. Sappers and Miners	No. 12 F.B. 2/3 Ben.A. (Lieutenant W. Dowell). Detachment: 3/8 R.A. (Lieutenant E. Maitland). 6/11 R.A. (Lieutenant H. A. Tracey).	one 18-pounder gun. one 8-inch howitzer. two 8-inch mortars. two 5-inch mortars.

O.C.A., 2nd Captain and Brev.-Major W. H. Goodenough, R.A.
S.O., 2nd Captain J. McN. Campbell, R.A.

Affair at Jamo.—On October 8 at daybreak Barker marched to Jamo and attacked the enemy, about 12,000 strong with 4 guns, driving them from the village after a feeble resistance with a loss of 1,000 men and 3 of their guns. Barker lost 4 men killed and 55 wounded, principally belonging to the Oudh police, who made a flank attack on the rebels.

Capture of Birwah, October 21.—After the action the enemy fled northwards and made their next stand at the strong fort of Birwah. Here there was a central citadel with quadrilateral outworks. The west face consisted of a strong wall with a ditch 40 feet wide and 30 deep; the other three faces were defended by two or more walls and entered by three massive gates. Barker determined to assault the west face. The approach on this side was covered by an entrenched village, but 2/3 Ben.A. speedily cleared it by enfilade fire, and the advance was continued up to the jungle which as usual surrounded the fort. It was possible, however, to see a small portion of the west wall through the thick thorny bushes, and Lieutenant W. S. Brown, R.A.,[1] opened a breaching fire upon it with an 18-pounder gun and an 8-inch howitzer, while Lieutenant H. A. Tracey, R.A., with $5\frac{1}{2}$-inch mortars and Captain A. Pearson, Ben.A., with field guns shelled the interior of Birwah. Barker then sent parties one to the east gate and one to the north gate with orders to blow them in [2] when the assault on the west wall was delivered. In a short time Brown made a practicable breach, and the infantry, crossing the formidable ditch by means of ladders, delivered the assault, Ensign H. E. Richards, Rifle Brigade, leading,

[1] In command of 3/8 R.A., Lieutenant E. Maitland being absent.
[2] Lieutenant H. A. L. Carnegie, Bengal Engineers, and Sergeant Charlton, R.A., superintended the blowing in of the two gates.

closely followed by Major W. H. Goodenough, R.A., who received a serious wound in the arm. When the gates were blown in the enemy, hastily leaving the outworks, took refuge in the citadel, whence they opened a galling small-arm fire on the British. A portion of the citadel was blown up, and the remainder set on fire, but the rebels held their ground till the night, when they made their escape, after losing 250 men and 8 guns. The conflict lasted eight hours, and Barker had 15 killed and 111 wounded.

Capture of Ruiya, October 28.—Meanwhile Lieut.-Colonel E. B. Hale, 82nd Foot, was marching from Futtehghur to reinforce Barker with the 8th Foot, some irregular cavalry, four guns of 7/14 R.A. (Major C. H. Smith), and a Sikh Company Ben.A. (Captain G. Holland).[1] On the 28th the united force took Ruiya, which had again been occupied by rebels, while troops from Shahjehanpore operated farther to the north to complete the cordon on the western frontier of Oudh.

Affair at Bunkagaon, October 8.—On October 8 Major-General Sir Thomas Seaton issued from Shahjehanpore and marched to the assistance of the Rajah of Powain,[2] who was threatened by rebels. He took with him a detachment of the Carabineers and some native cavalry, the 1st Battalion 60th **King's Royal Rifles** and 82nd Foot, and four guns of No. 7 F.B. 1/1 Ben.A. (Captain C. Cookworthy), and encountering the rebels at Bunkagaon easily dispersed them. He

---

[1] There were three Sikh companies at this period, but the author has been unable to identify their respective services.

[2] See p. 364 n.

then returned to Shahjehanpore, whence a column under Brigadier C. Troup[1] set out in an easterly direction on October 18. It was composed as follows:

6th D.Gs. (Carabineers) (1 squad.).	1st Batt. K.R.R.	3/3 Ben.H.A. (Captain and Brev.-Lieut.-Colonel F. F. Remmington).
Detachment 4th Ben.Irr.C.	93rd Highlanders. 66th Ben.N.I. (Gurkhas)	
Mooltani Horse.		No. 7 F.B. 1/1 Ben.A. (2 guns) (Lieut. A. J. Wake).
		4/4 Ben.A. ⎫
		Detachments: ⎰ Heavy
		1/4 Ben.A. ⎱ Battery.
		3/4 Ben.A. ⎭
		(2nd Captain F. W. Stubbs).

O.C.A., Lieutenant-Colonel J. Brind, Ben.A.
S.O., Major H. P. Bishop, Ben.A.

Brigadier Troup's Operations.—On November 8 Troup encountered and drove back to the east a large force of rebels under Khan Behadur Khan[2] and Feroz Shah, who were covering the Fort of Mithaulee. After a day's shelling the fort was found to be evacuated, and Brind, following up the enemy with the cavalry and horse artillery, fell upon them at the village of Mehndi and drove the greater part across the Chauka; but Feroz Shah and his followers doubled back and eventually escaped (December 7) into Central India.

The Eastern Frontier of Oudh.—After his success on November 26, Rowcroft pushed along the north bank of the Rapti, and on December 3 again defeated the "Nazim," whom he overtook at Bururiah, nine miles west of Dumurreagunge. He then held the line from Hir to Basti, which was continued to the

---

[1] Lieutenant-Colonel and Brevet-Colonel C. Troup, 68th Ben.N.I.
[2] See page 41.

south by Colonel Kelly, who with the 4th Madras Cavalry, the 34th Foot, and J Battery, 5/3 R.A. (Major G. C. Henry) advanced from Azimghur along the Tons River and occupied Dosta and Tunda. Thus when on November 2 the Commander-in-Chief quitted Allahabad part of his plan was already in execution.

VI. THE WINTER CAMPAIGN IN OUDH

The Force at the Disposal of the Commander-in-Chief.—The Commander-in-Chief, now Lord Clyde,[1] had the following forces at his disposal :—

Brigadier E. R. Wetherall[2] was at Saraon in command of the following troops :

1st Punjaub C.	79th Highlanders.	E Troop R.H.A. (Captain and
	9th Punjaub I.	Brev.-Major W. A. Middleton).
	(wing).	3/14 R.A. (Heavy Battery)
	Beluchi Batt.	(2nd Captain and Brev.-Major
		W. G. Le Mesurier).

Brigadier F. G. A. Pinckney[3] was at Pertabgurh on the Sye with the following troops :

6th D.Gs. (Carabineers).	5th Foot.	4/14 R.A. (Captain and Brev.-
6th Madras L.C.	54th Foot.	Lieut.-Colonel S. E. Gordon).
Oudh Mounted Police.	1st Sikh I.	H.F.B. Sikh Company Ben. A.
Pathan Horse.	Oudh Police.	
	4th Co. R.E.	
	Delhi Pioneers.	

Major-General Sir James Hope Grant was at Jug-

---

[1] On November 1 the Proclamation dissolving the East India Company and placing India under the Queen's Government was made, and Lord Canning became first Viceroy of India. Sir Colin Campbell was created Baron Clyde, and Major-General J. H. Grant was given the K.C.B.

[2] Colonel E. R. Wetherall, C.B., unattached.

[3] Lieutenant-Colonel and Brev.-Col. F. G. A. Pinckney, 73rd Foot.

dispore,[1] whither he had marched westwards from Sultanpore. His force consisted of the following troops :

7th Hussars.	32nd Foot.	F Troop R.H.A. (Captain and Brev.-Major the Hon. D.M'D. Fraser).
Hodson's Horse.	2nd Batt. Rifle Brig.	
	1st Mad.E.F.	
	5th Punjaub I.	5/12 R.A. (2 guns) (Lieutenant T. B. Strange).
		5/13 R.A. Heavy Battery (Capt. and Brev.-Major H. L. Talbot).

Capture of Rampore Kussia, November 3.—The half of the first part of the Commander-in-Chief's plan was already performed, for Brigadier Barker had cleared the rebels out of that portion of Oudh between the Ganges and the Gogra which lies west of Lucknow. The next step was to deal with the Baswarra district on the east, where powerful Talukdars were in revolt. The first object of attack was Rampore Kussia on the Sye, a stronghold three miles in circumference and surrounded by jungle except on the north-west. Wetherall and Grant were ordered to proceed against it; but Wetherall having discovered a weak spot attacked the place himself on November 3 without waiting for Grant's co-operation, and took it, inflicting a loss of 300 men on the rebels and capturing 23 guns. The British had 78 casualties, and Wetherall incurred the censure of the Commander-in-Chief.

Capture of Amithi and Shunkerpore, Nov. 10 and 16.—A combined movement was then made on Amithi, fifteen miles north-east of Rampore Kussia, where lay the stronghold of the Talukdar Lall Madho. Grant invested it on the north, Wetherall on the south, and

[1] Jugdispore in Oudh, must not be confused with the other Jugdispore in Behar. There are also two Amithis in Oudh. See p. 377.

Pinckney on the east, and on November 10 Lall Madho personally surrendered; but most of his followers made their escape to Shunkerpore, the fort of Beni Madho, another powerful Talukdar. On November 11 Clyde, with Pinckney's column, set out westwards for Shunkerpore, Grant on his right and Wetherall on his left, with the intention of surrounding Shunkerpore on three sides, while Brigadier Evelegh marched from Newabgunge (on the Lucknow—Cawnpore road) to complete the investment. But this part of the programme Evelegh was unable to carry out, and on the night of November 15 Beni Madho, with 4,000 men and 10 guns, escaped towards the jungles round Peroo,[1] lying to the westward of Roy Bareilly, whither Grant was ordered to follow in pursuit. At daybreak, November 16, Shunkerpore was occupied, but only two or three guns were found. The same day Brigadier R. C. H. Taylor,[2] who succeeded Wetherall in his command, was sent to Fyzabad, and on the 19th Clyde, with Pinckney's Brigade, marched into Roy Bareilly.

Brigadier Evelegh's Movements.—Meanwhile Brigadier Evelegh left Newabgunge for Shunkerpore on November 8, with the following force :

1st Sikh C.	Detachments :	1/1 Ben.H.A. (Captain and Brev.- Major H. Le G. Bruce).
Oudh Mounted Police.	20th Foot. 23rd Foot. 80th Foot. Oudh Police.	Detachments : 6/11 R.A. (Lieutenant E. C. Cuthbert). 3/5 Ben.A. (siege guns) (Lieutenant H. Latham).

---

[1] Peroo is in Oudh, and must not be confused with Peroo in Behar.

[2] Lieutenant-Colonel and Brev.-Col. R. C. H. Taylor, 79th Highlanders.

He at once encountered forces of the rebels at Poorwah, and after a successful skirmish at that place stormed the fort of Simri. On the 17th he reached Bera, where he was attacked by a large body of rebels whom he defeated and drove westward. Shunkerpore having been taken, Evelegh was ordered to keep the enemy in sight and to retrace his steps towards Simri, previously sending his siege guns (3/5 Ben.A.) and all his impedimenta into Roy Bareilly. On the 21st Clyde left Roy Bareilly with the bulk of Pinckney's Brigade, and marched to Buchraon; he then turned westward and joined Evelegh on the 23rd near Nuggur.

Affair at Dundeakhera, November 24.—Next day the Commander-in-Chief ordered Brigadier H. R. Jones, who had succeeded Pinckney in his command, to march on Buksur while Brigadier Evelegh made for Dundeakhera. It was soon evident that Grant's movement on Peroo had taken effect, and that Beni Madho had been forced back towards the Ganges, where he was now caught between the two advancing columns. Clyde sent a message to the Talukdar giving him an hour's grace in which to surrender, but no answer having been received, the rebel position was attacked by a line of skirmishers thrown out from both columns. The skirmishers forced back the enemy, who made off on both flanks up and down the river, and the British main body was not engaged. The cavalry were sent in pursuit, and 400 rebels were slain and their 7 guns captured.

Gordon's and Carmichael's Columns.—The main line of the rebel retreat being uncertain, Clyde halted during November 25, but it was soon apparent that the mass of the rebels had worked round towards Peroo, and Lieutenant-Colonel S. E. Gordon, R.A., was sent in pursuit with a small force including four guns of 4/14 R.A. (2nd Captain C. Johnston). Gordon reached Roy Bareilly without encountering the rebels, who, with Beni Madho, continued their retreat towards the Gumti. Gordon halted at Roy Bareilly, and Colonel J. D. Carmichael[1] continued the pursuit from that place on December 31 with the 32nd Foot, the 19th Punjaub Infantry, some Oudh Mounted Police, and two guns of E Troop Mad.H.A. (Lieutenant L. Bridge).

On November 29 Beni Madho was severely handled by some of Grant's cavalry,[2] and on December 5 was finally driven across the Gogra by Carmichael, who had briskly followed up his tracks.

Brigadier Horsford's Operations.—General Grant, having headed off the rebels from the Peroo jungles, reached Jugdispore on November 17, where he handed over his command to Brigadier Horsford and proceeded himself to Fyzabad to assume command of the trans-Gogra operations which were about to begin. Horsford, having ascertained that the fort of Rehora on the Gumti, north of Jugdispore, was occupied by rebels, sent Colonel Galwey on November 23 with the Madras Fusiliers, two guns of F Troop R.H.A. (Lieutenant H. T. Arbuthnot), and two mortars of 5/13 R.A. (Lieutenant F. G. E. Warren) to turn them out. Galwey carried out this operation with little resistance, and then went on to the neighbouring fort

---

[1] Major and Brevet-Lieutenant-Colonel J. D. Carmichael, 32nd Foot.
[2] See p. 392.

of Koeli, which also fell after Arbuthnot had blown in the gate. On the 29th Lieutenant-Colonel Sir William Russell, 7th Hussars, with the cavalry and four guns of F Troop R.H.A. (Major Fraser), intercepted a part of Beni Madho's following on their way to the Gumti and drove them across that river in great disorder.[1]

Completion of the First Part of Sir Colin's Plans.—At the end of November the Commander-in-Chief, with Brigadiers H. R. Jones and Evelegh, was at Lucknow, 4/14 R.A. and the 54th Foot being left at Roy Bareilly under Lieutenant-Colonel Gordon, R.A. On the last day of the month Brigadier Evelegh was sent to reduce the fort of Oomeriah (twenty-one miles north of Lucknow) with the following force :

1st Sikh C.	5th Foot.	1/1 Ben.H.A. (Captain and
	3rd Batt. Rifle Brig.	Brev.-Major H. Le G. Bruce).
	Oudh Police.	H.F.B. Sikh Company Ben.A.
	23rd Company R.E.	

---

[1] The enemy apparently consisted of a large body of sowars, who made off closely followed by F Troop, which came into action whenever opportunity occurred. During the pursuit a gun horse in Lieutenant A. K. Rideout's Division fell, or was shot. Rideout remained with the gun until the horse was replaced, and then when following up the column lost his way owing to the clouds of dust that hung over the land and went too far to the right. Finding out his mistake, he was pressing forward when, from a stretch of low dense jungle on his right, a brisk fire of musketry was opened at a range of 80 to 100 yards, wounding two of the drivers, who, however, did not quit their saddles till all danger was over. Rideout, who was without escort, sent back the trumpeter to hurry up the rear guard and immediately opened fire with case shot on the jungle. The sound of the firing was heard by the main body of F Troop, who were able to take in flank the rebels driven back by Rideout. After he had been in action a few minutes the rear guard—a troop of the 7th Hussars—came up ; but during this space of time the safety of the gun depended on its own fire. Subsequently it was found that the rebels had suffered considerable loss. This account was obtained by the author from the late Major-General

This was successfully accomplished on December 3, and a few days afterwards Brigadier Horsford marched into Lucknow. These operations cleared the southeast portion of Oudh; Brigadier Barker had already cleared the south-west, and the rebels were now concentrated beyond the Chauka and the Gogra.

Passage of the Gogra.—In the middle of November Brigadier Taylor's force at Fyzabad was thus composed :

2nd D.Gs.	53rd Foot.	E Troop R.H.A. (Captain and Brev.-Major W. A. Middleton).
9th Lancers (1 wing).[1]	79th Highlanders. 1st Sikh I.	3/1 Ben.H.A. (Captain and Brev.-Major G. Moir).
1st Punjaub C.		2/2 Ben.H.A.[1] (3 guns) (Lieutenant R. R. Franks).
Hodson's Horse.		5/12 Q Battery (2nd Captain A. C. Johnson).
4th Company R.E.		3/14 R.A. (Heavy Battery) (2nd Captain and Brev.-Major W. Le Mesurier).

The rebels were in force on the left bank of the Gogra opposite Fyzabad, but they took up a position some miles from the river and offered no opposition to the construction of a bridge of boats and of a covering earthwork at its outward end. On November 26 a detachment of 3/14 R.A. with 18-pounder guns and 8-inch mortars (Lieutenant J. R. Oliver) occupied this work, and the 1st Sikhs were sent across by boats higher up the river, to threaten the rebels' right flank during the subsequent advance. Grant had arrived a few days previously and assumed the

---

A. K. Rideout, who had kept a diary in the Mutiny. It differs slightly from that given in the " Records of the Royal Horse Artillery," p. 14.

[1] The 9th Lancers had marched from Rohilkhund; 2/2 Ben.H.A. from the Punjaub.

command, and on the 27th at daybreak he led his main body across the bridge of boats. But when Oliver opened fire and the Sikhs made their flank attack, the rebels at once fell back and were pursued through swamps and jungle for two miles. Then they made a stand and brought a heavy gun into action, but Middleton took E Troop R.H.A. up to short range,[1] and the rebels again fled, leaving their guns behind them. Continuing his northerly advance by Bungaon to Muchligaon, Grant,[2] on December 3, captured 2 guns after a short skirmish. On the 9th he took the fort of Gonda with 5 guns, and on the 10th, in accordance with orders received from Lord Clyde, he turned westward to Secrora. For a time the rebels retiring to the north-east were unmolested.

Clyde's March from Lucknow to Fyzabad.—On December 5 Lord Clyde set out from Lucknow for Byram Ghat on the Gogra with the following force:

6th D.Gs.	20th Foot.	F Troop R.H.A. (Captain and
7th Hussars.	2nd Batt. Rifle	Brev.-Major the Hon. D.
6th Mad.L.C.	Brig.	McD. Fraser).
Lahore Horse.	Beluchi Batt.	5/13 R.A. (Heavy Battery)
Oudh Mounted Police.	23rd Company R.E.	(2nd Captain and Brev.- Major H. L. Talbot).

Lord Clyde was followed by Brigadier Evelegh, who after he had taken Oomeriah made a short march to the north-east to clear the country, and then turned due south on Newabgunge-Barabanki. On December 7 the Commander-in-Chief reached Byram

---

[1] Middleton was on the enemy before they realised he was coming and fired over him (Knollys, p. 314).

[2] 3/1 Ben.H.A. remained at Fyzabad and joined Lord Clyde's column.

Ghat. It now appeared that there was a concentration of rebels a mile above the junction of the Chauka and the Gogra, and Lord Clyde had no means of crossing at Byram Ghat. He therefore made the following arrangements :

(1) He ordered Major-General Sir J. H. Grant to march on Secrora so as to turn the rebels' position.

(2) He posted Colonel W. P. Purnell [1] at Byram Ghat with orders to watch the rivers as far as Jehanghir, where he would get into touch with Brigadier Troup and in concert with him prevent any rebels crossing the Gogra and Chauka to the westward. He placed some irregular cavalry, the 23rd R.W. Fusiliers, the 90th Light Infantry, 6/13 R.A. (2nd Captain C. J. Tyler [2]), and the heavy battery manned by the Sikhs at his disposal, and directed him to detach Lieutenant-Colonel Pratt and a wing of the 23rd R. W. Fusiliers and two guns of 6/13 R.A. (Lieutenant W. Smith) to patrol the Mullapore Doab.

(3) The Engineers were to make a bridge at Byram Ghat.

Lord Clyde, followed by Brigadier Evelegh, then marched along the right bank of the Gogra to Fyzabad.[3]

Grant's Movements.—The mere menace of Grant's

---

[1] Lieutenant-Colonel and Brevet-Colonel W. P. Purnell, 90th Light Infantry.

[2] After the transfer of Major Middleton to E Troop R.H.A., 6/13 R.A. was commanded by 2nd Captain C. J. Tyler until the arrival of Captain F. J. Soady.

[3] At the same time Lieutenant-Colonel J. D. Carmichael's column, which was encamped near Byram Ghat, after it had driven the rebels across the Gogra was sent to Roy Bareilly.

advance was sufficient to cause the rebels to abandon their position above the junction of the Chauka and the Gogra, and when he reached Secrora on December 12 they had already fled to the north. The purpose of his march having been effected, he again resumed his advance in a north-easterly direction towards Bulrampore.

Lord Clyde's Movements.—The Commander-in-Chief crossed the Gogra at Fyzabad and marching up its left bank reached Secrora the same day that Grant left it; and, halting there till December 22, he despatched Lieutenant-Colonel S. T. Christie[1] to the westward to clear the Mullapore Doab with a force consisting of detachments of Carabineers, Hodson's Horse, Oudh Police, of the 20th and 80th Foot, of Madras Sappers and Miners and 3/1 Ben.H.A. (Major G. Moir). On December 23 communication with the rear was ensured by the completion of the bridge at Byram Ghat, and the same day Lord Clyde marched northwards in order to sweep the rebels across the Nepaul frontier.

Leaving two guns of 3/1 Ben.H.A. and some native cavalry and infantry at Secrora, Clyde set out towards Nunparah on the 25th, where he obtained intelligence that the enemy were in force near Burgeedia.

Grant's Movements.—When Grant marched eastwards from Secrora his object was to follow up the rebels he had already defeated near Fyzabad, and while on the march news reached him that they were to be found at Tolsipore. Thus the forces of the enemy beyond the Gogra were collected into two

[1] Major and Brev.-Lieutenant-Colonel S. T. Christie, 80th Foot.

main bodies, namely, at Tolsipore and Burgeedia, some sixty miles apart. But owing to his western march Grant was not within striking distance of Tolsipore; he therefore sent orders to Brigadier Rowcroft at Hir to attack the rebels, and, sending on the 53rd Foot by forced marches to reinforce him, he himself followed as speedily as possible with his main body.

Action at Tolsipore, December 23.—At Tolsipore a large body of rebels with eight guns were assembled under Bala Rao, the Nana's brother, and on December 23 Rowcroft defeated them and occupied Tolsipore, where a few days afterwards Grant arrived with all his troops except the 1st Punjaub Cavalry, which had been ordered to join the Commander-in-Chief.

Action at Burgeedia, December 26.—On December 26 Lord Clyde attacked the rebels at Burgeedia. His first line was formed with a squadron of the 6th Madras Cavalry on the right, four guns of F Troop R.H.A. in the centre, and the 7th Hussars on the left. The remainder of his force was on his left rear. On approaching the rebel position the horse artillery and cavalry went rapidly forward, and when within range of the enemy's guns suddenly took ground to their right for several hundred yards, and when opposite the enemy's left they again advanced to the front and F Troop came into action at 600 yards. The effect of these movements was instantaneous, the enemy broke in disorder, fled northwards, and were pursued till dark.

Capture of Musjeediah, December 27.—Next day, December 27, Clyde marched on the fort of Musjeediah, six miles to the north-west, whither a large portion of the rebels had fled. On approaching the fort he ordered an 8-inch howitzer and a 24-pounder gun of 5/13 R.A. to take up a sheltered position 350 yards from the east face, while two 8-inch mortars manned by the same company took post on the south-west and infantry skirmishers crept forward through the jungle to fire at the embrasures at a range of 300 yards. The horse artillery and cavalry watched the flanks. After three hours' firing, the enemy gradually ceased to reply, and it was then found that the bulk of them had escaped into the almost impenetrable jungle in rear of the fort.

The Rebels enclosed.—On the 29th Lord Clyde returned to Nunparah, having ordered Christie to continue his northern march and clear the country up to the Gurwa River. On December 31 Christie reached Pudnaha, and at the same time Brigadier Troup sent Colonel M. G. Dennis [1] with a small mixed force including two guns of 3/1 Ben.H.A. across the Chauka and Dour, to sweep the country up to the Kyseeghur jungle. Farther to the north Brigadier Walpole with 5/1 Ben.H.A. (Colonel J. H. Smyth), 180 Punjaubi Cavalry, half a company of the 42nd Highlanders, and a few Gurkhas, fell upon 2,000 rebels, and on January 5 drove them across the Upper Chauka at Sassaya.

Action at Banki, December 30.—On December 30 the rebels having again collected at Banki, Clyde, leaving a wing of the 20th Foot and 5/13 R.A. to hold Nunparah, resumed his northward march. At Banki there was a belt of jungle traversed by

[1] Lieutenant Colonel and Brevet-Colonel M. G. Dennis, 60th K.R.R.

two roads. The belt was about three miles from the River Rapti, and on December 31 the rebels were found occupying a position three-quarters of a mile to the south of it with their flanks resting on the roads. Against this position the horse artillery and cavalry were formed for attack, a squadron of the Carabineers and of the 1st Punjaub Cavalry on the right, F Troop R.H.A. in the centre, the 7th Hussars and three squadrons of the 1st Punjaub Cavalry on the left. The line went forward at a steady pace, F Troop came into action, when, after a few rounds, the rebels retired by both roads, leaving two guns behind them. Breaking up into two columns, the cavalry and horse artillery pursued, but once in the jungle the rebels made a stand and held their ground until driven out by the Rifle Brigade. When our troops emerged from the jungle it was perceived that the enemy had taken up another position about 800 yards distant, on rising ground and in a wood, with a deep nullah across their front. A frontal attack was at once made by the infantry, and the cavalry and horse artillery began a turning movement. The bulk of the cavalry and two guns of F Troop (Lieutenant H. T. Arbuthnot) found a way across the nullah, and soon a charge along the banks of the Rapti drove the enemy across the river. Two more guns were taken, and heavy loss was inflicted on the rebels. One officer and two men of the 7th Hussars were drowned while crossing the river in pursuit.

Subsequent Movements.—Clyde halted at Banki, and early in January, 1859, was joined there by the columns under Christie and Pratt, and by 5/13 R.A.

Meanwhile to the eastward Grant had given the rebels little rest after their defeat at Tolsipore. They were again encountered, after some marching and countermarching, at Kunda Koti close to the Nepaul Hills, and were scattered with the loss of their 15 guns.

By the defeats of Banki and Kunda Koti the rebels were cleared out of Oudh, and on January 8 Lord Clyde with a portion of his troops set out for Lucknow, leaving Brigadier Horsford to watch the passes of Nepaul with the following force :

7th Hussars.	3rd Batt. Rifle Brig.	3/1 Ben.H.A. (4 guns)
1st Punjaub C.	5th Punjaub I.	(Captain and Brev.-
	Beluchi Batt.	Major G. Moir).

He also placed Sir J. H. Grant in charge of the whole frontier, one hundred miles in extent.

During the spring of 1859 desultory engagements took place. In February Horsford entered Nepaul, where the remnant of the mutinous sepoys and the followers of the " Nazim " and Bala Rao still attempted to draw to a head. On the 9th Horsford encountered a strong force near the Soona Valley, and scattered them, capturing 15 guns. Shortly afterwards Colonel Kelly entered Nepaul and twice defeated another force in the neighbourhood of Butwul on March 25 and 28. A third force which had doubled back across the frontier was met by Horsford on March 31 at Tolsipore and was immediately put to flight, the troops employed being the 53rd Foot, the 1st Sikh Infantry, and 5/12 R.A. (Captain A. C. Johnson). Soon after this the " Nazim " surrendered. On April 1 the fort of Bungaon, a useful place of refuge for the rebels, was captured by a small force under Lieutenant-Colonel C. P. B. Walker, 2nd Dragoon Guards, and active operations may be said to have ended on May 21, at the Jerwa Pass, where the rebels were finally scattered. On this occasion Grant himself was present with a small force consisting of the 2nd Dragoon Guards, the Ferozepore Regiment, the 7th Punjaub Infantry, and two guns of 6/13 R.A. (Lieutenant W. Smith). Shortly afterwards Bala Rao and his infamous brother the Nana were both reported to be dead, and the frontier was left in peace.[1]

VII. Pursuit of Tantia Topi, 1858-59

Position of British Troops after Capture of Gwalior.—After the capture of Gwalior Sir Hugh Rose departed to assume command in Bombay ; the Central India Field Force was broken up, and a Gwalior Division under Brigadier R. Napier was formed, the garrisons of Jhansi and Kalpi (reinforced) being included in it. The bulk of the Gwalior Division remained in that city ; the Saugor Field Force was round about Banda and Kirwi ; a brigade of the Rajputana Field Force, with the Head-quarters at Nusseerabad, was under Major-General H. G. Roberts ; another was detached to Neemuch under Brigadier W. Parke,[2] and the third under Brigadier M. W. Smith returned to Sipri. A brigade of the Bengal Army was at Agra under Brigadier St. G. D. Showers,

---

[1] Nevertheless years afterwards the Nana was reported to be alive.
[2] Lieutenant-Colonel W. Parke, 72nd Highlanders.

the Malwa Division of the Bombay Army under Major-General J. Michel was at Mhow, and farther south troops of the Madras Army were in Nagpore. Such was the position of the British troops south of the Jumna when the rebel leaders Tantia Topi, the Rao Sahib, and the Nawab of Banda made their last desperate throw for victory.

Rebel Hopes.—They hoped to organise a Mahratta rising that would seriously embarrass the British Government, and they fondly imagined that the Peshwa himself might display his standard in the Deccan and regain southern India. The British were busy on the far side of the Ganges, the native nobles had risen from Shahjehanpore to Azimghur, and throughout Oudh and Behar the Government, beyond the ground covered by its garrisons, had little or no authority. Then even though the Rajputana princes personally held to their treaties, their adherents might be induced to rally to the Peshwa, while hordes of wandering mutineers would gladly hail a leader. The country was well suited to guerilla warfare, and if they maintained themselves in Rajputana till the Nana escaped from Oudh, and could cross the Nurbudda under his leadership, success might still be theirs. With these hopes the rebels after their defeat at Jaora Alipore (June 21) hurried towards the west.

Tantia Topi's March to Tonk.—It was Tantia Topi's intention to go to Jeypore, whither his agents had already gone, but Roberts marched towards that city on June 25, and Showers advanced from Agra. Their movements led Tantia Topi to change his course, and he marched southwards to Tonk, a small

Mohammedan principality. On his approach the ruler of Tonk shut himself up in his citadel with such of his adherents as he could trust, in the vain hope that the remnant outside its walls would resist the invader. But they did nothing of the kind, and Tantia Topi, sweeping them and their four guns into his own following, disappeared southwards just as Roberts was approaching the town. Sending Lieutenant-Colonel J. Holmes [1] with a light field force in pursuit, Roberts marched westward to the Neemuch—Nusseerabad road. Holmes pressed on with all speed, but saw nothing of the nimble rebels, who lived by loot and yet had the sympathy of the inhabitants, who freely supplied them with food and information.

His Flight along the Banas.—After winding through the hills about Indarghur, Tantia Topi made for Boondi, but the Rajah shut his gates against him. The rebels then crossed the Banas, and following its tributary, the Kotari, reached Sanganeer on August 7. Their presence being known to Roberts, who was ten miles to the north of them, he at once marched to attack them, and after a successful skirmish (without casualties in his own force) he drove them back and inflicted on them a loss of 60 men. Tanti Topi made for Oodeypore, where he prostrated himself before the sacred Hindu shrine of Nathdwara, while Holmes, after a most fatiguing and fruitless march, arrived near Sanganeer and rejoined Roberts on

---

[1] Major and Brev.-Lieutenant-Colonel J. Holmes, 12th Bom.N.I. His force consisted of detachments of the 8th Hussars, 1st Bombay Lancers, Beluchi Horse, 72nd Highlanders, 12th Bom.N.I., and 2nd Troop Bom.H.A.: 570 cavalry, 710 infantry, and 6 guns.

August 9. The force under that general now consisted of the following troops :

Detachments :	72nd Highlanders.	2nd Troop Bom.H.A. (Captain J. G. Petrie).
8th Hussars.	83rd Foot.	
1st Bom. Lancers.	12th Bom. N.I.	No. 8 F.B. 2/6 Ben.A. (3 guns) (Lieutenant R. A. Stevenson, Bom.A.).
Gujrat Irr. Horse.	13th Bom.N.I.	
Beluchi Horse.		
	R.E. and Pioneers.	

O.C.A., Lieutenant-Colonel E. Price, R.A.
S.O., Lieutenant W. Stirling, R.A.

Following up the tracks of the rebels, Roberts overtook them on the Banas, twenty-eight miles east of Nathdwara, on August 13.

Action on the Banas, August 13.—On the approach of the British, who were advancing from the north, Tantia Topi took up a position on the south bank of the river. His four guns were posted on his right flank and commanded the ground on the north bank, which was level opposite to Tantia Topi's position, but towards his right front was diversified by hillocks. Favoured by this accident of the ground, Roberts' horse artillery and cavalry got to within 400 yards of the rebel guns without being observed and No. 2 Troop Bom.H.A. opened a rapid fire, under the protection of which the infantry and 2/6 Ben.A. crossed the river, which was only knee deep, and driving the enemy from the ridge they captured the four guns. The horse artillery and cavalry then crossed the river and pursued the enemy, who lost about 1,000 men. The British casualties (for the most part in the 8th Hussars) were 3 killed and 19 wounded.

Tantia Topi's Flight to Jhalawar Territory.—Tantia Topi fled eastwards along the Banas to Poonah, and turning southward and skirting the high ground about Retonghur, he crossed the Chumbal at a point sixty miles east of Neemuch and made for Jhalra Patan in the territory of the Rajah of Jhalawar. Roberts followed, and at Poonah meeting Brigadier W. Parke,[1] who had marched up from Neemuch, he reinforced him with cavalry and ordered him to continue the pursuit. As the Chumbal was reported to be in flood, there was good hope of overtaking and even capturing Tantia Topi, but when Parke reached the river he only found a few disabled ponies, and could see the rear guard of the rebels retreating from the opposite bank.

His Attempts to reach Indore.—The Rajah of Jhalawar was loyal, but his troops deserted him at once. Tantia Topi forced him to disburse fifteen lakhs of rupees for the use of the rebels, and, being reinforced by the Jhalawar levies and 30 guns, he then marched out of Jhalra Patan with the intention of making a descent on Indore and raising Holkar's troops. By the second week in September he reached Rajghur on the Parwan, but this movement was foreseen, and on September 14 Major-General J. Michel [2] was close

---

[1] The 8th Hussars and Beluchi Horse were attached to Parke's Brigade for this operation. Lieutenant A. H. Bell and 21 N.C.O.'s and men of 3/4 R.A. were serving under Parke at this time and were also employed in some minor operations under Major-General Roberts.

[2] At the end of August, '58, Major-General Michel succeeded Major-General Roberts in command of the Rajputana Field Force.

to that town. He had advanced from Mhow with the following force :

Detachments :	71st Foot.	No. 8 F.B. 2/6 Ben.A. (4
17th Lancers.	92nd Highlanders.	guns) (2nd Captain C. S.
3rd Bom.L.C.	4th Bom. N.I.	Lemarchand).
	19th Bom.N.I.	

Action at Beora, September 15.—Tantia Topi, now made for the Grand Trunk Road, and took up a strong position at Beora with 10,000 men, where when attacked by Michel on the 15th he made but a poor resistance and fled eastward, having lost 300 men and 27 guns. At Sironge he found four guns, and after halting there for a week he went northwards to Isaghur, where he took five more from a local chief, and dividing his force, he himself, with the Nawab of Banda, made for Chanderi, while the Rao Sahib marched on Tal Bahat and Lalitpore. But the garrison of Jhansi was strong enough to hold the Rao Sahib in check, and, being reinforced with cavalry by Michel, Brigadier M. W. Smith marched from Sipri towards Chanderi, in front of which town Tantia Topi had spent three fruitless days. On Smith's approach his position became dangerous, and he retired thirty miles southwards to Mangaoli on the left bank of the Betwa.

Action at Mangaoli, October 10.—But here he was attacked and defeated on October 10 by Michel, who two days previously had learnt his position. After losing 300 men and 6 guns Tantia Topi crossed the Betwa and joined the Rao Sahib at Lalitpore.

Action at Sindwaho.—Tantia Topi now determined

to make a dash himself for the Deccan without waiting for the Nana Sahib, and as a preliminary the Rao Sahib and the Nawab of Banda marched with 10,000 men and 4 guns to Sindwaho, fifteen miles south-east of Lalitpore, in order to turn the protected fords of the Jumni. On October 18 Michel, reinforced from Smith's Brigade, was at Narhat with the following Troops :

Detachments :	71st Foot.	No. 3 Troop Bom.H.A.
8th Hussars.	92nd Highlanders.	(Captain and Brev.-
17th Lancers.	95th Foot (detach-	Lieutenant-Colonel E. S.
1st Bom.L.C.	ment).	Blake).[1]
3rd Bom.L.C.	19th Bom.N.I.	2/6 Ben.A. (2nd Captain
Mayne's Horse.		C. S. Lemarchand).

Cavalry, 500 ; Infantry, 1,000.

On the 19th he marched to Sindwaho and found the rebels drawn up on a hill north of that place. Realising that they would attempt to pass eastwards, he made a strong attack on their left flank with his cavalry and horse artillery, while the 72nd and 92nd Highlanders and 2/6 Ben.A. were sent against their centre, the 19th Bom.N.I. remaining in reserve. For a time the British cavalry made no impression on the rebel left, and it was only when reinforced by the 19th Bom.N.I., and after salvos of case shot had been fired by the horse artillery, that they obtained the mastery. The rebel cavalry fought well on both flanks, and the 92nd Highlanders, on the British left, were obliged to wheel back to resist a flank attack. When the left of the rebels was overpowered the

[1] Possibly the troop was commanded at Sindwaho by Lieutenant T. M. Harris. In Colonel Blake's services though Kurai is included Sindwaho is not. (Spring, p. 91.)

attack on their centre was pushed home, and they were forced back on Lalitpore, where Tantia Topi, with the remainder of the rebels, was in position. They lost 500 men and their 4 guns. The British casualties were 4 killed and 19 wounded.

Action at Kurai.—Tantia Topi now determined to cross the Betwa and march southwards along its left bank, but was prevented doing so by Colonel J. Liddell from Jhansi. He then took a circuitous route along the right bank, and although unhampered by guns he only reached Kurai, thirty-five miles north-west of Saugor, by October 25. Here he was overtaken by Michel, who by following a more direct course was enabled to cut off some 3,000 rebels from the main body. Of these 350 were killed and the rest driven northwards.

Affair at Bagrode, October 26.—Tantia Topi himself pressed on southwards with the bulk of his force, and next day as he passed Bagrode he was gallantly attacked by Lieutenant-Colonel Becher,[1] who was marching from the Deccan to join Michel with a newly raised cavalry regiment. Becher accounted for some 40 rebels, but was not strong enough to inflict further damage; and Tantia Topi got among the friendly mountains and hurried to the Nurbudda.

Tantia Topi crosses the Nurbudda.—Crossing the river forty miles above Hoshungabad, he went southwards to Multai, and entering the town with great pomp proclaimed his army to be the advanced guard of the Peshwa, who was marching to take possession

---

[1] Captain and Brevet-Lieut.-Colonel C. G. Becher, 5th Bengal Light Cavalry. (?) The author has been unable to verify this.

of the Deccan after great victories in Central India. But the presence of British troops in Nagpore forbade a further advance in that direction, and Tantia Topi fled along the valley of the Tapti in the vain hope of finding in that wild country an outlet through which he might pass to the south.[1] Anticipated at all points, he abandoned the idea of marching to the Deccan and determined to go westward to Baroda, where support might be forthcoming.

Skirmish at Rajpore, November 19.—On November 19, after he had crossed the Grand Trunk Road, he was attacked at Rajpore by a small force of infantry and cavalry under Major R. M. Sutherland, 92nd Highlanders,[2] which was one of several columns despatched from Mhow to watch the fords of the Nurbudda.[3] A running fight took place in which the rebels lost 70 men and 2 guns; but they crossed to the north bank of the Nurbudda without further misfortune, and Tantia Topi and the Rao Sahib reached Chota Udepur on November 30. It was about this time that the Nawab of Banda made submission to the Government.

Action at Chota Udepur, December 1.—Meanwhile Michel was following up the rebels, and on November 7

[1] When passing Kargaon in Nimar on November 19, Tantia forced two troops of cavalry and a company of infantry, with two guns belonging to Holkar's Army, to join his forces.

[2] Under Sutherland were 120 of Holkar's Cavalry, detachments of the 71st Highland Light Infantry, 92nd Highlanders, and 4th Bombay N.I.—194 in all—and also 150 men of the 92nd mounted on camels.

[3] Among other columns which were in movement about this time was one from Ahmednuggur which between November 1858 and March 1859 marched some 1,300 miles. It included the 4th Troop Bom.H.A. (Captain D. Gaye) and 1/4 Bom.A. with No. 6 Light Field Battery (Captain J. B. Hardy).

he reached Hoshungabad, where he was joined by Parke with the Neemuch Brigade. He then marched to Charwah. Here information of the rebel movements was obtained, and Michel, ordering Brigadier Parke to continue the pursuit, turned back himself to Mhow. After marching 240 miles in ten days, Parke reached Chota Udepur, and on December 1 attacked the rebels. He had the following force:

Detachments:	72nd Highlanders	No. 2 F.B. 1/2 Bom.A. (Lieu-
8th Hussars.	(94 men on camels).	tenant T. B. Heathorn).
2nd Bom.L.C.		
Gujerat Irr. Horse.		
Baroda Irr. Horse.		
Mahratta Horse.		
Sind Horse.		
	793 all ranks.	

The force opposed to Parke numbered about 3,500, composed for the most part of well-mounted cavalry who at first attempted themselves to assume the offensive. As Parke approached them their trumpets were heard sounding the advance; but the rifle fire of the 72nd Highlanders and the case shot of 1/2 Bom.A. speedily broke down all resistance, while the British cavalry harassed them on both flanks. Driven from position to position, the rebels at length fled with a loss of 300 men, the British casualties being 10 killed and 15 wounded. This success had the effect of turning Tantia Topi from Baroda: he fled across the Mye, and on December 10 was hidden in the Banswarra jungle. He remained in hiding for a fortnight and then marched towards the Parwan and Parbati rivers, hoping to join Maun Sing, a feudatory of Sindiah's in arms against the Government, and to aid Prince Feroz Shah, who, more lucky or more adventurous than the Nana, had escaped from Rohilkhund and had crossed the Jumna.[1]

Affairs at Pertabghur and Zirapore, December 25 and 29.—On December 25 Tantia Topi was near Pertabghur, where he encountered Major R. Rocke, 72nd Highlanders, with a column of the Rajputana field force composed as follows:

Native Cavalry (137 all ranks).	Detachments: 72nd Highlanders. 95th Foot. 13th Bom.N.I. Total force, 450 men.	3/4 R.A. (2 guns) (Captain J. L. Bolton).

A sharp skirmish ensued in which Captain J. L. Bolton, R.A., was wounded, but Rocke was not strong enough to break up the rebel force and Tantia Topi continued his eastward march to Zirapore, where, on December 29, he was attacked by a column from Mhow under Lieutenant-Colonel H. R. Benson, composed of 226 men of the 17th Lancers and two guns of D Troop R.H.A. under Sergeant-Major Butler.

Affair at Chubbra, December 31.—After a short skirmish Tantia Topi marched to Chubbra, and on December 31 was overtaken by another Mhow column under Brigadier C. H. Somerset,[2] composed of 100 men 17th Lancers, 100 men of the 92nd Highlanders on camels, and 4 guns of D Troop R.H.A. (Captain L. G. Paget). Somerset's attack was met by 3,000 rebel horse who showed a disposition to charge the

[1] See p. 386.
[2] Lieutenant-Colonel and Brevet-Colonel C. H. Somerset, 72nd Highlanders.

small British force, but the fire opened by D Troop R.H.A. was both rapid and effective, and they were soon in full retreat. There was little loss on either side in these cavalry actions.

The excellent marching powers displayed by the Mhow columns deserves to be specially noted. Benson covered fifty miles between noon of December 30 and 6 a.m. on January 1, and from December 27 to January 1 Somerset marched 170 miles.

Meeting of Tantia Topi with Maun Sing and Feroz Shah.—Tantia Topi next made for Naharghur, where he was joined by Maun Sing. When, however, the rebels approached the town fire was opened on them from the walls and they went on to Paron.[1] Tantia Topi then marched to Indarghur, where, early in January, he was joined by Prince Feroz Shah, but Maun Sing remained on the Parbati.

Maun Sing's Outbreak.—When in the previous August Maun Sing assembled his armed retainers, he only professed hostility against Sindiah, his overlord, and declared he had no quarrel with the British; but at once his ranks were swelled by mutineers of the Bengal Army and the contingents, and he seized the fort of Paori, twenty miles to the west of Sipri. Brigadier M. W. Smith forthwith marched against him, and finding his field guns insufficient for a siege when he arrived before Paori, Smith applied to Gwalior for assistance. Brigadier Napier speedily arrived with reinforcements, and on August 19 the combined

---

[1] Naharghur was frankly hostile to all comers, for when British troops approached it in January '59 they were opposed for a time. The fort was finally dismantled on the 21st of that month.

force under his orders consisted of the following troops :

8th Hussars.	86th Foot.	No. 3 Troop Bom.H.A. (Captain and Brev.-Lieutenant-Colonel E. S. Blake).
1st Bom. Lancers.	95th Foot.	
Meade's Horse.	10th Bom.N.I.	
	25th Bom.N.I.	5/14 R.A. {Siege Bhopal Art. guns (2nd Captain T. S. P. Field).
	Detachment R.E.	

Siege of Paori.—During August 20 and 21 Paori was bombarded by four 8-inch mortars and two 18-pounder guns, but it was found impossible to invest it completely, and on the night of the 21st Maun Sing and his followers escaped, leaving 17 guns behind them on the walls.

Affairs at Bijepur and Kundri.—Lieutenant-Colonel G. H. Robertson, 25th Bom.N.I., with a mixed force [1] went in pursuit, and pushing on himself at the head of 500 cavalry and infantry and two guns of No. 3 Troop (Lieutenant A. R. Hoskins), he overtook a number of the fugitives at Bijepur on September 5 and drove them across the Chumpet. Maun Sing was not heard of again until early in November, when, at the head of 3,000 men, he was approaching Mangaoli on the Betwa. Brigadier M. W. Smith at once moved against him, and on November 14 encountered him at Kundri. Smith's force was composed as follows :

8th Hussars.	95th Foot.	3rd Troop Bom.H.A. (Captain and Brev.-Lieutenant-Colonel E. S. Blake).
1st Bom. Lancers.	10th Bom.N.I.	
Meade's Horse.		

---

[1] Detachments of the 8th Hussars and Meade's Horse (200 men) and of the 86th, 95th, 10th, and 25th Bom. N.I. (340 men). Two guns of No. 3 Troop Bom.H.A. and two guns of 5/14 R.A.

Maun Sing made but a feeble resistance and, having lost 600 men, again disappeared until he was met by Tantia Topi before Naharghur at the end of December.

Reappearance of Feroz Shah, Ranode, December 17.—On December 12 Brigadier Napier, learning that Prince Feroz Shah was advancing through the jungles along the Sindh River, marched with a mixed force southwards from Gwalior to intercept him. Leaving his infantry and guns at Narwar and pushing on with a light force of cavalry and some forty of the 71st Highland Light Infantry on camels—200 men in all—he overtook Feroz Shah on December 17 at Ranode on the right bank of the Sindh. The rebels were moving in an irregular mass on a front of nearly a mile in extent, and fell an easy prey to the small disciplined force which took them completely by surprise. Having lost 450 men, Feroz Shah made for Chanderi, but British columns issuing from Jhansi and Lalitpore obliged him to turn aside, and, passing south of Isaghur and Pachor, he sought the jungles of Aroni.

Affair of Sarpore, December 22.—On the night of December 22 he was encamped in a deep glen at Sarpore, eleven miles south-west of Aroni, and was attacked by Captain W. Rice [1] with 200 infantry, part of a small column detached from Goonah. Rice, a noted tiger hunter, leaving his guns and handful of horsemen behind and favoured by a moonlight night, suddenly fell upon the encampment. The rebels fled in such haste that few were killed, but horses, camels, arms, and equipment were captured. Feroz Shah then

[1] Captain W. Rice, 25th Bom.N.I.

made his way to the north-west, and he joined Tantia Topi at Indarghur early in January, 1859.

Tantia Topi's Dash for Jeypore.—Aware that British columns were closing round him, Tantia Topi determined to make a second dash for Jeypore, and on January 13 he reached Dosa, thirty-five miles east of that city, accompanied by the Rao Sahib and Feroz Shah. This movement becoming known to Brigadier Showers at Kishenghur (whither he had gone from Agra), he marched at once to Dosa with 265 Irregular Cavalry, a wing of the 3rd Bengal European Regiment, and four guns of No. 21 F.B. 2/5 Ben.A. (Lieutenant R. S. Robinson).

Action at Dosa, January 14.—Early on the morning of January 14 Showers, who had marched fifty-seven miles in twenty-five hours, surprised a portion of the enemy, 3,000 strong, who were easily put to flight by the cavalry aided by a few rounds from the guns. Tantia Topi's whole force broke up into various bodies which were pursued for about five miles; 300 rebels were killed and elephants, horses, etc., captured. Showers had 7 men killed and 22 wounded. The rebels retreated by Ulwar to Sikar, where, on January 21, they were overtaken by Lieutenant-Colonel J. Holmes, with a column composed as follows :

Irregular Cavalry.	83rd Foot.	No. 8 F.B. 2/6 Ben.A. (2nd Cap-
	12th Bom.N.I.	tain J. Shekleton, Bom.A.).
	Detachment R.E.	

Surprise at Sikar, January 21.— Holmes left Nusseerabad on January 8 and reached Sikar on January 21 without arousing the enemy's suspicion. The surprise was complete; the guns were at once sent rapidly to the front, while the infantry and cavalry followed in support, and the rebels broke up and all who threw down their arms were spared.

Capture of Tantia Topi.—After this defeat the three rebel leaders quarrelled and Tantia Topi's force dissolved, some surrendering to the Rajah of Bikanir, others scattering in various directions, and 3,000 remaining under the leadership of Feroz Shah and the Rao Sahib. Tantia Topi himself with three or four servants fled to Maun Sing, who was hiding in the forests of Paron. Feroz Shah and the Rao set off southwards to Kushana; but their followers hearing of the approach of Brigadier R. W. Honner,[1] who, with a mobile force, arrived near that place on February 10, they broke up and fled.[2] Honner pursued them with his cavalry and Camel Corps and cut down some 200; while the fugitives, crossing the Aravelli range, traversed Rajputana and sought to hide themselves in the Sironge jungles, where they were finally broken up and scattered by the action of small mobile columns in March, 1859. Meanwhile Brigadier Napier was clearing the Paron forests, and in April Maun Sing surrendered and gave up Tantia Topi, who was hanged at Sipri on the 18th of that month.

[1] Lieutenant-Colonel and Brev.-Colonel R. W. Honner, 4th Bom.N.I. His force consisted of detachments of the 8th Hussars, 1st Bombay Lancers, Sikh Horse, 83rd Foot, 12th Bom.N.I., and a small Camel Corps. In all 300 cavalry, 200 infantry, and 70 Camel Corps. The Camel Corps was under Lieutenant R. A. Stevenson, Bom.A.

[2] Prince Feroz Shah escaped disguised as a pilgrim and was never heard of again. The Rao Sahib was arrested in 1862, and having been found guilty of murder, was hanged at Cawnpore, August 20, 1862.

VIII. The Final Operations of the Saugor Field Force

While the operations narrated in the last section were taking place in the western region of the theatre of war south of the Jumna, the eastern was the scene of sporadic outbreaks of revolt and disorder which had little or no connection with each other ; and for present purposes it will be sufficient to glance at the principal events that occurred in the various districts.

Jhansi District.—When the garrison of Jhansi was reinforced by Sir Hugh Rose, Colonel J. Liddell had the following troops under his orders :

14th L.Ds. (1 squad).	3rd Bom.E.R.	1st Troop Bom.H.A. (Captain
3rd Bom.L.C.	24th Bom.N.I.	J. G. Lightfoot).
Detachment:		Detach. Bom.A.
Hyderabad C.		Bhopal Artillery.

By the middle of August Liddell had pacified the country between the Betwa and the Dhasan and sent Captain W. Ashburner [1] towards Kunch with 140 men of the 3rd Bombay Light Cavalry, detachments of the 3rd Bombay Europeans, and 24th Bom.N.I., some Bombay Sappers and detachments of Bengal and Bhopal artillery ; in all 350 men. Scattering the enemy as he advanced, Ashburner marched down the Betwa, and then moved westward towards Daboh, where, on September 4, he defeated the rebels on the banks of the Pahuj, killing 100 of them and taking 3 guns. He then turned towards Jalaon, where

[1] Captain W. Ashburner, 3rd Bombay Light Cavalry.

Brigadier J. MacDuff was stationed with the following troops :

| Irregular Cavalry. | 43rd Foot. | 5/14 R.A. (4 guns) (Captain F. |
| Mounted Oudh Police. | Oudh Police. | M. M. Ommaney.) |

Jalaon District—Action at Sahao, September 5.— On September 5 MacDuff marched to Sahao—a village four miles north of Jalaon—where a force of 3,000 matchlock men and 500 sepoys, with a few guns, occupied a position on rising ground. Their left flank was protected by a village, their right—where the sepoys were posted—rested on houses and walled gardens. MacDuff formed his line with the guns and infantry in the centre and the cavalry on the flanks, and advanced to within 500 yards of the enemy. The rebel guns then opened fire; 5/14 R.A. replied, and in a few minutes drove the defenders from the village; and the rebel left, falling into confusion, was charged and ridden through by the cavalry. Their right, however, fell back with some steadiness and a second position was taken up. The guns, supported by the 43rd Light Infantry, were again pushed forward and opened fire while the cavalry harassed the enemy on both flanks. Some of the sepoys attempted to hold a walled enclosure, but they were surrounded and killed to a man by the 43rd, and the remainder of the rebels broke and fled. The cavalry pursued them for some six miles, killing 300. One gun— a 1-pounder—abandoned in their first position, was captured. The British casualties were slight. After this affair MacDuff joined Ashburner and together they cleared the Jalaon district.

Hamirpore District. — The Hamirpore District continued in a very disturbed state till September, when Brigadier Whitlock forced the rebels to retire south of Jaitpore. In November they were again causing trouble and two small columns under Captain G. T. Hilliard, 50th Mad.N.I., and Lieutenant-Colonel Oakes,[1] 12th Lancers, were sent against them. Hilliard's Column was composed as follows :

Irregular Cavalry. Detachments : No. 1, F.B. A/4 Mad.A. (2 guns)
                    43rd Foot.   (Lieutenant C. Johnson).
                    50th Mad.N.I.
                    Total, all ranks, 320.

On December 4 Hilliard caught and defeated the rebels at Tola Kanjaran, a village on the right bank of the Dhasan, driving them across the river, and on December 5 Oakes, who had under him a troop of the 12th Lancers and two guns of A/4 Mad.A. (Lieutenant R. Pope), encountered them on the left bank at Lucherra and killing 80 scattered the others in confusion.

Kirwi District. — The wild and hilly country south of Kirwi was long a hotbed of rebels who caused much disquiet by repeated descents from their mountain fastnesses into the plains of the district. About midday on August 13 Brigadier T. D. Carpenter,[2] in command at Kirwi, received intelligence of the advance of a body of 3,000 rebels upon the friendly villages of Nyagaon and Chikrata situated a few miles to the south of Kirwi at the foot of a range of wooded hills where a pass debouches into the plains.

[1] Major and Brevet-Lieutenant-Colonel T. G. A. Oakes.
[2] Lieutenant-Colonel and Brevet-Colonel T. D. Carpenter, 1st Mad.N.I.

He at once marched from Kirwi with the following force :

Mounted Police.	Detachments :	6/14 R.A. (Captain and Brev.-
	43rd Foot.	Major E. Palmer).
	1st Mad.N.I.	
	Total, all ranks, 365 men.	

Affair at Nyagaon.—As there were only thirty mounted police available for advance guard and reconnaissance, they were supplemented by a few mounted gunners of 6/14 R.A.,[1] and the guns of the force—two brass pieces of native manufacture—were carried concealed by a tarpaulin on the back of an elephant. After a march of two hours, during which one-third of his force was disabled by the heat, Carpenter reached the hills and was at once assailed with matchlock fire which ceased as soon as one of the small guns came into action. The rebels fled, losing 100 men. During the short engagement a thunderstorm broke over the hills and soon the dry nullahs became roaring torrents. Darkness set in rapidly and it was with great difficulty the force reached Kirwi at 1 a.m. on the 14th.

Carpenter's March to Kothi.—In order to clear the rebels away from the vicinity of Kirwi, Carpenter on August 27 set out for a march through the hills with the same force except that he had 50 irregular cavalry and 6/14 had 6-pounder guns and mortars.[2]

[1] Quartermaster-Sergeant R. Glasgow and Gunner John Watkins, 6/14 R.A., did good service while with the advanced guard. The former gallantly rescued a drowning comrade during the return march to Kirwi.

[2] Palmer had two 6-pounder guns drawn by bullocks, two 4⅘-inch mortars carried by an elephant, and a 3-pounder gun also carried by an elephant.

Carpenter's destination was Kothi, and on September 5 he was within fifteen miles of that place. In the vicinity were some 7,000 rebels, including 500 sepoys, who had evacuated all the hill villages before the British advance. Between the British and Kothi lay the Panghati Pass, and here the rebels determined to make a stand; but on September 6 they melted away before Carpenter's attack, and after losing 50 men they abandoned a position they could easily have held against a much greater force. The British, having suffered no casualties, reached Kothi that evening, and later in the month advanced as far as Nagode.

Attack on Kirwi Palace, December 25.—Towards the end of December Brigadier Whitlock was at Mahoba with a large part of the Saugor Field Force. Banda was occupied by native levies from the friendly state of Rewah, and on December 24 the British force left in Kirwi under Captain C. Woodland, 1st Mad.N.I., consisted of some 100 men of the 43rd Light Infantry and 1st Mad.N.I. and 11 men of 6/14 R.A. under Rough-riding Sergeant Smith. Intelligence having reached Woodland early that morning of the approach of 6,000 rebels, including 600 sepoys, he determined to hold the palace and offer its shelter to the inhabitants of the city. The palace consisted of a square enceinte surrounded by a wall 10 to 15 feet high with turrets at the four corners and in the middle of the sides, where one 10-pounder and four 3-pounder guns of native make were mounted.[1] On three sides the

---

[1] The Rough Rider, Sergeant Smith, disposed his men as follows: 10-pounder north gate, Sergeant Smith, Bombr. Rye, Gunner McGee;

palace was almost surrounded by houses some thirty yards distant, the fourth being protected by a cliff and a river. When they entered Kirwi about noon the rebels at once occupied these houses and opened a futile fire of small arms on the palace, for they had neither guns nor scaling ladders. About 5 p.m. they were dislodged by the cannon and rifles of the small garrison and retired with the intention of returning next day better provided for a siege. But on the night of the 23rd the rebel movements had been reported to Whitlock, who, at once setting out for Kirwi and pushing on with his guns and cavalry, reached the city on the morning of Christmas Day after a march of eighty-four miles in thirty-four hours, just in time to foil the efforts of the besiegers, who, with two guns wrested from a neighbouring chieftain, were about to renew their attack. They now retired to the hills south of Kirwi and took up a position at Panwari, where on December 29 Whitlock attacked them with the following force:[1]

Detachments:	Detachments:	
12th Lancers.	43rd Foot	A Troop Mad.H.A. (Captain and Brev.-Lieutenant Colonel J. D. Mien).
Hyderabad C.	3rd Mad.E.F.	
Rewah C.	1st Mad.N.I.	
	50th Mad.N.I.	6/14 R.A. (Captain and Brev.-Major E. Palmer).
	Rewah Inf.	
		Rewah Art.

3-pounder commanding river ford west side, Gunners Wilkinson and Phillips; 3-pounder gun south-east turret, Gunners Storey and Goff; 3-pounder north-east turret, Gunners Cawson and Rodgers; 3-pounder east gate, Gunner Widgate, Shoeing Smith Fox. ("R.A.I. Occ. Papers," i. p. 255.)

[1] The Rewah levies marched in from Banda, forty-eight miles in thirty-eight hours, and Whitlock's infantry came in from Mahoba on the night of December 27.

Action at Panwari, December 29.—The general divided his force into three columns. The centre column mounted the heights in front while the left column made a flank movement, and the combined attack forced the rebels back upon the right column. These operations were completely successful, and the enemy were scattered in all directions, losing their guns and about 300 men. Whitlock had 9 men wounded.

Affair on the Tons, March 4.—On March 4, 1859, news reached Brigadier P. P. Faddy,[1] commanding a column of the Saugor Field Force at Rewah, of a rebel gathering on the Tons River, and he at once marched to attack it with 168 men of the 2nd Sikh Cavalry and 2 guns of A. Troop Mad.H.A. (Lieutenant R. G. F. Henegan). The enemy, 1,500 strong, occupied a good position, but on the approach of the column they retired, and Faddy, following in pursuit, found himself in difficult ground where his guns could not keep up with the cavalry. Advancing with the Sikh horse, he himself led a vigorous charge which effectively broke up the rebels, of whom 100 were killed. The British casualties were 4 men killed and 12 wounded.

Chanderi District.—The Rajahs of Banpore and Sharghur for a time caused much disorder in the neighbourhood of Chanderi, but at the end of August 1858 they finally submitted, and one fruitful source of trouble came to an end. But this district generally, and those lying to the south, continued in a disturbed state, and when Tantia Topi marched up the Betwa

---

[1] Lieut.-Colonel P. P. Faddy, R.A.

and along the Nurbudda the country bordering these rivers was thrown into great excitement. Disorder continued for several months, and numerous engagements took place between small British columns and local freebooters and rebels, but during the summer of 1859 the country was finally pacified.

IX. REORGANISATION OF THE INDIAN ARMY

Changes in the Bengal, Madras, and Bombay Armies.—The Honourable East India Company was dissolved on November 1, 1858, and the Indian Armies became part of Her Majesty's Forces; but it was during the years which followed the Mutiny that certain drastic changes in organisation took place. Then all the European regiments in the three Presidencies became part of the British Army. The five cavalry regiments of Bengal became the 19th, 20th, and 21st Hussars; all the artillery and engineers were incorporated in the Royal Artillery and the Royal Engineers, the officers, however, remaining on their own lists for purposes of promotion; the 11 infantry battalions were the basis of the 100th and the nine following Regiments of the Line.[1]

A new native army was already in existence in Bengal. Of the seventy-four native infantry regiments few survived the Mutiny, but there were some notable exceptions. The identity of regiments or portions of regiments which like those at

[1] The German Legion was included in H.M. 109th Regiment of Foot.

Saugor and Lucknow had given proof of devoted loyalty was carefully preserved, and the steadfast 21st, whose good faith was never in question, took the right of the line under its new designation, the 1st Bengal Native Infantry.[1] The regiments of Kelat-i-Ghilzai, Ferozepore, and Loodianah, the Shekwatti Battalion, and the Calcutta Militia also received numbers, and in company with the loyal remnants of the Poorbeah Army constituted the first eighteen regiments of the new line. The remaining twenty-seven regiments were formed as follows: the 19th to the 32nd from Punjaubi infantry recruited during the Mutiny; the 33rd to the 40th and the 42nd to the 44th from various native levies; the 41st from the 1st Infantry, Gwalior Contingent, and the 45th from Rattray's Sikhs. The Gurkha Battalions were formed into five Gurkha Regiments, and the Punjaub Irregular Frontier Force remained practically unchanged.

The ten regiments of Light Cavalry had all disappeared, and eight regiments of Irregular Cavalry became the first eight regiments of the new Bengal Cavalry, eleven more being formed from Hodson's Horse, Wale's Horse, the Mooltani Horse, and other similar levies.

In the Madras and Bombay Armies no changes took place in the native infantry and cavalry except that the 21st and 27th Bom.N.I. had disappeared. The irregular troops in both Presidencies were loyal, and in company with the new levies raised during the Mutiny maintained for a time their original

[1] See p. 26 n.

status, though changes in nomenclature and organisation occurred at a later period.

While a native corps of sappers and miners was maintained in each presidency, it was determined that, with the exception of a few mountain batteries, there should be no native artillery. In Bengal the four native troops ceased to exist as such on September 8, 1859 : their designations were retained until the amalgamation, but their *personnel* was European; the native companies of Ben.A. all disappeared, and the artillery of the Punjaub Frontier Force [1] was converted into four mountain batteries.

In Madras and Bombay also the native artillery gradually disappeared with the exception of 1/4 Bom.A. and the Jacobabad Train, which were retained for mountain service in Bombay, while the artillery of the Hyderabad Contingent (which remained intact) fulfilled the same function in Madras.

The Record of the Native Artillery.—The record of the native artillery in the Mutiny compares favourably with that of the other arms, for the number of mutineers was proportionately smaller " and in no case did they take the lives of their own officers in cold blood." [2]

Two troops of Bengal Horse Artillery took part in the revolt. The 4th Troop 1st Brigade mutinied at Neemuch, and fought against us at Sassiah, at Delhi, and in Behar; the 4th Troop 3rd Brigade rose at Mooltan, but their action was more in the nature of desertion than of mutiny, and it

[1] The " Irregular " was dropped in 1862.
[2] Stubbs, iii. p. 595.

is claimed for them that there were extenuating circumstances.[1] The 4th Troop 2nd Brigade was at Peshawar when the Mutiny broke out, and betrayed no sign of insubordination; but for prudential reasons it was marched to Attock, where the native gunners were left; and the troop returned with a volunteer *personnel* from the 24th, 27th, and 70th Foot. The 5th Troop 1st Brigade did excellent service at Delhi and elsewhere. In a General Order dated July 14, 1858, special rewards for gallantry were conferred upon the men: 93 received the 3rd Class of the Order of Merit, one Havildar being given the 1st Class; two Jemadars got the 2nd Class of the Order of British India, and the Subadar was decorated with the 1st Class. The five seniors also received a step in rank. But in the following September the natives were all discharged or absorbed and their places were taken by Europeans.[2]

Of the Bengal Artillery six companies mutinied and fought against the British. Thus 2/7 Ben.A. mutinied at Nusseerabad; 3/7 Ben.A. at Delhi; 5/7 Ben.A. at Fyzabad; 6/7 at Cawnpore; 6/8 at Bareilly; and 4/9 at Jhansi and Nowgong. In the other twelve companies the men were allowed to take their discharge or were transferred to other native regiments, and one of them (1/7 Ben.A.), it will be remembered, did good service in 1857 at Umballa.[3]

In Bombay two native companies mutinied, namely: 3/4 Bom.A. at Shirkarpore and 5/4 at Hyderabad (Sind). In Madras there was no mutiny.

The Amalgamation of the Royal and Indian Artil-

[1] Stubbs, iii. p. 597.   [2] *Ibid.*, pp. 596, 597.   [3] Page 21.

lery.—In 1861 the official amalgamation of the Royal and Indian Artillery took place. At this period the Royal Regiment [1] consisted of the Horse Brigade, five Field Brigades (the 4th, 8th, 9th, 11th, and 14th), and ten Garrison Brigades (the 1st, 2nd, 3rd, 5th, 6th, 7th, 10th, 12th, 13th, and 15th [2]). It was now increased by four Horse Brigades, four Field Brigades (the 16th, 18th, 19th, and 23rd), three Garrison Brigades (17th, 24th, 25th), and three Mixed Brigades (the 20th, 21st, and 22nd). The new designation of the Indian Troops and Companies are set forth in the following tables : [3]

[1] The introduction of the Brigade System is described in "History of the Royal Artillery (Crimean Period)," by Colonel J. R. J. Jocelyn, p. 89, etc.

[2] The 15th Brigade was added to the Royal Artillery in 1858.

[3] In 1864 the five Horse Brigades were designated by the letters A, B, C, D, E. Field and Garrison Brigades retained their numbers, field batteries were designated by letters, and garrison companies by numbers, a practice only partially introduced in 1861. Immediately after the new nomenclature was published several changes amongst individual batteries took place.

THE KHURSHID MANZIL (MESS HOUSE), LUCKNOW.

THE AMALGAMATION OF THE INDIAN WITH THE ROYAL ARTILLERY

The Horse Artillery

Designation in 1860.	Designation in 1861.	Designation in 1912.[1]
1st Troop 1st Brigade Ben.H.A.	A Battery 2nd Horse Brigade R.A.	F Battery IV Brigade R.H.A.
2nd ,, ,, ,,	B ,, ,, ,,	56th Battery XLIV Brigade R.F.A.
3rd ,, ,, ,,	C ,, ,, ,,	58th Battery XXXV Brigade R.F.A.
4th ,, ,, ,,	D ,, ,, ,,	Reduced 1862.
5th ,, ,, ,,	E ,, ,, ,,	T Battery XI Brigade R.H.A.
1st Troop 2nd Brigade Ben.H.A.	A Battery 5th Horse Brigade R.A.	K Battery VI Brigade R.H.A.
2nd ,, ,, ,,	B ,, ,, ,,	52nd Battery XV Brigade R.F.A.
3rd ,, ,, ,,	C ,, ,, ,,	No. 1 Depot R.F.A.
4th ,, ,, ,,	D ,, ,, ,,	W Battery XII Brigade R.H.A.
1st Troop 3rd Brigade Ben.H.A.	F Battery 2nd Horse Brigade R.A.	L Battery VII Brigade R.H.A.
2nd ,, ,, ,,	G ,, ,, ,,	57th Battery XLIII Brigade R.H.A.
3rd ,, ,, ,,	E Battery 5th Horse Brigade R.A.	S Battery IX Brigade R.H.A.
4th ,, ,, ,,	F ,, ,, ,,	X Battery VIII Brigade R.H.A.
A Troop Mad.H.A.	A Battery 3rd Horse Brigade R.A.	J Battery IV Brigade R.H.A.
B ,, ,,	B ,, ,, ,,	M Battery VIII Brigade R.H.A.
C ,, ,,	C ,, ,, ,,	P Battery X Brigade R.H.A.
D ,, ,,	D ,, ,, ,,	R Battery X Brigade R.H.A.
1st Troop Bom.H.A.	A Battery 4th Horse Brigade R.A.	N Battery IX Brigade R.H.A.
2nd ,, ,,	B ,, ,, ,,	Y Battery XIII Brigade R.H.A.
3rd ,, ,,	C ,, ,, ,,	Q Battery VIII Brigade R.H.A.
4th ,, ,,	D ,, ,, ,,	Z Battery XIV Brigade R.H.A.

[1] Since 1912 some changes in the numbering of the Brigades have taken place; but batteries and companies have preserved their letters and numerals.

THE AMALGAMATION OF THE INDIAN WITH THE ROYAL ARTILLERY

Foot Artillery

Designation in 1860.	Designation in 1861–1863.	Designation in 1912.
No. 1 Company 1st Battalion Ben.A.	A Battery 16th Brigade R.A.	21st Battery II Brigade R.F.A.
,, 2 ,, ,, ,,	B ,, ,, ,,	10th Battery XVII Brigade R.F.A.
,, 3 ,, ,, ,,	C ,, ,, ,,	53rd Battery II Brigade R.F.A.
,, 4 ,, ,, ,,	D ,, ,, ,,	22nd Battery XXXIV Brigade R.F.A.
No. 1 Company 2nd Battalion Ben.A.	A Battery 19th Brigade R.A.	34th Battery XXXVIII Brigade R.F.A.
,, 2 ,, ,, ,,	B ,, ,, ,,	Reduced 1863
,, 3 ,, ,, ,,	C ,, ,, ,,	94th Company R.G.A.
,, 4 ,, ,, ,,	D ,, ,, ,,	35th Battery XXXVII Brigade R.A.
No. 1 Company 3rd Battalion Ben.A.	No. 1 Battery 22nd Brigade R.A.	28th Company R.G.A.
,, 2 ,, ,, ,,	,, 2 ,, ,, ,,	23rd Battery XL Brigade R.F.A.
,, 3 ,, ,, ,,	,, 3 ,, ,, ,,	97th Company R.G.A.
,, 4 ,, ,, ,,	,, 4 ,, ,, ,,	36th Battery XXXIII Brigade R.F.A.
No. 1 Company 4th Battalion Ben.A.	No. 1 Battery 24th Brigade R.A.	No. 2 Mountain Battery R.G.A.
,, 2 ,, ,, ,,	,, 2 ,, ,, ,,	No. 4 Mountain Battery R.G.A.
,, 3 ,, ,, ,,	,, 3 ,, ,, ,,	No. 6 Mountain Battery R.G.A.
,, 4 ,, ,, ,,	,, 4 ,, ,, ,,	Reduced 1905
No. 1 Company 5th Battalion Ben.A.	No. 1 Battery 25th Brigade R.A.	93rd Company R.G.A.
,, 2 ,, ,, ,,	,, 2 ,, ,, ,,	95th Company R.G.A.
,, 3 ,, ,, ,,	,, 3 ,, ,, ,,	Reduced 1907
,, 4 ,, ,, ,,	,, 4 ,, ,, ,,	54th Battery XXXIX Brigade R.F.A.
No. 1 Company 6th Battalion Ben.A.	No. 5 Battery 16th Brigade R.A.	29th Company R.G.A.
,, 2 ,, ,, ,,	,, 5 ,, 19th ,, ,,	59th Battery Brigade R.F.A.
,, 3 ,, ,, ,,	,, 5 ,, 22th ,, ,,	60th Battery Brigade R.F.A.
,, 4 ,, ,, ,,	,, 5 ,, 24th ,, ,,	16th Company R.G.A.
A Company 1st Battalion Mad.A.	No. 1 Battery 17th Brigade R.A.	71st Company R.G.A.
B ,, ,, ,,	,, 2 ,, ,, ,,	85th Company R.G.A.
C ,, ,, ,,	,, 3 ,, ,, ,,	24th Battery XXXIV Brigade R.F.A.
D ,, ,, ,,	,, 4 ,, ,, ,,	Reduced 1868

A Company 2nd Battalion Mad.A.	No. 1 Battery 20th Brigade R.A.	34th Company R.G.A.
B ,,	,, 2 ,,	50th Battery XXXIV Brigade R.F.A.
C ,,	,, 3 ,,	51st Battery XXXIX Brigade R.F.A.
D ,,	,, 4 ,,	49th Battery XL Brigade R.F.A.
A Company 3rd Battalion Mad.A.	A Battery 23rd Brigade R.A.	46th Battery XXXIX Brigade R.F.A.
B ,,	B ,,	47th Battery XLI Brigade R.F.A.
C ,,	C ,,	23rd Company R.G.A.
D ,,	D ,,	48th Battery XXXVI Brigade R.F.A.
A Company 4th Battalion Mad.A.	No. 5 Company 17th Brigade R.A.	79th Company R.G.A.
B ,,	,, 6 ,, ,,	44th Company R.G.A.
C ,,	No. 5 Company 20th Brigade R.A.	Reduced 1869
D ,,	E Battery 23rd Brigade R.A.	Reduced 1868
No. 1 Company 1st Battalion Bom.A.	A Battery 18th Brigade R.A.	86th Company R.G.A.
,, 2 ,,	No. 2 Battery 18th Brigade R.A.	15th Battery XXXVI Brigade R.F.A.
,, 3 ,,	C ,, ,,	30th Battery XLIV Brigade R.F.A.
,, 4 ,,	D ,, ,,	91st Company R.G.A.
No. 1 Company 2nd Battalion Bom.A.	No. 1 Battery 21st Brigade R.A.	81st Battery X Brigade R.F.A.
,, 2 ,,	,, 3 ,, ,,	90th Company R.G.A.
,, 3 ,,	B Battery 18th Brigade R.A.	31st Battery XXXVII Brigade R.F.A.
,, 4 ,,	No. 4 Battery 21st Brigade R.A.	55th Battery XXXVII Brigade R.F.A.
No. 1 Company 3rd [1] Battalion Bom.A.	E Battery 18th Brigade R.A.	70th Battery XXXIX Brigade R.F.A.
,, 2 ,,	F Battery 18th Brigade R.A.	71st Battery XXXVI Brigade R.F.A.
,, 3 ,,	G ,, ,,	90th Battery XVI Brigade R.F.A.
,, 4 ,,	6th Battery 21st Brigade R.A.	88th Company R.G.A.

[1] The Reserve Battalion Bombay Artillery was called the Third Battalion in 1859.

APPENDICES

Appendix I

THE INDIAN ARTILLERY

The Bengal Artillery

In the Bengal Artillery there were three brigades of horse artillery. In the First Brigade there were three European and two native troops (4/1 and 5/1 Ben.H.A.), and the Second and Third Brigades had each three European and one native troop (4/2 and 4/3 Ben.H.A.). In a European troop, with the exception of the two limber gunners, Nos. 6 and 7, all the combatant detail was European; in a native troop, with the exception of the officers and two or three non-commissioned officers, it was native. All troops had a detail of lascars, syces, and grass-cutters.

The usual armament for a troop of horse artillery consisted of five 6-pounder guns and one 12-pounder howitzer. There were teams of six horses for both gun and waggon, and pole draught was employed; there were no detachment horses, No. 1 and his horse-holder No. 14 being alone provided with saddle horses; but spare gun-teams accompanied the troop. The gun was driven by Nos. 8, 9, and 10; the off horses being ridden by Nos. 5, 2, and 4. The waggon was driven by Nos. 11, 12, and 13, No. 3 being mounted on the off leader. The waggon as a rule accompanied the gun into action, but if for any reason it did not do so, the leaders were unhooked and followed the gun. No. 2 usually carried the sponge, with the end of the stave in a lance bucket, when the troop was going into action.

The foot artillery consisted of six European battalions, numbered 1 to 6, each having four companies, and of three native battalions, numbered 7 to 9, each having six companies, which were designated by numerals. Twenty-one companies were employed as garrison, and twenty-one as field artillery, the equipment and extra *personnel* for twenty-one field batteries being provided; but there was no permanency in this allotment of duties, and any company, European or native, was liable to take its turn at either. The *personnel* of a field battery therefore consisted of the officers and men of the company allotted to it (a temporary *personnel*), and also of some sixty native drivers, a detail of lascars, syces, and grass-cutters with a few European non-commissioned officers in charge of them (a permanent *personnel*). The armament was usually five 9-pounder guns and one 24-pounder howitzer, with 130 horses or an equivalent number of bullocks, though sometimes 18-pounder guns and 8-inch mortars were employed. The twenty-one field batteries were designated by numerals, and such expressions as No. 1 Horse Field Battery, or No. 2 Bullock Field Battery were in common use. They were also called Light Field Batteries and Heavy Field Batteries according to the nature of the armament. When companies were detailed for field duties, they were said to " go into battery," and their full designations would be such as No. 1 Company 1st Battalion with No. 3 Horse Field Battery, or No. 2 Company 5th Battalion with No. 17 Bullock Field Battery. Pole draft was employed, and there were seats on the axle-tree boxes. There was also a mountain train manned by natives.

Madras Artillery

In Madras the Brigade of Horse Artillery comprised four European Troops (A, B, C, D) and two native troops (E and F). The detachments were mounted as in England, but pole draught was employed. In other respects they resembled the Bengal Horse Artillery. The foot artillery consisted of four European battalions, Nos. 1, 2, 3, and 4, and one native battalion, No. 5. The European battalions had each four companies, the native battalion had six companies; in each case designated by letters. Equipments for field batteries were maintained and companies of foot artillery attached thereto, in the same manner as in the Bengal Artillery.

Bombay Artillery

The Horse Brigades consisted of 4 troops, designated the First, Second, Third and Fourth. There were no native troops. In Bombay shafts were used and detachments mounted as in England. In other respects the Bombay resembled the Bengal Horse Artillery.

In the foot artillery there were 12 companies of Europeans and 12 companies of natives. The Europeans were organised in three battalions of four companies each, namely the First, Second and Reserve Battalions, the last being raised in 1857. The natives were organised in two battalions, designated the Third and Fourth, each having six companies.

In a similar manner to that of the other presidencies the equipment and *personnel* of some eighteen field batteries were maintained, and the companies of foot artillery took their turn of going into battery. Troops, battalions and companies were designated by numerals. There was also a mountain train manned by natives.

APPENDIX II

THE GARRISON OF INDIA IN 1857-9

In 1857 there was a Commander-in-Chief in each Presidency. In Bengal the Divisions and Brigades were as follows:

Divisions	1st and 2nd Class Brigades
Presidency	Barrackpore, 2nd Class
Dinapore	
Cawnpore	Cawnpore, 2nd Class Lucknow, 1st Class Saugor, 1st Class
Meerut	Meerut, 2nd Class Agra, 1st Class Muttra, 1st Class Delhi, 2nd Class Bareilly, 2nd Class
Sirhind	Umballa, 2nd Class Jullundur, 2nd Class
Lahore (Mian Mir)	Lahore, 2nd Class Mooltan, 2nd Class Ferozepore, 1st Class Sialkote, 1st Class Peshawar, 1st Class Sind Saugor (Rawal Pindi), 2nd Class

The Oudh Irregular Force was under the Chief Commissioner of Oudh, and the Punjaub Irregular Frontier Force under the Chief Commissioner of the Punjaub.

In Madras there were six divisions—namely, the Southern, Northern, Centre, Ceded Districts, Mysore, and Pegu.

In Bombay there were four divisions in 1857—namely, Poonah, Northern, Southern, and Sind. In 1858 the Malwa Division was added.

I. THE EUROPEAN FORCE IN INDIA IN APRIL 1857

The following regiments of Her Majesty's Army were quartered in the Bengal Presidency, Oudh, and the Punjaub:

CAVALRY

6th Regt. of Dragoon Guards (Carabineers), at Meerut.
9th (The Queen's Royal Regt. of Light Dragoons) Lancers, at Umballa.

INFANTRY

1st Battn. 8th (The King's) Regt. of Foot, at Jullundur.
1st Battn. 10th (The North Lancashire) Regt. of Foot, at Dinapore.
1st Battn. 24th (The 2nd Warwickshire) Regt. of Foot, at Rawal Pindi.
27th (or Inniskilling) Regt. of Foot, at Nowshera.
32nd (The Cornwall) Regt. of Foot, at Lucknow.
52nd (Oxfordshire) Regt. of Foot (Light Infantry), at Sialkote.
53rd (The Shropshire) Regt. of Foot, at Dum Dum.
1st Battn. 60th (The King's Royal Rifle Corps), at Meerut.
61st (The South Gloucestershire) Regt. of Foot, at Ferozepore.
70th (The Surrey) Regt. of Foot, at Peshawar.
75th Regt. of Foot, at Peshawar.
81st Regt. of Foot (Loyal Lincoln Volunteers), at Lahore.
87th Regt. of Foot (Royal Irish Fusiliers), at Peshawar.

The following European troops of H.E.I. Co.'s Service were in the Bengal Presidency, Oudh, and the Punjaub.

ARTILLERY

1st Troop 1st Brig. Ben.H.A., at Jullundur.
2nd Troop 1st Brig. Ben.H.A., at Meerut.
3rd Troop 1st Brig. Ben.H.A., at Sialkote.
1st Troop 2nd Brig. Ben.H.A., at Peshawar.
2nd Troop 2nd Brig. Ben.H.A., at Lahore.
3rd Troop 2nd Brig. Ben.H.A., at Lahore.
1st Troop 3rd Brig. Ben.H.A., at Peshawar.
2nd Troop 3rd Brig. Ben.H.A., at Umballa.
3rd Troop 3rd Brig. Ben.H.A., at Umballa.
1st Compy. 1st Battn. Ben.A. (No. 7 Field Batty.), at Peshawar.
2nd Compy. 1st Battn. Ben.A., at Govindghur.
3rd Compy. 1st Battn. Ben.A. (No. 17 Field Batty.), at Sialkote.
4th Compy. 1st Battn. Ben.A. (No. 9 Field Batty.), at Lucknow.
1st Compy. 2nd Battn. Ben.A. (No. 10 Field Batty.), at Peshawar.
2nd Compy. 2nd Battn. Ben.A., at Peshawar.
3rd Compy. 2nd Battn. Ben.A., at Peshawar.
4th Compy. 2nd Battn. Ben.A., at Peshawar.
1st Compy. 3rd Battn. Ben.A. (No. 4 Field Batty.), at Saugor.
2nd Compy. 3rd Battn. Ben.A. (No. 12 Field Batty.), at Benares.
3rd Compy. 3rd Battn. Ben.A. (No. 14 Field Batty.), at Meerut.
4th Compy. 3rd Battn. Ben.A., at Mooltan.
1st Compy. 4th Battn. Ben.A., at Mian Mir.
2nd Compy. 4th Battn. Ben.A., at Mian Mir.
3rd Compy. 4th Battn. Ben.A., at Mian Mir.
4th Compy. 4th Battn. Ben.A., at Mian Mir.

APPENDIX II

2nd *Compy. 5th Battn. Ben.A. (No.* 21 *Field Batty.*), at Agra.
3rd *Compy. 5th Battn. Ben.A. (No.* 20 *Field Batty.*), at Dum Dum.
4th *Compy. 5th Battn. Ben.A. (No.* 11 *Field Batty.*), at Dinapore.
1st *Compy. 6th Battn. Ben.A.*, at Cawnpore.
No. 2 *Compy. 6th Battn. Ben.A.*, at Mhow.
3rd *Compy. 6th Battn. Ben.A. (No.* 19 *Field Batty.*), at Ferozepore.
4th *Compy. 6th Battn. Ben.A.*, at Phillour.

ENGINEERS
The *Corps of Bengal Engineers.*

INFANTRY
1st *European Bengal Fusiliers*, at Dugshai.
2nd *European Bengal Fusiliers*, at Subathu.
3rd *European Regiment*, at Agra.

The following regiments of H.M. Army were quartered in the Madras Presidency:

12th *Prince of Wales' Royal Regt. of Lancers.*
74th *(Highland) Regt. of Foot.*
91st *(The Argyllshire) Regt. of Foot.*

The following European troops of H.E.I. Co.'s Service were also in Madras:

A, B, C, and D Troops Madras Horse Artillery.
A, B, C, and D Companies 1st, 2nd, 3rd, *and* 4th *Battns. Madras Artillery*
  (16 companies in all).
The *Corps of Madras Engineers.*
1st *Madras Fusiliers.*
2nd *European Light Infantry.*
3rd *Madras European Regt.*

The following regiments of H.M. Army were quartered in the Bombay Presidency:

14th *(The King's) Regt. of Light Dragoons.*
83rd *Regt. of Foot.*
86th *(The Royal County Down) Regt. of Foot.*

The following European troops of H.E.I. Co.'s Service were also in Bombay:

The 1st, 2nd, *and* 4th *Troops Bombay Horse Artillery.*[1]
The 2nd *and* 3rd *Companies* 1st *Battn. Bom.A.*[1]
The 1st, 2nd, 3rd, *and* 4th *Companies of the* 2nd *and Reserve Battn. Bom.A.*
The *Corps of Bombay Engineers.*
The 1st *European Regt. (Fusiliers) and the* 3rd *European Regt.*[1]

[1] The 3rd *Troop,* the 1st and 4th *Companies* 1st *Battn. Bombay Artillery,* and the 2nd *European Regiment* were in Persia.

II. THE NATIVE ARMY

The native troops in India were as follows:

(a) BENGAL

CAVALRY	INFANTRY	ARTILLERY
10 Regiments Light Cavalry.	74 Regiments of the Line.	3 Troops Horse Artillery.
18 Regiments Irregular Cavalry.	6 Local and Special Battalions.[1]	3 Battns. Foot Artillery (18 Companies, 10 of which were attached to Field Batteries).
	11 Regiments Local[2] Infantry.	
		Sappers and Miners.

The Oudh Irregular Force

3 Regiments Irregular Cavalry.	10 Regiments Irregular Infantry.	3 Field Batteries and 1 Reserve Company of Irregular Artillery.
Mounted Police.	Police.	

The Punjaub Irregular Frontier Force

Guide Corps

5 Regiments Punjaub Cavalry.	6 Regiments Punjaub Infantry.	3 Field Batteries.
	4 Regiments Sikh Infantry.	Mountain Train.

The Gwalior Contingent

2 Regiments Cavalry.	7 Regiments Infantry.	4 Companies Artillery.

(b) MADRAS

8 Regiments Light Cavalry.	52 Regiments of the Line.	Two Troops Horse Artillery and 1 Batt. Foot Artillery (6 Companies)
	Sappers and Miners.	

Local Levies.

---

[1] The Regiment of Kelat-i-Ghilzai, 2 Sikh Battns. (Ferozepore and Loodianah), and 3 Gurkha Battns. (Sirmur, Nusseeri, and Kumaon).

[2] Calcutta Native Militia, Ramgurh Light Infantry, Hill Rangers, 1st and 2nd Assam Regiments, Mhanwarrah Battn., Sylhet Light Infantry, Arracan Battn. Harriana Light Infantry, Pegu Light Infantry, and Shekwatti Battalion.

APPENDIX II

(c) BOMBAY

3 Regiments Light Cavalry.	29 Regiments of the Line.	Two Batts. Foot Artillery (12 Companies).
6 Regiments Irregular Horse.	2 Beluchi Battalions.	Sappers and Miners.

Hyderabad Contingent

4 Regiments Cavalry. 6 Regiments Infantry. 4 Companies Foot Artillery.

There were also the United Malwa, the Bhopal and the Kotah Contingents, the Jodpore Legion and other levies, each comprising a small force of the three arms.

III. THE REINFORCEMENTS

The following troops of H.M. Army reached India during the course of the Mutiny:

CAVALRY

From		Presidency or Port.	Date of Arrival.
England	1st (The King's) Regt. of Dragoon Guards	Madras	Nov. '57
,,	2nd (The Queen's) Regt. of Dragoon Guards	Bengal	Nov. '57
,,	3rd (The Prince of Wales') Regt. of Dragoon Guards	Bombay	Dec. '57
,,	7th (The Princess Royal's) Regt. of Dragoon Guards	Bengal	Jan. '58
,,	6th (Inniskilling) Regt. of Dragoons	Bombay	Oct. '58
,,	7th (The Queen's Own) Regt. of Light Dragoons (Hussars)	Bengal	Nov. '57
,,	8th (The King's Royal Irish) Regt. of Light Dragoons (Hussars)	Bombay	Dec. '57
,,	17th Regt. of Light Dragoons (Lancers)	,,	Dec. '57
,,	Military Train (acting as Cavalry)	Bengal	Aug. '57

APPENDICES

ARTILLERY

From		Presidency or Port.	Date of Arrival.
England	D Troop R.H.A.	Bombay	Dec. '57
,,	E Troop R.H.A.	Bengal	Nov. '57
,,	F Troop R.H.A.	,,	Nov. '57
,,	H Troop R.H.A.	Madras	Dec. '57
,,	8th Compy. 1st Battn. R.A. (X Batty.)	Karachi	Jan. '58
,,	8th Compy. 2nd Battn. R.A.	Calcutta	Dec. '57
,,	2nd Compy. 3rd Battn. R.A. (Y Batty.)	Bombay	Sept. '58
,,	3rd Compy. 3rd Battn. R.A.	Madras	Nov. '57
,,	5th Compy. 3rd Battn. R.A. (J Batty.)	Calcutta	Jan. '58
,,	3rd Compy. 4th Battn. R.A.	Bombay	Aug. '57
Cape of Good Hope	4th Compy. 5th Battn. R.A.	Calcutta	June '57
England	1st Compy. 6th Battn. (V Batty.) R.A.	Karachi	Mar. '65
,,	7th Compy. 6th Battn. (I Batty.) R.A.	Madras	Jan. '58
,,	6th Compy. 7th Battn. R.A.	Karachi	Jan. '58
Ceylon	3rd Compy. 8th Battn. R.A.	Calcutta	June '57
England	3rd Compy. 9th Battn. (T Batty.) R.A.	,,	Dec. '57
,,	6th Compy. 9th Battn. R.A.	Madras	Nov. '57
,,	2nd Compy. 11th Battn. R.A.	Karachi	Mar. '58
,,	6th Compy. 11th Battn. R.A.	Calcutta	Dec. '57
,,	7th Compy. 11th Battn. R.A.	Karachi	Mar. '58
,,	5th Compy. 12th Battn. (Q Batty.) R.A.	Calcutta	Dec. '57
China	5th Compy. 13th Battn. R.A.	,,	Sept. '57
,,	6th Compy. 13th Battn. R.A.	,,	Sept. '57
England	3rd Compy. 14th Battn. R.A.	,,	Sept. '57
,,	4th Compy. 14th Battn. R.A.	,,	Oct. '57
,,	5th Compy. 14th Battn. R.A.	Bombay	Dec. '57
,,	6th Compy. 14th Battn. R.A.	Madras	Nov. '57
,,	7th Compy. 14th Battn. R.A.	Calcutta	Oct. '57
,,	8th Compy. 14th Battn. R.A.	Bombay	Dec. '57
Burmah	1st Compy. 5th Battn. Ben.A.	Calcutta	June '57

ROYAL ENGINEERS

From		Presidency or Port.	Date of Arrival.
England	4th Compy. R.E.	Bengal	..
,,	11th Compy. R.E.	Bombay	Nov. '57
,,	21st Compy. R.E.	,,	Nov. '57
,,	23rd Compy. R.E.	Bengal	..

APPENDIX II

INFANTRY

From		Presidency or Port.	Date of Arrival.
England	1st Battn. 1st (The Royal) Regt. of Foot	Madras	Oct. '57
Mediterranean	1st Battn. 3rd (East Kent) Regt. of Foot (or The Buffs)	Bengal	Dec. '58
Mauritius	1st Battn. 4th (The King's Own) Regt. of Foot	Bombay	Jan. '59
,,	1st Battn. 5th Regt. of Foot (Northumberland Fusiliers)	Bengal	Aug. '57
Cape of Good Hope	1st Battn. 6th (The Royal 1st Warwickshire) Regt. of Foot	,,	Jan. '58
England	1st Battn. 7th Regt. of Foot (Royal Fusiliers)	,,	Nov. '57
Mediterranean	1st Battn. 13th or Prince Albert's Regt. of Light Infantry	,,	Oct. '57
England	1st Battn. 18th (Royal Irish) Regt. of Foot	Bombay	Dec. '57
,,	1st Battn. 19th (The 1st Yorkshire N. Riding) Regt. of Foot	Bengal	Dec. '57
,,	1st Battn. 20th (The East Devonshire) Regt. of Foot	,,	Nov. '57
China	1st Batt. 23rd (The Royal Welsh Fusiliers) Regt. of Foot	Bengal	Nov. '57
Mediterranean	28th (The 2nd Warwickshire) Regt. of Foot	Bombay	Dec. '58
Burmah	29th (The Worcestershire) Regt. of Foot	Bengal	Dec. '57
Mediterranean	31st (The Huntingdonshire) Regt. of Foot	Bombay	Dec. '58
Mauritius	33rd (The Duke of Wellington's) Regt. of Foot	,,	Aug. '57
England	34th (The Cumberland) Regt. of Foot	Bengal	Oct. '57
Burmah	35th (Royal Sussex) Regt. of Foot	,,	June '57
Ceylon	37th (The North Hampshire) Regt. of Foot	,,	June '57
England	38th (The 1st Staffordshire) Regt. of Foot		Nov. '57
,,	42nd (The Royal Highland) Regt. of Foot	,,	Nov. '57
Cape of Good Hope	43rd (Monmouthshire) Regt. of Foot (Light Infantry)	Madras	'57–58
England	44th (The East Essex) Regt. of Foot	Madras	Nov. '57
Mediterranean	46th (The South Devonshire) Regt. of Foot	Bombay	Oct. '58
,,	48th (The Northamptonshire) Regt. of Foot	Bengal	Oct. '58
England	51st (The 2nd Yorkshire West Riding) or The King's Own Light Infantry Regt.	Bengal	Dec. '57

442 APPENDICES

INFANTRY—continued

From		Presidency or Port.	Date of Arrival.
England	54th (The West Norfolk) Regt. of Foot	Bengal	Jan. '58
,,	56th (The West Essex) Regt. of Foot	Bombay	Dec. '57
Mediterranean	57th (The West Middlesex) Regt. of Foot	,,	Oct. '57
China	59th (The 2nd Nottinghamshire) Regt. of Foot	,,	Oct. '57
Cape of Good Hope	2nd Battn. 60th The King's Royal Rifle Corps	Bengal	May '58
England	3rd Battn. 60th The King's Royal Rifle Corps	Bengal	Oct. '57
Persia	64th (The 2nd Staffordshire) Regt. of Foot	,,	July '57
England	66th (The Berkshire) Regt. of Foot	Madras	Dec. '57
England	67th (The South Hampshire) Regt. of Foot	Bengal	Jan. '59
,,	68th (The Durham) Regt. of Foot Light Infantry	Madras	Mar. '58
,,	69th (The South Lincolnshire) Regt. of Foot	Madras	Dec. '57
Mediterranean	71st (Highland) Regt. of Foot, Light Infantry	Bombay	Feb. '58
England	72nd (The Duke of Albany's Own Highlanders) Regt. of Foot	,,	Dec. '57
Cape of Good Hope	73rd Regt. of Foot	Bengal	April '58
Australia	77th (The East Middlesex) Regt. of Foot		June '58
Persia	78th (Highland) Regt. of Foot (or Ross-shire Buffs)	,,	July '57
England	79th Regt. of Foot (Cameron Highlanders)	,,	Nov. '57
Cape of Good Hope	80th Regt. of Foot (Staffordshire Volunteers)	,,	Jan. '58
China	82nd Regt. of Foot (The Prince of Wales' Volunteers)	Bengal	Sept. '57
Burmah	84th (York and Lancaster) Regt. of Foot	,,	Jan. '57
England	88th Regt. of Foot (Connaught Rangers)	,,	Nov. '57
,,	89th Regt. of Foot	Bombay	Sept. '57
China	90th Regt. of Foot (Perthshire Volunteers), Light Infantry	Bengal	July '57
Mediterranean	91st (The Argyllshire) Regt. of Foot	Madras	Oct. '58
,,	92nd (Gordon Highlanders) Regt. of Foot	Bombay	Mar. '58
China	93rd (Sutherland Highlanders) Regt. of Foot	Bengal	Sept. '57
England	94th Regt. of Foot	,,	Jan. '58

APPENDIX II

INFANTRY—*Continued*

From		Presidency or Port.	Date of Arrival.
Cape of Good Hope	95th (*The Derbyshire*) Regt. of Foot	Bengal	
England	97th (*The Earl of Ulster's*) Regt. of Foot	,,	Nov. '57
,,	98th Regt. of Foot	,,	Jan. '58
,,	99th (*Lanarkshire*) Regt. of Foot	,,	Jan. '58
,,	2nd Battn. Rifle Brigade	,,	Nov. '57
,,	3rd Battn. Rifle Brigade	,,	Nov. '57
,,	The German Legion	Bombay	

IV. NEW REGIMENTS, ETC.

During the course of the Mutiny 3 regiments of European Cavalry and 3 regiments of European Infantry were raised in Bengal, and Volunteer Guards were formed at Calcutta, Madras, and other places. The Bengal Native Army was increased by 18 regiments of Sikhs and Punjaubis, 2 of Gurkhas, the East India Rifles, and 19 regiments of Irregular Horse. Twelve Infantry levies were also raised at Agra, Allahabad, and other important stations.

In Madras 3 new regiments of Native Infantry were raised and the Madras Rifles formed from picked men.

In Bombay 5 new battalions of Native Infantry were raised and the Irregular Horse increased.

Appendix III
CASUALTIES

DELHI—May 30–September 20

	Effective Strength of all ranks, Sept. 11.	Killed.						Wounded.					Total all ranks wounded.
		Officers.		N.C. Officers and Men.		Total killed all ranks.		Officers.		N.C. Officers and Men.			
		European.	Native.	European.	Native.			European.	Native.	European.	Native.		
STAFF		4	..	..	..	4		9	..	..	..		9
Artillery (including drivers, gun lascars, and Sikhs)	722	4	1	43	26	74		23	1	216	49		289
Engineers (including Engineers, Sappers, Miners, and Pioneers)	391	3	2	4	34	43		19	1	6	60		86
9th Lancers	123	1	..	26	..	27		2	..	64	..		66
6th Dragoon Guards (4 troops)	78	1	..	18	..	19		2	..	9	..		11
4th Ben. Irregular Cavalry	147	..	..	..	..	..		..	..	..	3		3
Detach. 1st Punjaub Cavalry	114	..	..	..	1	1		1	..	..	5		6
,, 2nd Punjaub Cavalry	107	..	..	..	..	..		..	..	..	3		3
,, 5th Punjaub Cavalry	462	..	..	..	..	..		..	1	..	3		4
Hodson's Horse	323	..	..	..	..	..		1	5	..	5		11
8th Foot	302	3	..	24	..	27		7	..	129	..		136
52nd Light Infantry	300	1	..	18	..	19		4	..	73	..		77
60th Rifles	402	4	..	109	..	113		10	..	266	..		276
61st Foot	459	2	..	30	..	32		7	..	112	..		119
75th Foot	427	5	..	79	..	84		14	..	184	..		198
1st Ben. E. Fusiliers	370	3	..	95	..	98		11	..	210	..		221
2nd Ben. E. Fusiliers	212	4	..	79	..	83		6	..	156	..		162
Sirmoor Battn.	312	1	..	..	85	86		6	8	..	219		233
Kumaru Battn.	585	1	..	..	20	21		2	3	..	33		38
Guide Corps	414	2	5	..	65	72		6	10	..	215		231
4th Sikh Infantry	664	1	2	..	43	46		3	7	..	106		116
1st Punjaub Infantry	650	3	3	..	71	77		5	5	..	141		151
2nd Punjaub Infantry	541	1	1	..	41	43		2	4	..	103		109
4th Punjaub Infantry	322	1	..	..	9	10		..	2	..	59		61
Beluch Battn. (wing)	No return	..	..	1	7	8		..	1	..	48		49
Pioneers (unarmed)		..	..	..	24	25		..	1	1	128		130
TOTAL	9,866	46	14	526	426	1,012		140	49	1,426	1,180		2,795

In addition 12 European rank and file and 18 natives missing = 30.

(Signed) H. W. NORMAN, A.A.G.

APPENDIX III

CHINHUT—JUNE 30

	Killed.					Wounded.				
	Officers.		N.C. Offrs. and Men.		Total all ranks.	Officers.		N.C. Officers and Men.		Total all ranks.
	European.	Native.	European.	Native.		European.	Native.	European.	Native.	
General and Brigade Staff	..	..	..	..	..	1	..	..	..	1
Artillery	..	1	1	51	53	2	..	8	1	11
32nd Foot	4	..	111	..	115	1	..	38	..	39
13th Ben.N.I.	..	..	..	35	35	1	3	..	10	14
48th Ben.N.I.	..	2	..	14	16	..	..	..	..	2
71st N.I.	..	..	1	3	4	..	..	..	2	..
Volunteer Cavalry	1	..	..	..	1	2	..	..	..	2
Oudh Irr. Cavalry	..	2	..	66	68	1	..	..	8	9
TOTAL	5	5	113	169	292	8	3	46	21	78

(Signed) J. INGLIS, Brigr.,
Commdg. Lucknow Garrison.

LUCKNOW GARRISON—JUNE 30–SEPTEMBER 26

	Killed.					Wounded.				
	Officers.		N.C. Offrs. and Men.		Total all ranks.	Officers.		N.C. Officers and Men.		Total all ranks.
	European.	Native.	European.	Native.		European.	Native.	European.	Native.	
General and Brigade Staff	1	..	..	..	1	3	..	..	..	3
Artillery	2	..	22	11	35	6	1	19	9	35
Engineers	2	..	2	..	4	..	..	1	..	1
Officers 7th Ben.L.C.	2	..	2	..	4	1	..	..	..	1
32nd Foot	3	..	81	..	84	6	..	133	..	139
84th Foot	..	..	12	..	12	1	..	2	..	3
13th Ben.N.I.	2	2	..	26	30	1	7	..	45	53
41st Ben.N.I.	..	..	1	5	6	3	..	..	..	3
48th Ben.N.I.	..	..	..	3	3	6	1	..	6	13
71st Ben.N.I.	..	1	1	8	10	2	1	..	11	14
Oudh Irr. Force	2	..	..	3	5	5	..	..	13	18
Pensioners and New Levies	..	1	..	10	11	..	..	..	35	35
Lucknow Magazine	..	..	3	2	5	..	..	..	2	2
Officers (unattached)	2	..	..	..	2	1	..	..	..	1
TOTAL	16	4	124	68	212	35	10	155	121	321

(Signed) J. INGLIS, Brigr.
Commdg. Lucknow Garrison.

HAVELOCK'S FORCE—JULY 11–AUGUST 16

| | Killed. | | | | | Wounded. | | | | |
	Officers.		N.C. Offrs. and Men.		Total all ranks.	Officers.		N.C. Officers and Men.		Total all ranks.
	European.	Native.	European.	Native.		European.	Native.	European.	Native.	
Artillery	..	..	2	1	3	..	..	16	3	19
64th Foot	..	..	..	2	2	3	..	77	..	80
84th Foot	..	..	..	5	5	1	..	23	..	24
78th Foot	..	..	..	11	11	4	..	58	..	62
1st Mad. Fusiliers	..	2	..	15	17	3	..	45	..	48
Ferozepore Regt.	..	..	..	3	3	..	..	..	40	40
Volunteer Cav.	..	..	1	..	1	3	..	..	..	3
Irr. Cav.	..	1	..	5	6	..	..	..	3	3
TOTAL	..	3	3	42	48	14	..	219	46	279

OUDH FIELD FORCE UNDER BRIGADIER-GENERAL HAVELOCK, C.B.—SEPTEMBER 25, 26

| | Killed. | | | | | Wounded. | | | | |
	Officers.		N.C. Officers and Men.		Total all ranks.	Officers.		N.C. Officers and Men.		Total all ranks.
	European.	Native.	European.	Native.		European.	Native.	European.	Native.	
Staff	2	..	..	..	2	8	..	..	..	8
Artillery	3	..	14	5	22	2	..	22	12	36
5th Foot	..	..	7	..	7	2	..	29	..	31
64th Foot	1	..	..	..	1	..	..	10	..	10
84th Foot	2	..	9	..	11	4	..	24	..	28
78th Foot	2	..	37	..	39	6	..	75	..	81
90th Foot	..	..	11	..	11	3	..	46	..	49
1st Mad. Fusiliers	..	..	13	..	13	2	..	35	..	37
Ferozepore Regt.	..	..	..	6	6	..	1	..	36	37
Volunteer Cav.	..	..	3	..	3	3	..	11	..	14
12th Ben. Irr. Cav.	1	1	..	2	4	..	..	..	8	8
TOTAL	11	1	94	13	119	30	1	252	56	339

In addition there were 76 European N.C.O. and men and 1 Sikh missing.

(Signed) J. HUDSON, Lieutenant,
D.A.A.G., Oudh F.F.

APPENDIX III

FIELD FORCE UNDER GENERAL SIR COLIN CAMPBELL
NOVEMBER 12-22

	Killed.					Wounded.				
	Officers.		N.C. Officers and Men.		Total all ranks.	Officers.		N.C. Officers and Men.		Total all ranks.
	European.	Native.	European.	Native.		European.	Native.	European.	Native.	
Staff	2	..	..	..	2	6	..	..	..	6
Naval Brigade	1	..	4	..	5	3	..	20	..	23
Engineers	..	..	3	..	3	1	..	19	..	20
Artillery	1	..	14	..	15	6	..	59	..	65
9th Lancers	..	..	..	..	..	..	..	..	..	..
Military Train	1	..	..	..	1	..	..	1	..	1
1st Punjaub Cav.	..	..	..	2	2	..	..	..	1	1
2nd Punjaub Cav.	..	..	..	1	1	..	..	..	..	..
5th Punjaub Cav.	..	..	..	..	..	..	..	..	1	1
Hodson's Horse	..	..	..	..	..	1	..	..	3	4
8th Foot	..	..	..	..	..	..	..	..	1	1
Detach. 5th Foot	..	..	5	..	5	..	..	3	..	3
,, 64th Foot	..	..	4	..	4	..	..	7	..	7
,, 78th Foot	..	..	..	..	..	..	..	..	..	..
2nd Punjaub Inf.	1	1	..	4	6	1	1	..	21	23
53rd Foot	..	..	10	..	10	3	..	63	..	66
Detach. 84th Foot.	1	..	1	..	2	..	..	8	..	8
,, 90th L.I.	..	..	6	..	6	3	..	22	..	25
93rd Horse	2	..	37	..	39	7	..	62	..	69
Detach. 1st Mad.Fus.	1	..	3	..	4	..	..	12	..	12
4th Punjaub Inf.	..	..	..	13	13	3	3	50	..	56
23rd R.W. Foot	..	..	3	..	3	1	..	23	..	24
82nd Foot	1	..	1	..	2	1	..	13	..	14
TOTAL	11	1	91	20	123	36	4	362	27	429

(Signed) H. W. NORMAN, Captain A.A.G.
("State Papers," ii. pp. 348, 353.)

ARMY UNDER COMMANDER-IN-CHIEF—MARCH 2–21

	Killed.					Wounded.				
	Officers.		N.C. Officers and Men.		Total all ranks.	Officers.		N.C. Officers and Men.		Total all ranks.
	European.	Native.	European.	Native.		European.	Native.	European.	Native.	
General Staff	..	..	..	..	..	1	..	..	..	1
H.M.S. *Shannon*	1	..	1	..	2	1	..	13	..	14
F. Troop R.H.A.	..	..	..	..	..	..	..	2	..	2
2/1 Ben.H.A.	..	..	..	..	..	..	..	1	..	1
3/3 Ben.H.A.	..	..	..	..	..	1	..	1	..	2
5/12 R.A.	..	..	..	..	..	..	..	9	..	9
Brigade Staff	1	..	..	..	1	..	..	..	..	..
8/2 R.A.	..	..	..	..	..	..	..	3	..	3
3/8 R.A.	..	..	1	..	1	1	..	2	..	2
6/11 R.A.	..	..	2	..	2	1	..	5	..	6
6/13 R.A.	..	..	1	..	1	..	..	3	..	3
3/14 R.A.	..	..	1	..	1	..	..	5	..	5
3/5 Ben.A.	..	..	..	..	..	..	..	1	..	1
23rd Compy. R.E.	1	..	13	..	14	..	..	1	..	1
Bengal Sappers & Miners	1	..	..	2	3	2	..	..	4	6
Punjaubi Dragoons	..	..	..	4	4	1	2	..	23	26
Delhi Pioneers	1	..	..	..	1	..	..	..	6	6
Brig. Staff	..	..	..	..	..	1	..	..	..	1
9th Lancers	..	..	1	..	1	1	..	4	..	5
2nd Punjaub Cav.	1	..	..	..	1	..	..	..	9	9
Detach. 5th Punjaub C.	1	..	..	..	1	..	..	..	6	6
1st Sikh Irr. C.	1	..	..	2	3	4	1	..	7	12
2nd Dragoon Guards	..	..	2	..	2	1	..	5	..	6
7th Hussars	..	..	..	..	..	3	..	2	..	5
Hodson's Horse	..	..	..	2	2	1	1	..	7	9
Pathan Horse	1	..	..	1	2	..	..	..	..	..
Divisional Staff	..	..	..	..	..	..	..	..	..	..
5th Foot	..	..	..	..	..	1	..	3	..	4
78th Foot	..	..	..	..	..	..	..	1	..	1
90th Foot	..	..	3	..	3	..	..	28	..	28
Ferozepore Regt.	1	2	..	5	8	1	..	..	45	46
Divisional Staff	..	..	..	..	..	1	..	..	..	1
34th Foot	..	..	..	..	..	..	..	4	..	4
38th Foot	..	..	1	..	1	3	..	22	..	25
53rd Foot	..	..	1	..	1	2	..	..	..	2
42nd Foot	..	..	5	..	5	1	..	39	..	40
93rd Foot	2	..	12	..	14	2	..	59	..	61
4th Punjaub Rifles	..	..	..	8	8	4	..	..	30	34
Brigade Staff	..	..	..	..	..	1	..	..	..	1
23rd Foot	..	..	4	..	4	3	..	25	..	28
79th Foot	..	..	7	..	7	2	..	21	..	23
1st E. Ben.F.	..	..	8	..	8	3	..	21	..	24
2nd Battn. R.B.	1	..	..	..	1	1	..	13	..	14
3rd Battn. R.B.	..	..	..	..	..	1	..	6	..	7
2nd Punjaub Infantry	1	1	..	7	9	1	..	32	..	33
Divisional Staff	1	..	..	..	1	2	..	..	..	2
10th Foot	..	..	4	..	4	1	..	23	..	24
20th Foot	..	..	7	..	7	2	..	28	..	30
97th Foot	1	..	2	..	2	..	..	21	..	21
TOTAL	16	3	77	31	127	51	4	403	137	595

In addition 5 European and 8 Native N.C.O. and men missing.

(Signed) H. W. NORMAN, Major, D.A.G.

APPENDIX III

GENERAL SIR HUGH ROSE'S CAMPAIGN IN CENTRAL INDIA

	Killed.					Wounded.				
	Officers.		N.C. Officers and Men.		Total all ranks.	Officers.		N.C. Officers and Men.		Total all ranks.
	European	Native	European	Native		European	Native	European	Native	
Staff	2	..	..	..	2	8	..	..	..	8
14th Light Dragoons	..	..	11	..	11	1	..	48	..	49
3rd Bom.L.C.	..	..	..	4	4	..	..	..	8	8
R.A.	1	..	2	..	3	..	..	6	..	6
Bom.A.	1	..	1	..	2	5	..	10	..	15
R.E.	..	..	..	..	..	1	..	..	..	1
Mad.E.	..	..	..	3	3	1	..	..	9	10
Bom.E.	2	..	..	..	2	..	..	..	6	6
71st Foot	..	..	..	..	..	..	..	1	..	1
86th Foot	..	..	4	..	4	8	..	91	..	99
Camel Corps	..	..	..	..	..	..	..	8	..	8
3rd Bom. Eur. Regt.	..	..	12	..	12	3	..	71	..	74
24th Bom. N.I.	2	..	..	8	10	1	2	..	19	22
25th Bom.N.I.	..	..	..	16	16	2	2	..	42	46
Hyderabad Contingent	1	2	..	29	32	6	13	..	87	106
TOTAL	9	2	30	60	101	36	17	235	171	459

SIR HUGH ROSE'S GWALIOR CAMPAIGN

	Killed.					Wounded.				
	Officers.		N.C. Officers and Men.		Total all ranks.	Officers.		N.C. Officers and Men.		Total all ranks.
	European	Native	European	Native		European	Native	European	Native	
8th Hussars	1	..	9	..	10	..	..	7	..	9
14th Light Dragoons	..	..	..	..	..	..	..	4	..	4
1st Bom.L.C.	1	..	..	1	2	..	..	..	6	7
Bom.A.	..	..	3	..	3	..	..	7	..	7
71st Foot	1	..	6	..	7	..	..	7	..	7
95th Foot	..	..	2	..	2	3	..	15	..	18
10th Bom.N.I.	..	..	..	..	..	..	..	..	3	3
25th Bom.N.I.	1	..	..	2	3	..	..	..	7	7
Hyderabad Contingent	..	..	..	2	2	..	..	..	3	3
TOTAL	4	..	20	5	29	6	..	40	19	65

Compiled from an official but unsigned return.

APPENDIX IV

SERVICES OF THE ROYAL AND INDIAN ARTILLERY IN THE MUTINY

N.B.—*The designations of troops and companies given in italics are those of 1912. The numbers in columns headed "Service" refer to Table XVII, p. 487, etc. The letters (a), (b), (c), etc., refer to Notes at the end of the Appendix, p. 491. Abbreviations, etc., are explained in Note (a).*

(*b*) ⊙ *Signifies a step of honorary rank on retirement.*

(*b*) I.—OFFICERS OF THE ROYAL ARTILLERY HOLDING COMMANDS OR STAFF APPOINTMENTS DURING THE INDIAN MUTINY

Number in Kane's List.	Names.	Employment.	Service.	Remarks.
1714	Major-Gen. J. E. Dupuis, C.B.	G.O.C.R.A. in India	56, 58	Gen. Sir John E. Dupuis, Col. Comdt. d. 1876.
1767	Lieut.-Col. and Brev.-Col. D. E. Wood, C.B.	O.C.R.H.A.	64, 76	Gen. Sir David E. Wood, G.C.B., Col. Comdt. d. 1894.
1836	Lieut.-Col. W. T. Crawford	O.C.R.A.	51, 58	Lieut.-Col. W. T. Crawford, C.B. d. 1862.
1846	Lieut.-Col. and Brev.-Col. G. R. Barker, C.B.	O.C.A, O.C. Troops	63, 76, 132, 137, 140, 142	Lieut.-Col. and Brev.-Col. Sir G. Barker, K.C.B. d. 1861.
1847	Lieut.-Col. P. P. Faddy	O.C. Troops	175	Lieut.-Col. and Brev.-Col. P. P. Faddy. r. 1862.
1855	Lieut.-Col. C. J. B. Riddell	O.C.A.	76, 91, 100	Major-Gen. C. J. B. Riddell, C.B., F.R.S. r. 1866.
1858	Lieut.-Col. E. Price	O.C.A.	85, 119, 121	Gen. E. Price, C.B., Col. Comdt. d. 1887.
1866	Lieut.-Col. E. Maberly	O.C.A.	73, 74, 76, 78, 83, 90	Major-Gen. E. Maberly, C.B. r. 1872.
1891	Lieut.-Col. J. M. Adye, C.B.	A.A.G.R.A.	56, 58	Gen. Sir J. M. Adye, G.C.B., Col. Comdt. d. 1900.
1911	Capt. and Brev.-Lieut.-Col. C. L. D'Aguilar, C.B.	O.C.A. and O.C.H.A.	65, 76	Gen. Sir C. L. D'Aguilar, G.C.B., Col. Comdt. d. 1912.
1928	Capt. F. J. Travers	O.C.R.A.	47, 51, 58	Lieut.-Col. and Brev.-Col. F. J. Travers. r. 1866

2033	Capt. F. M. M. Ommaney	O.C.A.	79, 81, 84, 86, 101, 106, 107, 125	Commanding 5/14 R.A. See p. 457.
2052	Capt. and Brev.-Lieut.-Col. S. E. Gordon	O.C. Troops		Commanding 4/14 R.A. See p. 457
2058	Capt. N. S. K. Bayly	O.C. Troops	87	Commanding 2/11 R.A. See p. 455.
2095	Capt. D. S. Greene	A.D.C. to G.O.C. R.A.	56, 58	⊙Major-Gen. D. S. Greene, C.B. r. 1884.
2097	Capt. and Brev.-Major W. W. Barry	Adj. R.A.	51, 58, 76	⊙Major-Gen. W. W. Barry, C.B. d. 1883.
2105	Capt. A. M. Calvert	D.A.Q.M.G.R.A. at Headquarters		⊙Col. A. M. Calvert. r. 1881
2119	2nd Capt. and Brev.-Major J. E. Michell	O.C. Troops	100	Major-Gen. J. E. Michell, C.B. d. 1883.
2137	2nd Capt. C.F. Young	Qrmr. R.A.	76, 91, 100	Lieut.-Col. C. F. Young. d. 1875.
2144	2nd Capt. and Brev.-Major N.O.S. Turner	Adj. R.A.	76, 91, 100	⊙Lieut.-Gen. N.O.S. Turner, C.B. r. 1883.
2206	2nd Capt. J. McC. Campbell	Qrmr. R.A.	63, 76, 132, 137, 160	Col. J. McC. Campbell. d. 1879.
2303	Lieut. G. J. Smart	Adj. R.A.	73, 74, 76, 90, 102, 111	Major-Gen. G. J. Smart. r. 1887.
2311	Lieut. T. B. Strange	Qrmr. R.A. and S.O.	73, 74, 76, 78, 83, 90, 102, 111	p. to 5/12 R.A. See p. 455.
2325	Lieut. A. T. G. Pearse	Qrmr. R.A.	85	⊙Major-Gen. A. T. G. Pearse. r. 1882.
2348	Lieut. H. C. S. Dyer	Qrmr. R.A.	47, 51, 58, 63, 76, 99, 103, 106, 109	Major and Brev.-Lieut.-Col.H.C.S. Dyer. r. 1876.
2365	Lieut. R. Biddulph	Adj., B.M., D.A.Q.M.G. and A.A.G.	63, 76, 132, 136, 139A, 144, 149, 150, 155A, 170, 179	Gen Sir R. Biddulph, G.C.B., G.C.M.G., Col. Comdt. Master Gunner, St. James's Park
2366	Lieut. W. Stirling	Adj. and B.M. R.A.	85	Lieut.-Gen. Sir W. Stirling, K.C.B., Col. Comdt. d. 1906.
2404	Lieut. W. J. Hall	Qrmr. R.H.A.	76, 141, 143, 144	Major W. J. Hall. r. 1877.
2416	Lieut. J.C.F. Ramsden	Adj. R.H.A.	64, 76	2nd Capt. J. C. F. Ramsden. r. 1862.
2437	Lieut. W. J. Smith-Neill	Adj. R.A.	57, 76, 111	2nd Capt. J. Smith-Neill. r. 1864.

D Troop, R.H.A. (E Batty. IIIrd Brig. R.H.A.)

Service of Troop, 166, 169

Number in Kane's List.	Name.	Service.	Remarks.
1902	Capt. and Brev.-Major H. L. Gardiner		General Sir H. L. Gardiner, K.C.V.O., C.B. (Civil), Col. Comdt. d. 1897.
2037	Capt. L. G. Paget (from 8/14 R.A.). See p. 458.	169	Lieut.-Col. L. G. Paget. r. 1866.
2367	Lieut. P. E. Hill		Lieut.-Col. and Brev.-Col. P. E. Hill. r. 1884.
2373	Lieut. H. P. Tillard		Lieut. H. P. Tillard. d. 1858.

E Troop, R.H.A. (D Batty. IIIrd Brig. R.H.A.)

Service of Troop, 64, 75, 76, 91, 100, 116, 117, 141, 143, 144, 150, 155a, 170

Number in Kane's List.	Name.	Service.	Remarks.
1945	Capt. and Brev.-Major W. A. Middleton (from 6/13 R.A.). See p. 456	141, 143, 144, 150, 155a, 170	Col. W. A. Middleton, C.B. d. 1875.
1950	Capt. and Brev.-Major J. R. Anderson, C.B.	64, 75, 76	⊙ Major-Gen. J. R. Anderson, C.B. r. 1871.
2119	2nd Capt. and Brev.-Major J. E. Michell	75, 76, 91, 100, 116, 117	Major-Gen. J. E. Michell, C.B. d. 1883.
2225	2nd Capt. S. M. Grylls (from 5/3 R.A.). See p. 453	141, 143, 144, 150, 155a, 170	Major and Brev.-Lieut.-Col. S. M. Grylls. r. 1862.
2300	Lieut. C. E. Torriano	64, 75, 76, 116, 117	Col. C. E. Torriano. r. 1884.
2334	Lieut. F. L. H. Lyon	64, 75, 76, 90, 91, 100, 141, 143, 144	Lieut.-Col. F. L. H. Lyon. r. 1877.
2369	Lieut. S. J. M. Maxwell	64, 75, 76, 91, 100, 141, 143, 144	d. 1860.

F TROOP, R.H.A. (G Battry, Vth Big. R.H.A.)
Service of Troop, 65, 76, 78, 118, 123, 136, 139a, 143, 144, 148, 152, 164, 165, 168

1911	Capt. and Brev.-Lieut.-Col. C. L. D'Aguilar		See Table I, p. 450.
1999	Capt. and Brev.-Major the Hon. D. McD. Fraser	148, 152, 164, 165, 168	⊙ Gen. the Hon. D. McD. Fraser, G.C.B, Col. Comdt. d. 1906.
2131	2nd Capt. and Brev.-Major H. P. Yates	65, 76, 118, 123 136, 139a	⊙ Gen. H. P. Yates, C.B., Col. Comdt. d 1896.
2292	Lieut. E. Markham	65	Lieut.-Gen. Sir E. Markham, K.C.B., Col. Comdt.
2312	Lieut. F. Lyon	76, 123, 136, 139a, 143, 144, 152, 164, 165, 168	Col. F. Lyon. Killed at Shoeburyness, 1885.
2364	Lieut. H. T. Arbuthnot	65, 76, 78, 118, 123, 139a, 143, 144, 152, 164, 165, 168	⊙ Major-Gen. H. T. Arbuthnot, C.B.
2371	Lieut. A. K. Rideout	123, 139a, 143, 144, 152, 164, 165, 168	⊙ Major-Gen. A. K. Rideout, C.B. r. 1884.

8TH COMPANY, 2ND BATT. R.A. (20th Batty. IXth Brig. R.F.A.)
Service of Company, 65, 73, 74, 76, 82, 91, 100

2063	Capt. J. E. Thring	65, 73, 74, 76, 91, 100	⊙ Major-Gen. J. E. Thring. r. 1877.
2266	2nd Capt. W. N. Waller	65, 73, 74, 76, 91, 100	⊙ Major-Gen. W. N. Waller. r. 1881.
2508	Lieut. R. C. Smith	62, 91, 100	Major R. C. Smith. d. 1880.
2516	Lieut. J. Robertson	82, 91, 100	2nd Capt. J. Robertson. d. 1866.
2584	Lieut. S. C. Kyle	65, 73, 74, 76, 91, 100	⊙ Col. S. C. Kyle. r. 1884.

2ND COMPANY, 3RD BATT. R.A., Y Battery (No. 54 Company R.G.A.)
Service of Company, 179a

2651	Lieut. A. J. Raitt	179a	Capt. A. J. Raitt, C.B. r. 1877.
2656	Lieut. H. L. Gwyn	179a	Lieut.-Col. H. L. Gwyn. r. 1886.

5TH COMPANY, 3RD BATT. R.A., J. Battery (13th Batty. 1st Brig. R.F.A.)
Service of Company, 100, 176

2120	Capt. and Bv.-Maj. G. C. Henry	176	Col. G. C. Henry, r. 1885. t. to E. Troop, R.H.A. (See p. 452).
2225	2nd Capt. S M. Grylls		
2307	2nd Capt. R. H. Newbolt	176	Lieut.-Col. R. H. Newbolt. r. 1880.
2419	Lieut. C. R. Franklen	100, 176	Lieut.-Col. C. R. Franklen. r. 1879- d. 1858.
2459	Lieut. T. G. Poulden		
2586	Lieut. H. L. Mitchell	100, 176	Col. H. L. Mitchell. r. 1890.

DETAIL OF R.A. TROOPS AND COMPANIES—continued.

3RD COMPANY, 4TH BATT. R.A. (9th Compy. R.G.A.)
Service of Company, 162

Number in Kane's List.	Name.	Service.	Remarks.
2087	Capt. J. L. Bolton	162	⊙ Major-Gen. J. L. Bolton. r. 1878.
2435	Lieut. H. B. Maule	162	⊙ Major-Gen. H. B. Maule. r. 1886.
2526	Lieut. A. H. Bell	162	Lieut. A. H. Bell. d. 1865.

4TH COMPANY, 5TH BATT. R.A. (No. 35 Compy. R.G.A.)
Service of Company, 47, 51

2086	Capt. W. N. Hardy	47, 51	k. Nov. 16, 1857.
2499	Lieut. A. Ford	47, 51	t. to 3/8 R.A. See below.
2602	Lieut. W. S. Brown	47, 51	t. to 3/8 R.A. See below.

1ST COMPANY, 6TH BATT. R.A., V Battery (25th Batty. XLVIth Brig. R.F.A.)
Service of Company, 124

2080	Capt. and Brev.-Major J. Singleton	124	Major-Gen. J. Singleton. r. 1875.
2249	2nd Capt. A. H. W. Williams	124	Major-Gen. Sir A. H. W. Williams, K.C.V.O., Col. Comdt.
2391	Lieut. M. Tweedie	124	Major-Gen. M. Tweedie. r. 1885.
2422	Lieut. F. A. Anley	124	Lieut.-Col. F. A. Anley. r. 1887.
2522	Lieut. the Hon. R. Hare	124	Major the Hon. R. Hare. r. 1877.

3RD COMPANY, 8TH BATT. R.A. (No. 18 Compy. R.G.A.)
Service of Company, 22, 23, 24, 25, 27, 28, 30, 32, 33, 37, 38, 39, 40, 57, 76, 129, 132, 137, 140

2153	2nd Capt. F. C. Maude	22, 23, 24, 25, 27, 28, 30, 32, 33, 37, 38, 39, 40, 57, 76	Capt. and Brev.-Col. F. C. Maude, C.B., V.C. r. 1866.
2319	Lieut. E. Maitland	22, 23, 24, 25, 27, 28, 30, 32, 33, 37, 38, 39, 40, 57, 76, 129, 132	Col. E. Maitland, C.B., r. 1888.
2499	Lieut. A. Ford (from 4/5 R.A.) See above.	57, 76	t. to 6/14 R.A. See p. 457.
2602	Lieut. W. S. Brown (from 4/5 R.A.) See above.	57, 76, 132, 137, 140	r. 1866.

2ND COMPANY, 11TH BRIG. R.A. (*No. 42 Compy. R.G.A.*)

Service of Company, 87

2058	Capt. N. S. K. Bayly	87	Capt. and Brev.-Major N. S. K. Bayly. d. 1865.
2274	2nd Capt. A. W. Johnson	87	Capt. A. W. Johnson. r. 1866.
2418	Lieut. F. C. Elton	87	Colonel F. C. Elton, C.B. r. 1894.
2434	Lieut. H. Y. Wortham	87	☉ Major-Gen. H. Y. Wortham. r. 1886.
2530	Lieut. F. S. Stoney	87	Major and Brev.-Lieut.-Col. F. S. Stoney. r. 1881.

6TH COMPANY, 11TH BATT. R.A. (*65th Batty. VIIth Brig. R.F.A.*)

Service of Company, 76, 129, 132, 137, 140, 146, 149

2248	2nd Capt. W. H. Goodenough	76, 132, 137, 140	Lieut.-Gen. Sir W. H. Goodenough, K.C.B. d. 1898.
2386	Lieut. E. C. Cuthbert (*c*)	57, 64, 76, 129, 132, 137, 146, 149	2nd Capt. E. C. Cuthbert. d. 1864.
2560	Lieut. H. A. Tracey	76, 129, 132, 137, 140	Lieut.-Col. and Brev.-Col. H. A. Tracey. d. 1886.
2576	Lieut. E. D. Tarleton	76, 129, 132, 137, 140	☉ Col. E. D. Tarleton. r. 1883.

5TH COMPANY, 12TH BATT., Q Battery (*8th Batty. XIIIth Brig. R.F.A.*)

Service of Company, 76, 102, 111, 123, 136, 143, 144, 150, 155a, 170, 177

1949	Capt. J. R. Gibbon	76, 102	☉ Lieut.-Gen. J. R. Gibbon, C.B. r. 1880.
2268	2nd Capt. A. C. Johnson	76, 102, 111, 150, 155a, 170, 177	Major-Gen. A. C. Johnson, C.B.
2311	Lieut. T. B. Strange (from Staff). See p. 451	123, 136, 143, 144, 150	☉ Major-Gen. T. B. Strange. r. 1881.
2408	Lieut. A. D. Burnaby	76, 102, 111, 143, 144, 150	Major A. D. Burnaby. t. to Control Department, 1871.
2412	Lieut. V. D. Majendie	76, 102, 143, 144, 150	Col. Sir V. D. Majendie, K.C.B. (Civil). r. 1881.
2438	Lieut. H. Le G. Geary		Lieut.-Gen. Sir H. Le G. Geary, K.C.B. Col. Comdt.
2607	Lieut. C. E. Bethune	76, 102, 111, 150, 155a, 170, 177	d. 1863.

DETAIL OF R.A. TROOPS AND COMPANIES—continued.

5TH COMPANY, 13TH BATT. R.A. (66th Batty. IVth Brig. R.F.A.)
Service of Company, 51, 58, 70, 75, 76, 90, 91, 100, 102, 123, 143, 144, 148, 164, 165

Number in Kane's List.	Name.	Service.	Remarks.
1944	Capt. C. S. Longden	51, 58	⊙Major-Gen. C. S. Longden. r. 1869.
2184	2nd Capt. H. L. Talbot	51, 70, 58, 76, 90, 102, 143, 144	Lieut.-Col. H. L. Talbot. d. 1876.
2259	2nd Capt. F. C. Griffin	51, 58, 75, 76, 90, 102, 123, 143, 144, 148, 164, 165	Capt. F. C. Griffin. k. 1864 (Bhootan).
2347	Lieut. F. G. E. Warren		Lieut.-Col. F. G. E. Warren, C.B., C.M.G. r. 1889.
2355	Lieut. M. A. Fitzmaurice	51, 58, 76, 91, 100, 143, 144	2nd Capt. M. A. Fitzmaurice. d. 1865.
2490	Lieut. E. S. Burnett	51, 58, 76, 91, 100, 143, 144	Major E. S. Burnett. d. 1880.

6TH COMPANY, 13TH BATT. R.A. (67th Batty. 1st Brig. R.F.A.)
Service of Company, 51, 58, 59, 73, 74, 76, 90, 178, 179

Number in Kane's List.	Name.	Service.	Remarks.
1945	Capt. W. A. Middleton	51, 58, 59, 73, 74, 76, 90	t. to E. Troop, R.H.A. See p. 452.
2091	Capt. F. J. Soady	178	Lieut.-Col. F. J. Soady. d. 1872.
2140	2nd Capt. and Brev.-Major J. F. Pennycuick	51, 58, 59, 73, 74, 76, 90	⊙Gen. J. F. Pennycuick, C.B. r. 1886.
2302	2nd Capt. C. J. Tyler	178	⊙Major-Gen. C. J. Tyler. r. 1881.
2356	Lieut. W. D. Milman	51, 58, 59, 73, 74, 76, 90	2nd Capt. W. D. Milman. d. 1860.
2474	Lieut. W. Smith	51, 58, 59, 73, 74, 76, 90, 178, 179	⊙Major-Gen. W. Smith. r 1887.
2514	Lieut. C. H. Pickering	51, 58, 59, 73, 74, 76, 90, 178	Major C. H. Pickering. d. 1876.

3RD COMPANY, 14TH BATT. R.A. (83rd Batty. XIth Brig. R.F.A.)
Service of Company, 76, 99, 108, 109, 141, 143, 144

Number in Kane's List.	Name.	Service.	Remarks.
2238	2nd Capt. and Brev.-Major W. G. Le Mesurier	76, 99, 108, 109, 141, 143, 144	Capt. and Brev.-Lieut.-Col. W. G. Le Mesurier, C.B. d. 1864.
2467	Lieut. E. Staveley	76, 99, 108, 109	⊙Major-Gen. E. Staveley. r. 1887.
2493	Lieut. J. R. Oliver (d)	56, 76, 99, 108, 109, 141, 143, 144, 150, 156	⊙Major-Gen. J. R. Oliver, C.M.G. r. 1887.

2512	Lieut. C. D. Chalmers	76, 99, 108, 109, 141, 143. | Col. C. D. Chalmers. r. 1887. 144

4TH COMPANY, 14TH BATT. R.A. (68th Batty. XIVth Brig. R.F.A.)
Service of Company, 58, 64, 116, 117, 143, 144, 149

2052	Capt. and Brev.-Lieut.-Col. S.	64, 143, 144, 149 ☉ Gen. S. E. Gordon, C.B. r. 1882.
	E. Gordon	
2203	2nd Capt. C. Johnston	58 ☉ Col. C. Johnston. r. 1881.
2468	Lieut. W. Gilmour	58, 64 2nd Capt. W. Gilmour. r. 1871.
2494	Lieut. E. Egan	2nd Capt. E. Egan. r. 1871.
2578	Lieut. E. R. Cottingham	64, 116, 117 Lieut.-Col. and Brev.-Col. E. R. Cottingham. r. 1889.

5TH COMPANY, 14TH BATT. R.A. (69th Batty. 1st Brig. R.F.A.)
Service of Company, 79, 81, 84, 86, 98, 101, 104, 106, 107, 122, 125

2033	Capt. F. M. M. Ommaney. See	79, 81, 84, 86, 101, 104, ☉ Major-Gen. F. M. M. Ommaney. r. 1875.
	p. 451	106, 107, 125
2204	2nd Capt. T. S. P. Field	98, 101, 104, 106, 107, 122 ☉ Col. T. S. P. Field. r. 1881.
2495	Lieut. J. C. J. Lowry	79, 81, 84, 98, 101, 104, 2nd Capt. J. C. J. Lowry. r. 1871.
		106, 107, 125
2518	Lieut. L. C. A. A. de Cetto	79, 81, 84, 86 ☉ Lieut.-Col. L. C. A. A. de Cetto. r. 1881.
2573	Lieut. R. Moresby	79, 81 k. March 11, 1858.
2505	Lieut. G. Arbuthnot (e)	66, 77, 84, 101 ☉ Col. G. Arbuthnot. r. 1877.

6TH COMPANY, 14TH BATT. R.A. (No. 33 Compy. R.G.A.)
Service of Company, 93a, 94, 120, 126, 163, 167

2034	Capt. E. Palmer	93a, 94, 120, 126, 166 Capt. and Brev.-Lieut.-Col. E. Palmer. r. 1877.
2496	Lieut. J. P. Morgan	93a, 94, 120, 126, 168 ☉ Major-Gen. J. P. Morgan. r. 1887.
2499	Lieut. A. Ford (from 3/8 R.A.	Col. A. Ford, C.B. r. 1883.
	See p. 454	
2544	Lieut. T. H. Redhead	93a, 94 d. May 20, 1858.
2550	Lieut. H. Brackenbury	93a, 94, 126 Gen. the Rt. Hon. Sir H. Brackenbury, G.C.B.,
		K.C.S.I., P.C. Col. Comdt. d. 1914.
2624	Lieut. C. H. S. Pasley	120, 126, 167 r. 1861.

DETAIL OF R.A. TROOPS AND COMPANIES—*continued.*

7TH COMPANY, 14TH BATT. R.A. (84*th Batty. XIth Brig. R.F.A.*)
Service of Company, 58, 63, 88, 105, 140

Number in Kane's List.	Name.	Service.	Remarks.
2083	Capt. and Brev.-Major C. H. Smith	58, 63, 88, 140	⊙Major-Gen. C. H. Smith, C.B. r. 1877.
2174	2nd Capt. and Brev.-Major C. E. Walcott	58, 63, 105	Capt. and Brev.-Lieut.-Col. C. E. Walcott. d. 1868.
2417	Lieut. F. A. Whinyates	58, 63, 88, 140	⊙Colonel F. A. Whinyates. r. 1881.
2471	Lieut. H. Edmeades	58, 63, 105	⊙Major-Gen. H. Edmeades. r. 1886.
2582	Lieut. J. C. Auchinleck	58, 63, 88, 140	Lieut.-Col. and Brev.-Col. J. C. Auchinleck. r. 1890.

8TH COMPANY, 14TH BATT. R.A. (*No.* 64 *Compy. R.G.A.*)
Service of Company, 110

Number in Kane's List.	Name.	Service.	Remarks.
2037	Capt. L. G. Paget	110	t. to D Troop, R.H.A. See p. 452.
2472	Lieut. J. M. Burn	110	⊙Major-Gen. J. M. Burn. r. 1887.

(*b*) III.—OFFICERS OF THE BENGAL ARTILLERY HOLDING COMMANDS AND IN STAFF EMPLOYMENT

Number in Stubbs's List.	Name.	Employment.	Service.	Remarks.
483	Lieut.-Col. and Brev.-Col. A. Wilson	Brigr. and G.O.C. Delhi and Lucknow	2, 8, 9, 10, 11, 12, 13, 76	Lieut.-Gen. Sir A. Wilson, Bt, K.C.B. d. 1874.
496	Lieut.-Col. and Brev.-Col. H. Garbett	Brigr. of A. Delhi	10, 11, 12	w. Sept. 1857; d. Jan. 1858.
511	Lieut.-Col. and Brev.-Col. F. Brind, C.B.	O.C. Troops, Sialkote		k. July 1857.
512	Lieut.-Col. and Brev.-Col. J. L. Mowatt	O.C.A. Umballa	10, 11, 12	d. May 1857.
550	Major F. Gaitskell	Brigr. of A. Delhi	13	Col. F. Gaitskell, C.B. r. 1862.
580	Major G. Larkins	O.C.A. Cawnpore	7	k. June 1857.

		O.C. Foot Artillery; O.C.A.		
588	Major J. Brind		10, 11, 12, 13, 92, 96, 99, 103, 108, 109, 142, 146a	Gen. Sir J. Brind, K.C.B. d. 1888.
594	Capt. and Brev.-Lieut.-Col. R. R. Kinleside	O.C.A.	60, 61, 62	Major-Gen. R. R. Kinleside. r. 1871.
601	Capt. and Brev.-Major J. H. Campbell	O.C.A. No. 2 Siege Batt. Delhi	13	Lieut.-Col. J. H. Campbell. r. 1858.
606	Capt. and Brev.-Major G. L. Cooper	O.C.A.	38, 39	k. Sept. 26, 1857.
620	Capt. and Brev.-Major V. Eyre	O.C. Troops	29, 31, 36	Commanding 1/5 Ben.A. See p. 470.
624	Capt. and Brev.-Lieut.-Col. M. Mackenzie	O.C.H.A.	8, 9	w. July 1857. d. Oct. 1857.
639	Capt. and Brev.-Major F. Turner (from 3/3 Ben.H.A. See p. 466	O.C. Ben.A.; O.C.A.	42, 45, 51, 58, 63, 75, 76	Gen. Sir F. Turner, K.C.B., Col. Comdt. d. 1890.
653	Capt. J. H. Smyth (f).	B.M.; O.C. Troops	43, 58, 95 Delhi, June 17	Commanding 5/1 Ben.H.A. See p.468.
706	Capt. and Brev.-Lieut.-Col. H. Tombs	O.C. Column		Commanding 2/1 Ben.H.A.
732	Capt. H. Le G. Bruce	A.A.G. of A.	51, 58	p. to 2/3 Ben.A. See p. 464.
736	Capt. and Brev.-Major E. B. Johnson	A.A.G. of A.	2, 8, 9, 10, 11, 12, 13, 76	Gen. Sir E. B. Johnson, G.C.B., C.I.E. d. 1893.
758	1st Lieut. J. A. Angelo	Adjt. Ben.A.	10	Major-Gen. J. A. Angelo. r. 1882.
762	1st Lieut. A. Bunny	Adjt. and B.M. Ben. A.	8, 9, 10, 11, 12, 13, 42, 45, 51, 58, 63, 76, 92, 99, 109, 109a	Lieut.-Gen. A. Bunny, C.B. r. 1883.
764	1st Lieut. W. J. Gray	O.C. Ferozepore Siege Train	12, 13	Major-Gen. W. J. Gray. r. 1882.
780	1st Lieut. C. Dempster	Adjt. Ben.A.	7	k. June 1857.
789	1st Lieut. H. P. Bishop	Adjt. Ben.A.	9, 10, 11, 12, 13	t. to 3/3 Ben.H.A. See p. 466.
800	1st Lieut. J. S. Frith	B.M. of A.	2, 9, 10, 11, 12, 13, 76	Lieut.-Col. J. S. Frith. d. 1872.
812	1st Lieut. A. O. Mayne	A.Q.M.G.	51	k. Nov. 1857.
833	1st Lieut. M. C. Sankey	Adjt. Ben.A.	1, 9, 10, 11, 12, 13, 35	Capt. and Brev.-Major M. C. Sankey. d. 1868.
887	1st Lieut. J. A. H. Eckford	Qrmr. Ben.A.	7	k. June 1857.
895	2nd Lieut. F. S. Roberts	Artillery and General Staff	1, 9, 10, 11, 12, 13, 42, 45, 51, 58, 63, 75, 76, 83	Field Marshal Earl Roberts, V.C., K.G., K.P., G.C.B., G.C.S.I., Col. Comdt. d. 1914.

IV.—OFFICERS OF BENGAL ARTILLERY UNPOSTED OR ABSENT FROM THEIR UNITS AT OUTBREAK OF MUTINY

Number in Stubbs's List.	Name.	Battalion, Troop, or Company.	Service.	Remarks.
502	Lieut.-Col. and Brev.-Col. R. Horsford	3rd Batt. Ben.A		Major-Gen. R. Horsford. d. 1869.
740	1st Lieut. T. E. Kennion	5th Batt. Ben.A.	9, 10, 11	Lieut.-Col. T. E. Kennion. r. 1867.
761	1st Lieut. J. E. Watson	4th Batt. Ben.A.		Lieut.-Col. J. E. Watson. r. 1868.
806	1st Lieut. T. H. Salt	3rd Batt. Ben.A.	10, 11, 12, 13	Captain T. H. Salt. r. 1869.
834	1st Lieut. J. Y. Worthington	2/2 Ben.H.A.		2nd Capt. J. Y. Worthington. r. 1864.
858	1st Lieut. B. Ashburner	2nd Batt. Ben.A.	7	k. June 1857.
907	1st Lieut. E. Salwey	4/3 Ben.H.A.	9, 10, 11, 12	d. Oct. 1857.
929	2nd Lieut. C. G. Robinson	4/7 Ben.A.	2, 9, 10, 11, 12, 13	p. to 3/1 Ben.H.A. See p. 465.
934	2nd Lieut. W. T. Somerville	3/8 Ben.A.	9, 10, 11, 12	d. Sept. 1857.
939	2nd Lieut. J. H. Shuldham	6/8 Ben.A.	9, 10, 11, 12	1st Lieut. J. H. Shuldham. d. Sept. 1857.
951	2nd Lieut. Sir R. de L. St. George, Bt.	Unposted	9, 10, 11, 12, 13	1st Lieut. Sir R. de L. St. George, Bt. d. 1861.
952	2nd Lieut. H. Chichester	Unposted	2, 9, 10, 11, 12, 13	p. to 3/3 Ben.A. See p. 469.
953	2nd Lieut. D. S. Pemberton	Unposted	9, 10, 11, 12, 13	p. to 3/1 Ben.A. See p. 467.
954	2nd Lieut. W. R. Craster	Unposted	2, 8, 9, 10, 11, 12, 13	Lieut.-Col. and Brev.-Col. W. R. Craster. r. 1890.
955	2nd Lieut. E. T. Hume	Unposted	9	p. to 2/4 Ben.A. See p. 470.
956	2nd Lieut. J. N. Martin	Unposted	7	k. June 1857.
957	2nd Lieut. T. Ryan	Unposted	9, 10, 11, 12, 13	p. to Sikh Company, Ben.A. See p. 473.
958	2nd Lieut. G. M. W. Sotheby	Unposted	7	k. June 1857.
960	2nd Lieut. F. W. Burney	Unposted	7	k. June 1857.
961	2nd Lieut. W. O'Brien	Unposted	9, 10, 11, 12, 13	Capt. W. O'Brien. d. 1867.
963	2nd Lieut. H. Girardot	Unposted	9	p. to 2/4 Ben.A. See p. 470.
967	2nd Lieut. E. H. Dyke	Unposted	57	Lieut.-Col. and Brev.-Col. E. H. Dyke. r. 1890.

Number in Stubbs's List	Name	Troop, Company, etc.	Service	Remarks
623	Capt. and Brev.-Lieut.-Col. M. Mackenzie	4/1 Ben.H.A.		See p. 459.
645	Capt. G. Kirby	6/8 Ben.A.	Outbreak at Bareilly	Lieut.-Col. G. Kirby. r. 1860.
649	Capt. and Brev.-Lieut.-Col. J. Abercrombie (g)	4/3 Ben.H.A.		Lieut.-Col. J. Abercrombie. r. 1860.
657	Capt. A. W. Hawkins	Gwalior Contingent	Outbreak at Gwalior	k. June 1857.
670	Capt. and Brev.-Major J. Mill	5/7 Ben.A.	Outbreak at Fyzabad	Drowned in Gogra, June 1857. [d. 1895.
673	Capt. H. P. de Teissier	3/7 Ben.A.	Outbreak at Delhi	Gen. H. P. de Teissier, Col. Comdt.
704	Capt. A. P. Simons	2/8 Ben.A.	16, 19 [abad.	w. June 30, d. Sept. 1857.
710	Capt. C. W. Timbrell	2/7 Ben.A.	Outbreak at Nusseer-	r. 1860.
753	1st Lieut. A. Pearson	Gwalior Contingent	Outbreak at Gwalior	p. to 2/5 Ben.A. See p. 471.
767	1st Lieut. W. Stewart	Gwalior Contingent	Outbreak at Gwalior	k. June 1857.
778	1st Lieut. E. W. E. Walker	4/1 Ben.H.A.	Outbreak at Neemuch	p. to 1/5 Ben.A. See p. 470.
788	1st Lieut. W. D. Couchman	4/1 Ben.H.A.	Outbreak at Neemuch	p. to 1/1 Ben.H.A. See p. 464.
792	1st Lieut. F. R. de Bude	4/3 Ben.H.A.	1, 13a.	Major F. R. de Bude. r. 1872.
794	1st Lieut. C. S. Le Marchand	Gwalior Contingent	Outbreak at Gwalior	p. to 2/6 Ben.A. See p. 472.
839	1st Lieut. D. C. Alexander	Oudh Irr. A.	16, 19	k. Sept. 1857.
849	1st Lieut. W. Wilson	3/7 Ben.A.	Outbreak at Delhi	p. to 2/1 Ben.H.A. See p. 464.
855	1st Lieut. F. H. Turnbull	4/9 Ben.A.	Outbreak at Jhansi, 4	k. June 1857.
869	1st Lieut. St. G. Ashe	Oudh Irr. A.	7	k. June 1857.
884	1st Lieut. J. Percival (h)	5/7 Ben.A.	74, 76	p. to 2/3 Ben.A. See p. 469.
896	2nd Lieut. H. O. Hitchens	2/7 Ben.A.	45, 57, 60, 61, 76	p. to 4/6 Ben.A. See p. 472.
903	2nd Lieut. J. H. Bryce	Oudh Irr. A.	16, 19	d. Aug. 1857.
907	2nd Lieut. E. Salwey	4/3 Ben.H.A.	9, 10, 11, 12, 13	d. Oct. 1857.
914	2nd Lieut. C. H. Barnes	4/1 Ben.H.A.	44	p. to 1/1 Ben.H.A. See p. 464.
915	2nd Lieut. J. Bonham	Oudh Irr. A.	16, 19	Capt. and Brev.-Lieut.-Col. J. Bonham, C.B. r. 1874. [1889.
919	2nd Lieut. D. MacFarlan	Oudh Irr. A.	16, 19	Lieut.-Gen. D. MacFarlan, C.B. r.
925	2nd Lieut. R. Currie	5/7 Ben.A.	Outbreak at Fyzabad	Drowned in Gogra, June 1857.
928	2nd Lieut. R. Aislabie	3/7 Ben.A.	2, 8, 9, 10, 11, 12, 13	Lieut.-Col. and Brev.-Col. R. Aislabie. r. 1885.
936	2nd Lieut. S. E. Townsend	4/9 Ben.A.	Outbreak at Nowgong	k. June 1857.
937	2nd Lieut. H. L. Jones	2/7 Ben.A.	Outbreak at Nusseerabad.	p. to 2/5 Ben.A. See p. 471.

VI.—OFFICERS OF BENGAL ARTILLERY IN THE ORDNANCE DEPARTMENT AND VARIOUS MILITARY APPOINTMENTS

Number in Stubbs's List.	Name	Employment.	Service.	Remarks.
547	Lieut.-Col. F. R. Bazeley	C. of O.	39	k. Sept. 1857.
578	Major G. F. C. Fitzgerald	Gurkha F.F.	East Oudh	Lieut.-Col. G. F. C. Fitzgerald. r. 1858.
584	Major G. H. MacGregor	Gurkha F.F.	76, East Oudh	Lieut.-Col. and Brev.-Col. Sir G. H. MacGregor, K.C.B. r. 1858.
638	Major and Brev.-Lieut.-Col. C. Hogge	D. of O.	9, 10, 11, 12, 13, 76	Col. C. Hogge, C.B. d. 1865.
662	Capt. C. Douglas	O.C.A. Hyderabad F.F.	84, 101, 104, 106, 107	Gen. C. Douglas. d. 1885.
680	Capt. and Brev.-Maj. A. Robertson	Gun Carriage Agency	15	w. July. d. Sept. 1857, a fugitive in Oudh.
724	1st Lieut. G. Holland	Gurkha F.F.	76, East Oudh	t. to Sikh Companies Ben.A. See p. 473.
735	1st Lieut. D. McNeill	Gurkha F.F.	76, East Oudh	Lieut.-Col. and Brev.-Col. D. McNeill. r. 1872.
737	1st Lieut. H. H. Maxwell	Attached to H.M.S. *Shannon*, Naval Brigade	51, 58, 63	Gen. H. H. Maxwell, C.B. d. 1892.
746	1st Lieut. W. C. Russell	C. of O.	5	Col. W. C. Russell. d. 1878.
777	1st Lieut. J. C. Griffiths	C. of O.	9, 10, 11, 12, 13	Capt. and Brev.-Major J. C. Griffiths. d. 1871.
779	1st Lieut. W. A. Mylne	Gurkha F.F.	E. Oudh	2nd Capt. W. A. Mylne. d. 1859.
796	1st Lieut. L. Machell	C. of O.	Agra and Doab	Captain L. Machell. d. 1864.
828	1st Lieut. W. T. Brown	C. of O.	51, 58, 63, 76, 99, 103	Lieut.-Col. and Brev.-Col. W. T. Brown, C.B. r. 1878.
832	1st Lieut. W. Dowell	Post guns	Outbreak at Dacca	p. to 2/3 Ben.A. See p. 469.
841	1st Lieut. G. D. Willoughby	C. of O.	Magazine at Delhi	k. May 1857.
844	1st Lieut. J. S. Gibb	Gurkha F.F.	76, 100	Major J. S. Gibb. r. 1872.
853	1st Lieut. T. N. Harward	Attached to 3/8 R.A.	5, 22, 23, 24, 25, 33, 37, 39, 58	Lieut.-Gen. T. N. Harward. r. 1885.
854	1st Lieut. A. H. Bogle	Gurkha F.F.	76	Major-Gen. A. H. Bogle. r. 1882.
868	1st Lieut. T. Nicholl	C. of O.	18	Lieut.-Gen. T. Nicholl, 1890.
888	1st Lieut. G. R. Manderson	Kashmir Contingent	13	p. to 1/1 Ben.H.A. See p. 464.
906	1st Lieut. J. R. Pearson	C. of O.	99, 103, 109, 109a	2nd Capt. J. R. Pearson. r 1866.

VII.—OFFICERS OF THE BENGAL ARTILLERY IN CIVIL EMPLOYMENT

Number in Stubbs's List.	Name.	Employment.	Service.	Remarks.
442	Lieut.-Col. and Brev.-Col. C. G. Dixon	Commissioner of Ajmere		d. June 1857.
533	Lieut.-Col. and Brev.-Col. Sir H. M. Lawrence, K.C.B.	Chief Commissioner of Oudh	16, 19	Major-Gen. Sir H. M. Lawrence, K.C.B. w. July 4. d. July 6, 1857.
613	Capt. and Brev.-Major R. C. Shakespear	Political		Lieut.-Col. and Brev.-Col. Sir R. C. Shakespear, C.B. d. 1861.
655	Capt. W. Maxwell	D.P.W.		⊙ Major-Gen. W. Maxwell. r. 1872.
678	Capt. C. H. Dickens	D.P.W.		Lieut.-Gen. C. H. Dickens, C.S.I. r. 1878.
686	Capt. J. Young	D.P.W.	8, 9, 10, 11, 12, 13	Col. J. Young. d. 1872.
695	Capt. R. C. H. B. Fagan	D.P.W.	9, 10, 11, 12, 13	k. Sept. 1857.
755	1st Lieut. H. M. Boddam	D.P.W.		Capt. H. M. Boddam. t. to Ben.S.C. 1865.
786	1st Lieut. A. H. Heath	D.P.W.	9, 10, 11, 12, 13	⊙ Major-Gen. A. H. Heath. r. 1878.
803	1st Lieut. G. Baillie	Political		2nd Capt. Baillie. r. 1862.
804	1st Lieut. J. Fulton	D.P.W.	9, 10, 11, 12, 13	⊙ Lieut.-Gen. J. Fulton. r. 1883.
811	1st Lieut. G. O. Rybot	D.P.W.	9, 10, 11, 12, 13	⊙ Lieut.-Col. G. O. Rybot. r. 1872.
831	1st Lieut. D. Thomson	D.P.W.	9, 10, 11, 12, 13	Major D. Thomson. r. 1869.
837	1st Lieut. J. H. Lamb	Political		p. to 2/5 Ben.A. See p. 471.
878	1st Lieut. E. L. Earle	D.P.W.	9, 10, 11, 12, 13	Capt. E. L. Earle was t. to Ben.S.C. 1868.
879	1st Lieut. E. H. Willoughby	Survey	18	w. June 1857. d. when 2nd Captain, 1860.
883	1st Lieut. T. E. Dickins	D.P.W.		k. July 1857.
905	1st Lieut. R. A. Smith	D.P.W.	9, 10	d. Sept. 1857.

VIII.—DETAIL OF TROOPS AND COMPANIES

1st TROOP, 1st BRIG. BEN.H.A. (*F. Batty. IVth Brig. R.H.A.*)

Service of Troop, 1, 6, 10, 11, 12, 13, 35, 42, 45, 51, 57, 58, 59, 67, 76, 83, 130, 146, 149, 153

Number in Stubbs's List.	Name.	Service.	Remarks.
689	Capt. and Brev.-Major H. A. Olpherts	6, 10, 11	Capt. and Brev.-Lieut.-Col. H. A. Olpherts. r. 1860.
722	Capt. F. F. Remmington	6, 10, 11, 12, 13, 35, 42, 45, 51, 57, 58, 59, 67, 76, 83	t. to 3/3 Ben.H.A. (q.v.).
732	Capt. and Brev.-Major H. Le G. Bruce (from 2/3 Ben.A.). See p. 468	130, 146, 149, 153	Major-Gen. H. Le G. Bruce, C.B. r. 1887.
788	1st Lieut. W. D. Couchman (from 4/1 Ben.H.A.). See p. 461	130, 146, 149, 153	Major W. D. Couchman. r. 1872.
863	1st Lieut. H. Murray	8, 9, 10, 11, 12, 13, 35, 42, 45, 51, 58, 59, 63, 67	Major H. Murray. r. 1872.
873	1st Lieut. A. H. Lindsay	9, 10, 11, 12, 13, 35, 67, 76	Lieut.-Col. A. H. Lindsay, C.B. r. 1880.
881	1st Lieut. G. M. Dobbin	1	Major G. M. Dobbin. r. 1874.
888	1st Lieut. G. R. Manderson (from Kashmir Contingent). See p. 462	45, 51, 76	Lieut.-Col. and Brev.-Col. G. R. Manderson, C.B. r. 1885.
912	2nd Lieut. G. B. Traill	1, 9, 10, 11, 12, 13, 45, 51, 58, 59, 76	Lieut.-Col. and Brev.-Col. G. B. Traill. r. 1887.
914	2nd Lieut. C. H. Barnes (from 4/1 Ben.H.A.). See p. 461	130, 146, 149, 153	Lieut.-Col. and Brev.-Col. C. H. Barnes. r. 1884.
965	2nd Lieut. A. Dixon (from 3/1 Ben.A.). See p. 467	57, 130	Major A. Dixon. d. 1882.

2nd TROOP, 1st BRIG. BEN.H.A. (*56th Batty. L Brig. R.F.A.*)

Service of Troop, 2, 8, 9, 10, 11, 12, 13, 35, 76, 92, 96, 99, 109, 109a

Number in Stubbs's List.	Name.	Service.	Remarks.
706	Capt. and Brev.-Major H. Tombs	2, 8, 9, 10, 11, 12, 13, 35, 76, 92, 96, 99, 109, 109a	Major-Gen. Sir H. Tombs, K.C.B., V.C. d. 1874.
842	1st Lieut. H. G. Perkins	2	k. May 1857.
849	1st Lieut. W. Wilson (from 3/7 Ben.A.). See p. 461	2, 8, 9, 10, 11, 12, 13, 35, 76, 92, 96, 99, 109, 109a	Lieut.-Col. W. Wilson. r. 1878.
876	1st Lieut. T. P. Smith	76, 92, 96, 99, 108, 109, 109a	Major-Gen. T. P. Smith. r. 1889.
917	2nd Lieut. J. Hills	2, 8, 9, 10, 11, 12, 13, 37, 76, 92, 96, 99, 109, 109a	Lieut.-Gen. Sir J. Hills-Johnes, V.C., G.C.B. r. 1888.

3RD TROOP, 1ST BRIG. BEN.H.A. (58th Batty. XLVIth Brig. R.F.A.)
Service of Troop, 1, 14, 52, 16oa

633	Capt. and Brev.-Lieut.-Col. M. Dawes	1, 14	Capt. and Brev.-Lieut.-Col. M. Dawes, C.S.I. r. 1858.
682	Capt. and Brev.-Major G. Moir (from 3/5 Ben.A.). See p. 471	16oa	Lieut.-Col. and Brev.-Col. G. Moir, C.B. r. 1870.
744	1st Lieut. C. Cookworthy	1, 14, 52	t. to 4/6 Ben.A. See p. 472.
814	1st Lieut. A. Fraser	52, 16oa	Capt. A. Fraser. d. 1870.
929	2nd Lieut. C. G. Robinson (from 4/7 Ben.A.). See p. 460	16oa	⊙ Major-Gen. C. G. Robinson. r. 1885.
943	2nd Lieut. E. Fraser	76, 16oa	Capt. E. Fraser. r. 1871.
979	2nd Lieut. F. W. Ward (from 3/1 Ben.A.). See p. 467	16oa	Lieut.-Col. and Brev.-Col. 1886.

5TH TROOP, 1ST BRIG. BEN.H.A. (T Batty. XIth Brig. R.H.A.)
Service of Troop, 6, 10, 11, 12, 13, 60, 61, 95, 172

653	Capt. J. H. Smyth. See p. 459	172	Col. J. H. Smyth, C.B. r. 1872.
774	1st Lieut. G. A. Renny	6, 10, 11, 12, 13, 60, 61, 95, 172	⊙ Maj.-Gen. G. A. Renny, V.C. r. 1878.
860	1st Lieut. H. J. Evans	6, 10, 11, 12, 13, 60, 61, 95	2nd Capt. H. J. Evans. d. 1860.
870	1st Lieut. C. W. Maynard	6, 60, 61, 95	Capt. C. W. Maynard. r. 1867.
938	1st Lieut. E. C. Griffin (from 2/5 Ben.A.). See p. 471	60, 61	Lieut.-Gen. E. C. Griffin, Col. Comdt.

1ST TROOP, 2ND BRIG. BEN.H.A. (K Batty. VIth Brig. R.H.A.)
Service of Troop, 1

769	1st Lieut. G. R. Brown	1	⊙ Major-Gen. G. R. Brown. r. 1878.

2ND TROOP, 2ND BRIG. BEN.H.A. (52nd Batty. XVth Brig. R.F.A.)
Service of Troop, 150, 155a, 158, 170

927	Lieut. R. R. Franks	93, 95, 98a, 99, 155a, 156, 158, 170	Lieut.-Col. and Brev.-Col. R. R. Franks. r. 1884.

1ST TROOP, 3RD BRIG. BEN.H.A. (L Batty. VIIth Brig. R.H.A.)
Service of Troop, 21

783	1st Lieut. C. H. Cookes	21	⊙ Major-Gen. C. H. Cookes. r. 1878.
892	1st Lieut. C. E. Lewes	21	d. Nov. 1857.

DETAIL OF TROOPS AND COMPANIES—continued.

2ND TROOP, 3RD BRIG. BEN.H.A. (57th Batty. XLIVth Brig. R.F.A.)
Service of Troop, 8, 9, 10, 11, 12, 13, 35, 42, 45, 47, 51, 58, 76, 83, 90, 102, 111

Number in Stubbs's List.	Name.	Service.	Remarks.
654	Capt. E. K. Money	8, 9, 10, 11	☉ Major-General E. K. Money. r. 1872.
715	Capt. W. A. Mackinnon	76, 83, 90, 102, 111	Lieut.-Col. W. A. Mackinnon, C.B. d. 1867.
742	1st Lieut. C. H. Blunt	8, 9, 10, 11, 12, 13, 35, 42, 45, 47, 51, 58	t. to 3/1 Ben.A. See p. 467.
856	1st Lieut. F. H. McLeod	8, 9, 10, 11, 12, 13, 37, 42, 45, 51, 58, 76, 90, 102, 111	☉ Major-General F. H. McLeod. r. 1880.
899	1st Lieut. G. Cracklow (from 3/1 Ben.A.). See p. 467	76, 90, 102, 111	2nd Capt. G. Cracklow. r. 1864.
909	2nd Lieut. C. Hunter	8, 9, 10, 11, 12, 13, 37, 42, 45, 51, 58, 76, 90, 102, 111	Lieut.-Col. and Brev.-Col. C. Hunter. r. 1888.

3RD TROOP, 3RD BRIG. BEN.H.A. (S Batty. IXth Brig. R.H.A.)
Service of Troop, 8, 9, 10, 11, 12, 13, 60, 61, 75, 76, 92, 96, 99, 109, 142, 146a

Number in Stubbs's List.	Name.	Service.	Remarks.
639	Capt. and Brev.-Major F. Turner	8, 9, 10, 11, 12, 13	To Staff. See p. 459.
722	Capt. and Brev.-Lieut.-Col. F. F. Remmington (from 1/1 Ben.H.A.). See p. 464	92, 96, 99, 142	Captain and Brev.-Lieut.-Col. F. F. Remmington, C.B. d. 1861.
789	1st Lieut. H. P. Bishop	60, 61, 62, 75, 76, 92, 96, 146a	☉ Major-General H. P. Bishop. r. 1878.
829	2nd Capt. C. McW. Mercer	142, 146a	☉ Major-General C. McW. Mercer. r. 1878.
852	1st Lieut. T. A. Dirom	8, 9, 10, 11, 12, 13, 75, 99	Lieut.-Col. T. A. Pasley-Dirom. r. 1872.
904	1st Lieut. H. E. Harington (from 3/1 Ben.A.). See p. 467	92, 99, 141	2nd Capt. H. E. Harington, V.C. d. 1861.
930	1st Lieut. C. S. S. Taylor	92, 96, 99, 109, 146a	Lieut.-Col. and Brev.-Col. C. S. S. Taylor. d. 1877.
933	2nd Lieut. P. Thompson	8, 9, 10, 11, 12, 13, 76, 99	Capt. P. Thompson. r. 1868.
955	2nd Lieut. E. T. Hume (from 2/4 Ben.A.). See p. 470	92, 99, 109, 142, 146a	Lieut.-Col. and Brev.-Col. E. T. Hume. r. 1890.
971	2nd Lieut. H. de G. Warter (from 1/5 Ben.A.). See p. 470	146a	Lieut.-Col. H. de G. Warter. d. 1889.

NO. 1 COMPANY, 1ST BATT. BEN.A. NO. 7 F.B. (21st Batty. IInd Brig. R.F.A.)
Service of Company, 93, 95, 98a, 99, 103, 108, 109, 133, 142

707	Capt. A. G. Austen (i)	56, 58, 93, 95, 98a, 99, 103, 108, 109	⊙ Major-Gen. A. G. Austen. r. 1868.
920	2nd Lieut. J. C. G. Price	93, 95, 98a, 99, 103, 108, 109, 133	⊙ Lieut.-Col. J. C. G. Price. r. 1872.
940	2nd Lieut. A. J. Wake	142	Lieut.-Col. and Brev.-Col. A. J. Wake. r. 1885.
963	2nd Lieut. H. Girardot (from 2/4 Ben.A.). See p. 470	93, 95, 98a, 99, 103, 108, 109, 133	Capt. H. Girardot. d. 1871.
964	2nd Lieut. W. J. Stewart	57, 93, 95, 98a, 99, 103, 108, 109, 133	1st Lieut. W. J. Stewart was t. to Ben.S.C. 1865.
1005	2nd Lieut. G. G. Nelson	93, 95, 98a, 99, 103, 108, 109, 133	Capt. G. G. Nelson. d. 1878.

NO. 3 COMPANY, 1ST BATT. BEN.A. NO. 17 F.B. (53rd Batty. IInd Brig. R.F.A.)
Service of Company, 1, 12, 13, 14, 42, 45, 47, 51, 58, 67, 106, 107, 132, 137, 140

681	Capt. G. Bourchier	1, 12, 13, 14, 42, 45, 51, 58, 67, 106, 107	⊙ Major-Gen. Sir G. Bourchier, K.C.B. r. 1872.
742	Capt. C. H. Blunt (from 2/3 Ben.H.A.). See p. 466		⊙ Major-Gen. C. H. Blunt, C.B. r. 1867.
753	Capt. A. Pearson (from 2/5 Ben.A.). See p. 471	132, 137, 140	Major A. Pearson. r. 1861.
899	2nd Lieut. G. Cracklow	1, 12, 13, 14, 42, 45, 51, 58, 67	To 2/3 Ben.H.A. See p. 466.
902	1st Lieut. E. L. Hawkins (from 1/5 Ben.A.). See p. 470	42, 45, 58, 106, 107	⊙ Major-Gen. E. L. Hawkins. r. 1886.
904	2nd Lieut. H. E. Harington	1, 12, 13, 14, 45, 47, 51, 58	p. to 3/3 Ben.H.A. See p. 466.
923	2nd Lieut. J. Sconce	1, 12, 13, 14, 42, 45, 47, 51, 58	Capt. J. Sconce was t. to Ben.S.C. 1870.
947	2nd Lieut. R. T. Hare (from 4/6 Ben.A.). See p. 472	42, 45, 47, 51, 58, 67, 106, 107	Lieut. R. T. Hare was t. to Ben.S.C. 1865.
950	1st Lieut. W. Gully (from 3/5 Ben.A.). See p. 471	107	⊙ Lieut.-Col. W. Gully. r. 1875.
953	2nd Lieut. D. S. Pemberton (from unposted list). See p. 460	42, 45, 47, 106, 107	Lieut.-Col. and Brev.-Col. D. S. Pemberton. d. 1889.
965	2nd Lieut. A. Dixon	42, 45	t. to 1/1 Ben.H.A. See p. 464.
979	2nd Lieut. F. W. Ward	45, 47	t. to 3/1 Ben.H.A. See p. 465.
1006	2nd Lieut. J. V. Wylie	137	Lieut. J. V. Wylie. d. 1860.

DETAIL OF TROOPS AND COMPANIES—continued.

No. 4 Company, 1st Batt. Ben.A. No. 9 F.B. (22nd Batty. XXXIVth Brig. R.F.A.)

Service of Company, 16, 19, 57, 76, 92, 96, 99

Number in Stubbs's List.	Name.	Service.	Remarks.
730	Capt. H. Francis	57, 76, 92, 96, 99	☉ Major-Gen. H. Francis. r. 1878. k. July 1857.
900	2nd Lieut. E. P. Lewin	16, 19	
916	2nd Lieut. J. Alexander	16, 19, 57, 76	2nd Capt. J. Alexander. d. 1867.
921	2nd Lieut. F. J. Cunliffe	16, 19	d. Sept. 1857.
964	2nd Lieut. W. J. Stewart	57, 76, 92, 96, 99	t. to 1/1 Ben.A. See p. 467.
968	2nd Lieut. J. R. Macleay	76, 99	Capt. J. R. Macleay. r. 1868.
987	2nd Lieut. J. A. Low	57, 76, 92, 96, 99	Capt. J. A. Low. Half-pay 1867.

No. 1 Company, 3rd Batt. Ben.A. No. 4 F.B. (No. 28 Compy. R.G.A.)

Service of Company, 18

717	Capt. W. B. Marshall	18	☉ Major-Gen. R. Marshall. r. 1872.
775	1st Lieut. H. V. Timbrell (?)	57	t. to 2/3 Ben. A. See below.
941	2nd Lieut. F. V. Eyre	18	Lieut.-Col. F. V. Eyre. r. 1880.

No. 2 Company, 3rd Batt. Ben.A. No. 12 F.B. (23rd Batty. XLth Brig. R.F.A.)

Service of Company, 3, 30, 32, 33, 37, 38, 39, 40, 51, 57, 76, 111, 128, 131, 132, 137, 140

640	Capt. and Brev.-Major H .A. Carleton	76, 111	Gen. H. A. Carleton, C.B., Col. Comdt. d. 1900.
687	Capt. W. Olpherts	3, 30, 32, 33, 37, 38, 39, 40, 51, 57, 76	Gen. Sir W. Olpherts, V.C., G.C.B., Col. Comdt. d. 1902.
732	Capt. H. Le G. Bruce (from the Staff). See p. 459.	128	t. to 1/1 Ben.H.A. See p. 464.
775	1st Lieut. H. V. Timbrell (from 1/3 Ben.A.). See above	76	☉ Major-Gen. H. V. Timbrell. r. 1878.

832	1st Lieut. W. Dowell (from Dacca). See p. 462	76, 99, 132, 137, 140	☉ Major-Gen. W. Dowell. r. 1878.
884	1st Lieut. J. Percival (from 5/7 Ben.A.). See p. 461	111, 128, 132, 137, 140	2nd Capt. J. Percival. d. 1866.
945	2nd Lieut. H. Smithett	30, 32, 37, 38, 39, 40, 51, 57, 76	☉ Major-Gen. H. Smithett. r. 1885.
966	2nd Lieut. F. Coddington	57	Lieut. F. Coddington was t. to Ben.S.C. 1865.
970	2nd Lieut. A. Swinton	111, 128, 131, 132, 137, 140	☉ Col. A. Swinton. r. 1891.
992	2nd Lieut. A. S. Heyland	111, 128, 132, 137, 140	Capt. A. S. Heyland. r. 1879.

No. 3 COMPANY, 3RD BATT. BEN.A. No. 14 F.B. (*No. 97 Compy. R.G.A.*)

Service of Company, 2, 8, 9, 10, 11, 12, 13, 99, 103, 108, 109

617	Capt. and Brev.-Major E. W. S. Scott	2, 8, 9, 10, 11, 12, 13	☉ Major-Gen. E. W. S. Scott. r. 1863.
679	Capt. H. Hammond	51, 63, 99, 103, 108, 109	☉ Major-Gen. H. Hammond. r. 1872.
886	1st Lieut. M. M. Fitzgerald	8, 9, 10, 11, 12, 13	☉ Major-Gen. M. M. Fitzgerald. r. 1884.
918	2nd Lieut. M. Elliot	2, 8, 9, 10, 11, 12, 13	☉ Major-Gen. M. Elliot. r. 1887. t. to 3/3 Ben.H.A. See p. 466.
930	2nd Lieut. C. S. S. Taylor	99, 103, 108, 109	
942	2nd Lieut. A. H. Davidson	2, 8, 99	☉ Major-Gen. A. H. Davidson. r. 1886.
952	2nd Lieut. H. Chichester (from unposted list). See p. 460	99, 103, 108, 109	☉ Major-Gen. H. Chichester. r. 1886.

No. 4 COMPANY, 3RD BATT. BEN.A. (36*th Batty. XXXIIIrd Brig. R.F.A.*)

Service of Company, 13a

850	2nd Capt. F. E. Smalpage	13a	2nd Capt. F. E. Smalpage. d. 1863.

No. 1 COMPANY, 4TH BATT. BEN.A. (*2nd Mountain Batty. R.G.A.*)

Service of Company, 10, 11, 12, 13, 93, 95, 99, 103, 108, 109, 142

950	2nd Lieut. W. Gully	10, 11, 12, 13	t. to 3/5 Ben.A. See p. 471.

DETAIL OF TROOPS AND COMPANIES—*continued.*

No. 2 Company, 4th Batt. Ben.A. (*No. 4 Mountain Batty. R.G.A.*)
Service of Company, 10, 11, 12, 13, 99, 108, 109

Number in Stubbs's List.	Name.	Service.	Remarks.
889	1st Lieut. H. M. Cadell	99, 108, 109	2nd Capt. H. M. Cadell. r. 1866.
910	2nd Lieut. H. D. Jackson	10, 11, 12, 13	Major H. D. Jackson. r. 1876.
963	2nd Lieut. H. Girardot (from unposted list). See p. 460	10, 11, 12, 13	t. to 1/1 Ben.A. See p. 467.
955	2nd Lieut. E. T. Hume (from unposted list). See p. 460	10, 11, 12, 13	t. to 3/3 Ben.H.A. See p. 466.

No. 3 Company, 4th Batt. Ben.A. (*No. 6 Mountain Batty. R.G.A.*)[1]
Service of Company, 10, 11, 12, 13, 93, 95, 99, 103, 108, 109, 142

793	1st Lieut. E. H. Hildebrand	10, 11, 12, 13	k. Sept. 1857.

No. 4 Company, 4th Batt. Ben.A. (*No. 103 Compy. R.G.A.; reduced 1905*)
Service of Company, 10, 11, 12, 13, 93, 95, 99, 103, 108, 109, 109a, 142

773	1st Lieut. W. F. Cox	103, 108	⊙ Lieut.-Col. W. F. Cox. r. 1868.
845	1st Lieut. F. W. Stubbs	93, 95, 99, 103, 109, 109a, 142	⊙ Major-Gen. F. W. Stubbs. r. 1878.

No. 1 Company, 5th Batt. Ben.A. No. 13 F.B. (*No. 93 Compy. R.G.A.*)
Service of Company, 29, 31, 36, 37, 38, 39, 40, 41, 51, 57, 76, 92, 96, 99, 108, 109

620	Capt. and Brev.-Major V. Eyre	29, 31, 36, 37, 38, 39, 40, 51, 57	Col. Sir V. Eyre, K.C.B. r. 1863.
778	1st Lieut. E. W. E. Walker (from 4/1 Ben.H.A.)	41, 57, 76, 92, 96, 99, 108, 109	Major-Gen. E. W. E. Walker. d. 1880.
890	1st Lieut. J. McK. Fraser	40, 57	t. to 3/5 Ben.A. See p. 471.
902	2nd Lieut. E. L. Hawkins	40, 57	t. to 3/1 Ben.A. See p. 467.
913	2nd Lieut. D. W. Gordon	31, 36, 41, 57	k. Jan. 1858.
971	2nd Lieut. H. de G. Warter		p. to 3/3 Ben.H.A. See p. 466.

[1] The Companies of the 4th Battn. were very weak and were sometimes grouped together.

No. 2 Company, 5th Batt. Ben.A. No. 21 F.B. (*No. 95 Compy. R.G.A.*)

Service of Company, 20, 45, 171

712	Capt. E. A. C. D'Oyly	20
753	1st Lieut. A. Pearson (from Gwalior Contingent) (*k*). See p. 461	20, 45, 76 — w. July 5, d. July 7, 1857. t. to 3/1 Ben.A. See p. 467.
795	Lieut. A. R. Fuller	20, 45 — Capt. A. R. Fuller was drowned in a torrent at Rawal Pindi, 1867.
837	1st Lieut. J. H. Lamb (from political employment). See p. 463	20 — w. July 5, d. Aug. 24, 1857.
937	2nd Lieut. H. L. Jones	45 — ⊙ Major-Gen. H. L. Jones. r. 1885.
938	2nd Lieut. E. C. Griffin	45 — t. to 5/1 Ben.H.A. See p. 465.
969	2nd Lieut. R. S. Robinson	171 — ⊙ Lieut.-Col. R. S. Robinson. r. 1879.

3rd Company, 5th Batt. No. 20 F.B. (*No. 25 Compy. R.G.A.; reduced* 1902)

Service of Company, 41, 57, 76, 83, 146

682	Capt. G. Moir	57, 76, 83 — t. to 3/1 Ben.H.A. See p. 465.
885	1st Lieut. P. C. Anderson	49 — Lieut.-Col. P. C. Anderson. d. 1882.
890	1st Lieut. J. McK.Fraser (from 1/5 Ben.A.). See p. 470	76, 83 — d. July 1858.
950	2nd Lieut. W. Gully (from 1/4 Ben.A.). See p. 469	41, 57, 76 — t. to 3/1 Ben.A. See p. 467.
1016	2nd Lieut. H. Latham	146 — Lieut.-Col. and Brev.-Col. H. Latham. r. 1890.

4th Company, 5th Batt. Ben.A. No. 11 F.B. (*54th Batt. XXXIXth Brig. R.F.A.*)

Service of Company, 73, 74, 76, 82, 151, 155

604	Capt. and Brev.-Lieut.-Col. A. Huyshe	General A. Huyshe. r. 1880.
821	1st Lieut. E. Simeon (*l*)	73, 74, 76, 83, 100 — Major E. Simeon. r. 1872.
893	2nd Lieut. D. J. Welsh	82, 151, 155 — Lieut.-Col. and Brev.-Col. D. J. Welsh. r. 1883.
993	2nd Lieut. N. D. Garrett	Capt. N. D. Garrett. r. 1872.

1st Company, 6th Batt. Ben.A. (*No. 29 Compy. R.G.A.*)

Service of Company, 7 (*m*)

DETAIL OF TROOPS AND COMPANIES—*continued*.

2ND COMPANY, 6TH BATT. BEN.A. NO. 8 F.B. (*59th Batty. XVIIIth Brig. R.F.A.*) (*n*)

Service of Company, 17, 46, 48, 54, 55, 85, 119, 121, 127, 134, 135, 138, 174

Number in Stubbs's List.	Name.	Service.	Remarks.
647	Capt. T. J. W. Hungerford	17, 46, 48, 54, 55	Lieut.-Col. T. J. W. Hungerford, C.B. r. 1859.
794	2nd Capt. C. S. Lemarchand (from Gwalior Contingent). See p. 461	127, 134, 135	Lieut.-Col. C. S. Lemarchand. r. 1873.
924	2nd Lieut. H. A. Mallock	17, 46	Lieut. H. A. Mallock was t. to Ben.S.C. 1869.

4TH COMPANY, 6TH BATT. BEN.A. (*No. 16 Compy. R.G.A.*)

Service of Company, 8, 9, 10, 11, 12, 13, 99, 103, 108, 109

665	Capt. and Brev.-Major E. Kaye	8, 9, 10, 11, 12, 13, 103, 108	Lieut.-Gen. E. Kaye, C.B. r. 1872.
744	Capt. C. Cookworthy (from 3/1 Ben.H.A.). See p. 465 (*o*)	99, 103, 108, 109, 133	Lieut.-Col. and Brev.-Col. C. Cookworthy. r. 1872.
896	2nd Lieut. H. O. Hitchens (from 2/7 Ben.A. See p. 461	99, 103, 108, 109	⊙ Col. H. O. Hitchens. r. 1882.
947	2nd Lieut. R. T. Hare	8, 9, 10, 11, 12, 13	t. to 3/1 Ben.A. See p. 467.

1ST COMPANY, 7TH BATT. (Ben. A.). (*Absorbed* 1862)

Service, Escort to Siege Train from Phillour to Umballa

718	Capt. W. K. Fooks		⊙ Major-Gen. W. K. Fooks. r. 1872.

RECRUIT DEPOT BEN.A.

741	1st Lieut. A. Light (*p*)	2, 8, 9, 10, 11, 12, 13, 60, 61	Major-Gen. A. Light. r. 1878.

SIKH COMPANIES, BEN.A. (*q*). (*Absorbed* 1862)

Service of Companies, 12, 13, 52, 60, 61, 140

724	Capt. G. Holland (from Gurkha F.F.). See p. 462.	140	Capt. G. Holland was cashiered, 1866.
847	1st Lieut. A. Gillespie	12, 13, 52, 60, 61	☉ Lieut.-Col. A. Gillespie. r. 1872.
857	1st Lieut. Sir W. Stirling-Hamilton, Bt.	12, 13	Gen. Sir W. Stirling-Hamilton, Bt., C.B., Col. Comdt. d. 1913.
957	2nd Lieut. T. Ryan (from un-posted list). See p. 460	52, 60, 61	Major-Gen. T. Ryan. r. 1893.

MOUNTAIN TRAIN. (*Absorbed in Indian Mountain Batteries*)

Service, 1

667	Capt. T. Brougham	1	☉ Major-Gen. T. Brougham. r. 1872.

(b) IX.—OFFICERS OF THE MADRAS ARTILLERY HOLDING COMMANDS, APPOINTMENTS, ETC., IN THE INDIAN MUTINY

Number in Leslie's List.	Name.	Employment.	Service.	Remarks.
393	Lieut.-Col. W. H. Miller	O.C.A.	93a, 94	Major-Gen. W. H. Miller, C.B., A.D.C. r. 1861.
406	Major T. Lavie	O.C. Mad.A.	93a, 94	Lieut.-Col. T. Lavie. r. 1859.
436	Capt. and Brev.-Major W. A. Orr	O.C. Hyderabad Contingent	46, 48, 53–55, 77, 84, 86, 101, 104, 106, 107, 114	Col. W. A. Orr, C.B., A.D.C. d. 1869.
463	Capt. and Brev.-Major J. L. Barrow	C. of O. Saugor F.F.	93a, 94	⊙ Major-Gen. J. L. Barrow, C.B. r. 1872.
484	Capt. A. W. Macintire	Hyderabad Contingent Cavalry	93a, 94, 163, 167	Lieut.-Gen. A. W. Macintire, C.B. Col. Comdt. d. 1895.
506	Capt. A. N. Scott	Hyderabad Artillery	Outbreak, Hyderabad	⊙ Col. A. N. Scott. r. 1866.
507	Capt. W. C. F. Gosling	B.M. Saugor F.F.		Gen. W. C. F. Gosling. Col. Comdt. d. 1885.
514	Lieut. and Brev.-Capt. C. H. Harrison	B.M. (acting) Saugor F.F.	93a, 94, 163, 167	⊙ Col. C. H. Harrison. r. 1865.
529	Lieut. and Brev. 2nd Capt. J. de C. Sinclair	Hyderabad Contingent Artillery	46, 48, 54, 55, 77, 84, 86, 101	Capt. J. de C. Sinclair. d. 1862.
540	Lieut. C. W. Crump	Attached to 3/8 R.A.	22–25, 27–29, 30, 32, 33, 37–39	k. Sept. 26, 1857.
554	Lieut. G. G. Pearse	O.C. Cav.; Harriana F.F. and 3rd Sikh Cav.	52, 76, 91, 100	Gen. G. G. Pearse, C.B. Col. Comdt. d. 1905.
568	Lieut. L. F. C. Thomas	C. of O. Lucknow	19	⊙ Major-Gen. L. F. C. Thomas. r. 1878.
571	Lieut. E. M. Playfair	O.C. Nagpore Irr. A.		Gen. E. M. Playfair. Col. Comdt. d. 1899.
582	Lieut. R. C. Henchy	Orderly Officer to O.C. Troops	173	⊙ Lieut.-Col. R. C. Henchy. r. 1872.
637	2nd Lieut. S. Penny	A.D.C. to Gen. N. Penny		p. to C/5 Mad.A. See p. 477.
657	Lieut. St. J. C. Shawe	Staff employment	175	d. 1866 (injured by a panther).

A.—DETAIL OF TROOPS AND COMPANIES

A Troop Mad.H.A. (J Batty. IVth Brig. R.H.A.)
Services of Troop, 93a, 94, 163, 167, 175

Number in Leslie's List.	Name.	Service.	Remarks.
466	Capt. and Brev.-Major J.D. Mein	93a, 94, 163, 167	⊙ Major-Gen. J. D. Mein. r. 1872.
544	Lieut. R. G. F. Henegan	93a, 94, 163, 167, 175	Lieut.-Col. and Brev.-Col. R. G. F. Henegan. d. 1876.
591	Lieut. J. Stewart	93a, 94, 163, 167	2nd Capt. J. Stewart. r. 1860.
606	Lieut. D. D. Anderson	93a, 94, 163, 167, 175	Capt. D. D. Anderson d. 1868.

B Troop Mad.H.A. (M Batty. VIIIth Brig. R.II.A.)
Service of Troop, 173

496	Capt. A. T. Cadell	173	⊙ Lieut.-Gen. A. T. Cadell. r. 1878.
577	2nd Capt. J. B. Swete	173	⊙ Lieut.-Col. J. B. Swete. r. 1872.
585	Lieut. G. C. M. Martin	173	Capt. G. C. M. Martin. d. 1868.

D Troop Mad.H.A. (R Batty. Xth Brig. R.H.A.)
Service of Troop, 180

438	Capt. and Brev.-Major J. E. Mawdsley	180	Lieut.-Col. and Brev.-Col. J. E. Mawdsley. r. 1801.
583	Lieut. A. J. Ogilvie	180	⊙ Lieut.-Col. A. J. Ogilvie. r. 1872.
586	Lieut. E. S. Milman	180	Lieut.-Col. E. S. Milman. r. 1872.

E Troop, Mad.H.A. (Native)
(*Absorbed* 1862)
Service of Troop, 51, 58, 180

495	Capt. F. G. Nuthall	150	Major-Gen. F. G. Nuthall. r. 1872.
503	Capt. H. T. Molesworth	180	⊙Major-Gen. H. T. Molesworth. r. 1872.
564	Lieut. L. Bridge	51, 58	Capt. and Brev.-Major L. Bridge. d. 1866.
609	2nd Lieut. B. L. Gordon	51, 58	Gen. Sir B. L. Gordon, K.C.B. r. 1890.

DETAIL OF TROOPS AND COMPANIES—continued.

F Troop, Mad.H.A. (Native)
(Absorbed 1862)
Service of Troop, 93a, 94, 156

Number in Leslie's List.	Name.	Service.	Remarks.
427	Capt. and Brev.-Major E. Brice	93a, 94	Col. E. Brice, C.B. d. 1868.
523	Lieut. and Brev.-Capt. G. B. B. Holmes	93a, 94	☉ Col. G. B. B. Holmes. r. 1869.
610	Lieut. W. R. D. Sewell	93a, 94	d. 1859.

C Company, 1st Batt. Mad.A. (24th Batty. XXXVIIIth Brig. R.F.A.)
Service of Company, 180

561	Lieut. M. C. Lawson	180	Major and Brev.-Lieut.-Col. M. C. Lawson. r. 1872.
580	Lieut. H. D. Gloag	180	2nd Capt. H. D. Gloag. r. 1864.

A Company, 3rd Batt. Mad.A. (r) (46th Batty. XXXIXth Brig. R.F.A.)
Service of Company, 65, 73, 74, 76, 78, 91, 100, 151, 155, 161

443	Capt. and Brev.-Major G. S. Cotter	65, 73, 74, 76, 91, 100	Lieut.-Col. and Brev.-Col. G. S. Cotter, C.B. r. 1861.
531	2nd Capt. and Brev.-Major R. Cadell (from C/5 Mad. A.). See P. 477 (r)	151, 155, 161	Gen. Sir R. Cadell, K.C.B. Col. Comdt. d. 1897.
543	2nd Capt. I. R. J. Robertson	100	Capt. J. R. J. Robertson. r. 1864.
548	Lieut. G. G. J. Campbell	65, 73, 74, 76, 91, 100	Capt. G. G. J. Campbell. d. 1865.
549	Lieut. W. J. Bradford	73, 74, 76, 78, 91, 100	Lieut.-Col. and Brev.-Col. W. J. Bradford. d. 1877.
560	2nd Capt. R. C. B. Highmore	151, 155, 161	d. 1862.
615	2nd Lieut. S. H. E. Chamier (from C/5 Mad.A. See p. 477)	65, 73, 74, 76, 91, 100	☉ Lieut.-Gen. S. H. E. Chamier, C.B. r. 1886.
659	Lieut. The Hon. A. B. De Montmorency		Lieut.-Col. and Brev.-Col. The Hon. A. B. De Montmorency. r. 1896.
660	Lieut. W. H. Montgomerie	151, 155a, 161	Major W. H. Montgomerie. d. 1878.

Service of Company, 29a, 180

494	Capt. C.H. Hutchinson [Jones	29a	☉ Major-Gen. C. H. Hutchinson. r. 1870.
528	Lieut. and Brev. 2nd Capt. A. C.	29a	Capt. A. C. Jones. d. 1860.
597	Lieut. R. A. Baker	29a	☉ Lieut.-Col. R. A. Baker. r. 1872.
602	Lieut. T. I. M. Hog	180	☉ Lieut.-Gen. T. I. M. Hog. r. 1885.
627	Lieut. H. P. Lane	29a	Lieut-Col. and Brev.-Col. H. P. Lane. d. 1884.
628	Lieut. W. H. Caine	29a	Lieut.-Gen. W. H. Caine. r. 1892.

A COMPANY, 4TH BATT. MAD.A. (*No. 79 Company R.G.A.*)
Service of Company, 93a, 94, 156, 157

469	Capt. R. Kinkhead		Capt. and Brev.-Major R. Kinkhead. d. 1860.
600	Lieut. R. Pope	93a, 94, 157	Capt. R. Pope. d. 1869.
601	Lieut. C. Johnson	93a, 94, 156	Capt. C. Johnson. r. 1867.
617	Lieut. J. C. Tayler	94	☉ Major-Gen. J. C. Tayler. r. 1889.

B COMPANY, 4TH BATT. MAD.A. (*No. 44 Compy. R.G.A.*)
Service of Company, 93a, 94, 163, 167

513	Capt. H. E. Hicks	163, 167	☉ Major-Gen. H. E. Hicks. r. 1872.
624	Lieut. F. H. Thompson	93a, 94, 163, 167	2nd Capt. F. H. Thompson was t. to Mad. Staff Corps, 1869.
647	Lieut. A. M. Rawlins		Maj.-Gen. A. M. Rawlins. r. 1894.
655	Lieut. W. Bryden	163, 167	r. 1867.

C COMPANY, 5TH BATT. MAD.A. (Native) (*Absorbed* 1862)
Service of company, 56 (s)

518	Lieut. and Brev.-Capt. S. Rippon		Capt. S. Rippon. r. 1860.
531	Lieut. and Brev. 2nd Capt. R. Cadell		t. to A/3 Mad.A.
538	Lieut. C. Desborough		Capt. C. Desborough. d. 1861.
612	Lieut. G. A. Goldingham		2nd Capt. G. A. Goldingham. r. 1864.
615	2nd Lieut. S. H. E. Chamier		t. to A/3 Mad.A.
687	2nd Lieut. S. Penny (from the Staff). See p. 474	56	Lieut.-Col. and Brev.-Col. S. Penny. d. 1887.

E COMPANY, 5TH BATT. MAD.A. (Native) (*Absorbed* 1862)
Service of Company, 180

517	Capt. J. H. Elwyn	180	☉ Major-Gen. J. H. Elwyn. r. 1870.
623	Lieut. F. E. Hadow	180	Major-Gen. F. E. Hadow. r. 1882.

(b) XI.—OFFICERS OF THE BOMBAY ARTILLERY HOLDING COMMANDS, APPOINTMENTS, ETC., IN THE INDIAN MUTINY

Number in Spring's List.	Name.	Employment.	Service.	Remarks.
286	Lieut.-Col. T. W. Hicks	O.C.A.	113, 114	Col. T. W. Hicks, C.B. r. 1858.
349	Capt. E. Wray	C. of Siege Train	85	⊙Lieut.-Gen. E. Wray, C.B. r. 1878.
368	Capt. A. M. Murray	Ordnance Dept.		⊙Major-Gen. A. M. Murray.
373	Capt. R. H. Keatinge	Political Agent, Mulwa	46, 48, 54, 55, 81	Brev.-Major R. H. Keatinge, V.C., was t. to Bom.S.C. 1861.
387	Lieut. T. T. Haggard	C. of O.	72, 84, 114, 115	Lieut.-Col. and Brev.-Col. T. T. Haggard. d. 1877.
388	Lieut. R. A. Morse	C. of O.	180	⊙Major-Gen. R. A. Morse. r. 1872.
390	Lieut. J. Shekleton	Attached to 2/6 Ben.A.	170	⊙Major-Gen. J. Shekleton. r. 1872.
413	Lieut. T. J. Maclachlan	Adj. Central India F.F.	66, 68, 72, 84, 85, 86, 101, 106, 107, 112, 114, 115	⊙Major-Gen. T. J. Maclachlan. r. 1878.
419	Lieut. C. S. de N. Lucas	Quartermaster and Interpreter 4th Batt. Bom.A.	180	⊙Major-Gen. C. S. de N. Lucas. d. 1887.
426	Lieut. G. Twiss	B.M. Cav.	180	Lieut.-Col. G. Twiss. r. 1872.
445	Lieut. R. A. Stevenson	Attached to 2/6 Ben.A.	119, 121	Lieut.-Col. R. A. Stevenson. d. 1879.
449	Lieut. J. Ritchie	Civil Employ, Neemuch		⊙Major-Gen. J. Ritchie. r. 1888.
459	Lieut. C. E. Newport	Adj.4th Batt.Bom.A.	180	Col. C. E. Newport. r. 1888.
464	Lieut. J. T. Leishman	Hyderabad Artillery	180	⊙Major-Gen. J. T. Leishman. r. 1888.

DETAIL OF TROOPS AND COMPANIES

1st Troop, Bom.H.A. (N Batty. IXth Brig. R.H.A.) (†)

Service of Troop, 66, 68, 69, 72, 77, 84, 86, 101, 106, 107, 112, 114, 115

Number in Spring's List.	Name.	Service.	Remarks.
304	Capt. and Brev.-Lieut.-Col. S. Turnbull	66, 68, 69, 72, 77, 84, 86	k. April 4, 1857.
350	Capt. J. G. Lightfoot	101, 106, 107, 112, 114, 115	Lieut.-Col. J. G. Lightfoot, C.B. r. 1861.
416	Lieut. R. Pittman	66, 68, 69, 72, 77, 84, 86, 101, 106, 107	2nd Capt. and Brev.-Major R. Pittman. d.1867.
417	Lieut. T. C. Crowe	66, 68, 69, 72, 77, 84, 86, 101, 106, 107, 112, 114, 115	⊙ Major-Gen. T. C. Crowe. r. 1878.
440	Lieut. C. P. Roberts		⊙ Col. C. P. Roberts. r. 1877.

2nd Troop, Bom.H.A. (Y Batty. XIIIth Brig. R.II.A.)

Service of Troop, 64a, 85, 119, 121

351	Capt. J. G. Petrie	64a, 85, 121	Lieut.-Gen. J. G. Petrie. r. 1878.
423	Lieut. J. H. P. Malcolmson	64a, 85, 121	Major-Gen. J. H. P. Malcolmson. r. 1882.
428	Lieut. F. Swanson	64a, 85	Lieut.-Col. F. Swanson. d. 1877.
433	Lieut. F. Hemming	64a, 121	d. 1860.
460	Lieut. J. Vibart	85, 119, 121	2nd Capt. J. Vibart. r. 1864.

3rd Troop, Bom.H.A. (Q Batty. VIIIth Brig. R.H.A.)

Service of Troop, 85, 113, 114, 122, 125a, 135, 138, 145

318	Capt. and Brev.-Major E. S. Blake	85, 113, 114, 122, 138, 145	Col E. S. Blake, C.B. d. 1862.
411	Lieut. T. M. Harris	85, 113, 114, 122, 135, 138, 145	Col. T. M. Harris. r. 1882.
435	Lieut. H. le Cocq	85, 113, 114, 144	Gen. H. le Cocq, Col. Comdt. d. 1911.
436	Lieut. A. R. Hoskins	85, 113, 114, 122, 125a, 135, 145	Col. A. R. Hoskins. r. 1877.

DETAIL OF TROOPS AND COMPANIES—*continued*.

4TH TROOP, BOM.H.A. (*Z Batty. XIVth Brig. R.H.A.*)
Service of Troop, 34, 180

Number in Spring's List.	Name.	Service.	Remarks.
362	Capt. D. Gaye	180	⊙ Major-Gen. D. Gaye. r. 1878.
441	Lieut. T. N. Holberton	34, 180	Col. T. N. Holberton. r. 1889.

No. 1 COMPANY, 1ST BATT. BOM.A., No. 3 L.F.B. (*No. 86 Compy. R.G.A.*)
Service of company, 180

356	Capt. W. S. Hatch	180	⊙ Gen. W. S. Hatch. r. 1858.
447	Lieut. W. W. Woodward	180	Col. W. W. Woodward. r. 1889.

3RD COMPANY, 1ST BATT. BOM.A., No. 1 L.F.B. (*30th Batty. XLIVth Brig. R.F.A.*)
Service of Company, 71

366	Capt. H. M. Douglas		d. 1862.
461	2nd Lieut. H. T. Vachell	71	Major-Gen. H. T. Vachell. r. 1885.

1ST COMPANY, 2ND BATT. BOM.A., No. 2 L.F.B. (*81st Batty. Xth Brig. R.F.A.*)
Service of Company, 64a, 85, 154

346	Capt. W. D. Aitken	64a	Gen. W. D. Aitken, Col. Comdt. d. 1897.
431	Lieut. T. B. Heathorn	64a, 85, 154	Capt. T. B. Heathorn. r. 1866.
457	Lieut. H. S. K. Pechell	64a, 85	2nd Capt. H. S. K. Pechell. r. 1863.

4TH COMPANY, 2ND BATT. BOM.A., No. 4 L.F.B. (*55th Batty. XXXVIIth Brig. R.F.A.*)
Service of Company, 46, 48, 54, 55, 66, 69, 72, 81, 84, 86, 101, 106, 107, 112, 114

352	Capt. J. D. Woollcombe	46, 48, 54, 55, 66, 69, 72, 81, 84, 107	Col. J. D. Woollcombe, C.B. d. 1875.
450	Lieut. C. H. Strutt	46, 48, 54, 55, 66, 69, 72, 81, 84, 86, 101, 106, 107, 112, 114	2nd Capt. and Brev.-Major C. H. Strutt. r. 1866.
468	Lieut. B. Christie	46, 48, 107	Lieut. B. Christie. d. 1860.

		Service of Company, 65, 66, 69, 72, 77, 84, 86, 101, 106, 107, 112, 114	
347	Capt. C. B. Fuller (*l*)	112, 113, 114	Gen. C. B. Fuller, Col. Comdt. t. to 1st Troop, Bom.H.A. See p. 479.
350	Capt. J. G. Lightfoot	66, 68, 69, 72, 77, 84	
454	Lieut. P. H. Harcourt	66, 68, 69, 72, 77, 84, 86, 101, 106, 107, 112, 114	Major-Gen. P. H. Harcourt. r. 1884.

3RD COMPANY, 3RD BATT. BOM.A. (*Reduced* 1870)
(Native)

Service of Company, 180

382	Capt. F. Conybeare	180	○ Major-Gen. F. Conybeare.

4TH COMPANY, 3RD BATT. BOM.A., No. 8 L.F.B. (*Distributed* 1860)
(Native)

Service of Company, 64a, 85

381	Lieut. G. G. Brown	64a, 85, 119	○ Major-Gen. G. G. Brown. r. 1878.

1ST COMPANY 4TH BATT. BOM.A. No. 6 L.F.B. (*Quetta Mountain Battery*)

Service of Company, 180

424	2nd Capt. J. B. Hardy	180	○ Lieut.-Col. J. B. Hardy. r. 1872.

2ND AND 3RD COMPANIES, 4TH BATT. BOM.A. (*Reduced* 1863 *and* 1870)
(Native)

Service of Companies, 34, 89

331	Capt. G. P. Sealy	34, 89	Major-Gen. G. P. Sealy, Col. Comdt. d. 1892.
466	2nd Lieut. C. E. Basevi	89	Hon. Col. C. E. Basevi. r. 1881.

XIII.—OFFICERS OF CAVALRY AND INFANTRY ATTACHED TO ARTILLERY

Lieut. J. Evans	9th Lancers	Mortar Battery at Delhi. See p. 116 n.
Cornet F. Trench	7th Ben.L.C.	No. 4 Siege Battery at Delhi. See p. 117 n.
Lieut. M. G. Clerk	1st Ben.N.I.	Alam Bagh. Nov. '57—March '58.
Lieut. R. A. Dobbin	7th Ben.N.I.	With 4/5 Ben.A.
Lieut. W. E. D. Broughton	8th Ben.N.I.	With 4/5 Ben.A.
Lieut. G. B. C. Simpson	23rd Ben.N.I.	With Central India Field Force. See p. 314.
Lieut. E. H. C. Simpson	39th Ben.N.I.	Alam Bagh with 3/8th R.A. See p. 216.
Capt. R. Ouseley	48th Ben.N.I.	Alam Bagh. Nov. '57—March '58.
Lieut. E. B. Ward	48th Ben.N.I.	Alam Bagh with 3/8th R.A. See p. 216.
Lieut. H. G. Delafosse	53rd Ben.N.I.	Alam Bagh. Nov. '57—March '58.
Capt. H. L. Evans	17th Bom.N.I. Dept. Comm. Oudh	Residency, p. 161. Alam Bagh.
Local Capt. E. O. Bradford	Oudh Commission	Stubbs, iii. p. 464.

XIV.—KEY TO THE SERVICES OF EXISTING BATTERIES AND COMPANIES

Royal Horse Artillery

Designation, 1912.	Designation, 1857-9.	Service. PAGE
D Battery III Brigade	E Troop R.H.A.	452
E ,, III ,,	D ,, ,,	452
F ,, IV ,,	1st ,, 1st Brig. Ben.H.A.	464
G ,, V ,,	F ,, R.H.A.	453
J ,, IV ,,	A ,, Mad.H.A.	475
K ,, VI ,,	1st ,, 2nd Brig. Ben.H.A.	465
L ,, VII ,,	1st ,, 3rd ,, ,,	465
M ,, VIII ,,	B ,, Mad.H.A.	475
N ,, IX ,,	1st ,, Bom.H.A.	479
Q ,, VIII ,,	3rd ,, ,,	479
R ,, X ,,	D ,, Mad.H.A.	475
S ,, IX ,,	3rd ,, 3rd Brig. Ben.H.A.	466
T ,, XI ,,	5th ,, 1st ,, ,,	479
Y ,, XIII ,,	2nd ,, Bom.H.A.	
Z ,, XIV ,,	4th ,, ,,	480

Royal Field Artillery

8th Battery XIII Brigade	5th Compy. 12th Battn. R.A.	455
13th ,, I ,,	5th ,, 3rd Battn. R.A.	453
20th ,, IX ,,	8th Compy. 2nd Battn. R.A.	453

KEY TO THE SERVICES—*continued*.

Designation, 1912.			Designation, 1857-8.				Service. PAGE
21st Battery	II	Brigade	1st Compy.	1st Battn. Ben.A.			467
22nd ,,	XXXIV	,,	4th ,,	,,	,,	,,	468
23rd ,,	XL	,,	2nd ,,	3rd ,,	,,		468
24th ,,	XXXVIII	,,	C ,,	1st Battn. Mad.A.			476
25th ,,	XLVI	,,	1st ,,	6th Battn. R.A.			454
30th ,,	XXX	,,	3rd Compy.	1st Battn. Bom.A.			480
36th ,,	XXIII	,,	4th ,,	3rd Battn. Ben.A.			469
46th ,,	XXXIX	,,	A ,,	3rd Battn. Mad.A.			476
48th ,,	XXXVI	,,	D ,,	3rd ,,	,,		477
52nd ,,	XV.	,,	2nd Troop	2nd Brig. Ben.H.A.			465
53rd ,,	II	,,	3rd ,,	1st Battn. Ben.A.			467
54th ,,	XXXIX	,,	4th ,,	5th ,,	,,		471
55th ,,	XXXVII	,,	4th ,,	2nd ,,	Bom.A.		480
56th ,,	L	,,	2nd ,,	1st Brig. Ben.H.A.			464
57th ,,	XLIV	,,	2nd ,,	3rd ,,	,,		466
58th ,,	XLVI	,,	3rd ,,	1st ,,	,,		465
59th ,,	XVIII	,,	2nd Compy.	6th Battn.Ben.A.			472
65th ,,	VII	,,	6th ,,	11th Battn. R.A.			455
66th ,,	IV.	,,	5th ,,	13th ,,	,,		456
67th ,,	I	,,	6th ,,	13th ,,	,,		456
68th ,,	XIV	,,	4th ,,	14th ,,	,,		457
69th ,,	I	,,	5th ,,	14th ,,	,,		457
71st ,,	XXXVI	,,	2nd ,,	Reserve Battn. Bom.A.			481
81st ,,	X	,,	1st Compy.	2nd Battn. Bom.A.			480
83rd ,,	XI	,,	3rd Compy.	14th Battn. R.A.			456
84th ,,	XI	,,	7th ,,	14th ,,	,,		458

ROYAL GARRISON ARTILLERY

No. 2 Mountain Battery	1st Compy.	4th Battn. Ben.A.		469
No. 4 ,, ,,	2nd ,,	4th ,,	,,	470
No. 6 ,, ,,	3rd ,,	4th ,,	,,	470
No. 9 Company	3rd ,,	4th Battn. R.A.		454
No. 16 ,,	4th ,,	6th Battn. Ben.A.		472
No. 18 ,,	3rd ,,	8th Battn. R.A.		454
No. 28 ,,	1st ,,	3rd Battn. Ben.A.		468
No. 29 ,,	1st ,,	6th Battn.Ben.A.		471
No. 33 ,,	6th ,,	14th Battn. R.A.		457
No. 35 ,,	4th ,,	5th Battn. R.A.		454
No. 42 ,,	2nd ,,	11th Battn. R.A.		455
No. 44 ,,	B ,,	4th Battn.Mad.A.		477
No. 54 ,,	2nd ,,	3rd Battn. R.A.		453
No. 64 ,,	8th ,,	14th Battn. R.A.		458
No. 79 ,,	A ,,	4th Battn. Mad.A.		477
No. 86 ,,	1st ,,	1st Battn. Bom.A.		480
No. 93 ,,	1st ,,	5th Battn. Ben.A.		470
No. 95 ,,	2nd ,,	5th ,,	,,	471
No. 97 ,,	3rd ,,	3rd ,,	,,	469
Quetta Mountain Battery	1st ,,	4th Battn. Bom.A.		481

(n) XV.—SURGEONS, ASSISTANT SURGEONS, AND VETERINARY SURGEONS

ROYAL ARTILLERY

Kane's List	Name.	Service.	Remarks.
262	Surgeon G. T. Ferris	With 5/13 R.A.	Senior Surgeon G. T. Ferris. t. to Staff, 1864.
265	Surgeon R. C. Elliot, C.B.	Head-qr. R.A.	Senior Surgeon R. C. Elliot, C.B. t. to Staff 1860.
281	Staff-Surgeon E. D. Allinson	With 8/14 R.A.	Surgeon G. D. Allinson. d. 1859.
284	Surgeon T. Rhys	With F Troop R.H.A.	d. 1863.
294	Asst. Surg. G. P. M. Woodward, M.D.	With D Troop R.H.A.	Surg. G. P. M. Woodward. t. to Staff, 1858.
297	Asst. Surg. J. A. M'Munn, M.D.	With F Troop R.H.A.	Surg. J. A. M'Munn. t. to Staff, 1860.
301	Asst. Surg. J. Barker	With 5/12 R.A.	Surgeon J. Barker. t. to 22nd Foot, 1872.
304	Asst. Surg. W. Houghton, M.B.	With 3/14 R.A.	r. 1861.
310	Asst. Surg. R. Webb	With 6/14 R.A.	D. Surgeon-General, 1873.
311	Asst. Surg. E. Bubb	With 5/13 R.A.	d. 1859.
314	Asst. Surgeon J. H. Hearn	With 3/8 R.A.	Surg. J. H. Hearn. r. 1869.
315	Asst. Surg. T. Tarrant, M.D.	With 7/14 R.A.	Dep.-Surg.-General, 1883.
317	Asst. Surg. G. Sharp	With 6/11 R.A.	r. 1862.
322	Asst. Surgeon T. J. Orton	With 6/13 R.A.	Surg. T. J. Orton. t. to 10th Foot, 1868.
326	Asst. Surgeon W. Y. Jeeves	With 6/13 R.A.	Surg. W. Y. Jeeves. t. to Staff, 1872.
329	Asst. Surg. A. R. Smith	With 8/2 R.A.	Surg. A. R. Smith. t. to Staff, 1868.
331	Asst. Surg. J. W. Rimmer	With 1/6 R.A.	d. 1867.
333	Asst. Surg. A. R. L. Veale, M.D.	With 6/13 R.A.	Brig.-Surg. H. R. L. Veale, M.D. r. 1883.
343	Asst. Surg. H. B. Franklyn, M.D.	With E Troop	Dep.-Surg.-General H. B. Franklyn, M.D. r. 1877.
351	Asst. Surg. G. D. Milne, M.D.	With 4/14 R.A.	r. 1865.
381	Asst. Surg. H. C. Miles	With 83rd Foot	Surg. H. C. Miles. d. 1871.
390	Asst. Surg. W. H. Pollard		t. to Staff, 1872.
16	Vet. Surg. J. S. Stockley	With Head-qr. R.A.	r. 1863.
17	Vet. Surg. H. Withers	With 6/13 R.A.	t. 3rd Dragoons, 1861.
18	Vet. Surg. M. J. Harpley	With E Troop R.H.A.	t. to Royal Horse Guards, 1860.
21	Vet. Surg. W. B. Lord	With D Troop R.H.A.	r. 1864.
23	Vet. Surg. G. I. Rollings	With F Troop R.H.A.	d. 1868.

27	Vet. Surg. C. Saunderson	With 4/14 and 7/14 R.A.	d. 1886.
34	Vet. Surg. J. B. W. Skoulding	With 8/14 R.A.	r. 1888.
36	Vet. Surg. E. Kelly	With 1/6 R.A.	r. 1881.
37	Vet. Surg. J. B. Hall	With 5/3 R.A.	d. 1870.
	Staff Asst. Surg. R. R. Scott	With F Troop R.H.A.	This officer belonged in 1858 to the 8th Hussars

(u) XVI.—SURGEONS, ASSISTANT SURGEONS, AND VETERINARY SURGEONS OF THE INDIAN ESTABLISHMENT SERVING WITH ARTILLERY

Bengal

Surg. C. Mackinnon	With 1st Brig. Ben.H.A.
Surg. J. C. Brown	With 5/1 Ben.H.A.
Surg. J. Bowhill	With 3/1 Ben.A. and 23 Ben.H.A.
Surg. D. J. O'Callaghan	With Ben.A., Delhi
Asst. Surg. J. Irving	Alam Bagh, Nov. '57—Mar. '58
Asst. Surg. E. J. Vivian	With 1/3 Ben.H.A.
Asst. Surg. J. J. Clarke	With 2/3 Ben.A.
Asst. Surg. A. Eteson	With 1/5 Ben.A.
Asst. Surg. F. Carter	With 3/1 Ben.H.A.
Asst. Surg. W. Walker	With 2/5 Ben.A.
Asst. Surg. G. A. Watson	With 2/6 Ben.A.
Asst. Surg. D. B. Smith	With 2/1 and 5/1 Ben.H.A.
Asst. Surg. N. B. Baillie	With Ben.A. at Delhi
Asst. Surg. C. J. Gayer	With 3/8 R.A.
Asst. Surg. W. W. Ireland	With 2/3 Ben.H.A.
Asst. Surg. C. E. Raddock	With 4/5 and 6/13 R.A.
Asst. Surg. J. H. White	With 5/1 Ben.H.A.
Asst. Surg. G. S. Sutherland	With 1/1 Ben.H.A.
Asst. Surg. H. Potter	With 1/5 Ben.A.
Vet. Surg. I. Bicknell	With 1st Brig. Ben.H.A.
Vet. Surg. J. R. Hoey	
Vet. Surg. G. Kettlewell	With 1/1 Ben.H.A.
Vet. Surg. J. Siddall	With 1/5 Ben.A.

SURGEONS, ASSISTANT SURGEONS, AND VETERINARY SURGEONS—*continued*

Madras

Name.	Service.	Remarks.
Asst. Surg. H. E. Busteed	With E Troop Mad.H.A.	
Asst. Surg. J. S. Morton, M.D.		
Asst. Surg. H. Adam		
Asst. Surg. W. H. Morgan		
Asst. Surg. P. Allen		
Asst. Surg. G. Dunman		
Asst. Surg. W Doyle		
Asst. Surg. J. M. Miller		

Bombay

Spring's List.	Name.	Service.	Remarks.
104	Asst. Surg. G. C. Bell	With 4th Troop Bom.H.A.	d. 1883.
109	Asst. Surg. W. C. Brown, M.D.	With 2nd Troop Bom.H.A.	d. 1863.
113	Asst. Surg. E. R. Butler, M.D.	With 4/2 Bom.A.	r. 1877.
117	Asst. Surg. J. Lumsdaine	With 1st Troop Bom.A.	r. 1883.

XVII. TABLE OF BATTLES, SIEGES, ACTIONS, ETC.

No.	Date.	Description.	Place.	Page.
	1857			
1	May 11–Sept. 20	Operations	Punjaub	21–31
2	May 30 and 31	Actions	Hindun River	68, 69
3	June 4	Disarming	Benares	55, 56
4	June 5–7	Defence	Jhansi Fort	36
5	Jan. 6	Defence	Allahabad Fort	58
6	June 7	Disarming	Jullundur	28
7	June 7–27	Defence	Cawnpore Entrenchment	51–4, 145–53
8	June 8	Action	Badli-ki-Serai	70, 71
9	June 1–22	Siege	Delhi	72–87
10	June 23–July 24	Siege	Delhi	87–95
11	July 25–Aug. 14	Siege	Delhi	95–8
12	Aug. 14–Sept. 6	Siege	Delhi	96–99
13	Sept. 6–21	Siege	Delhi	99–131
13a	June 11	Disarming	Mooltan	27
14	June 12	Affair	Trimoo Ghat	31
15	June 18–July 4	Defence	Futtehghur Fort	44, 45
16	June 30	Battle	Chinhut	154–59
17	July 1	Outbreak	Mhow	38, 39
18	July 1	Outbreak & Operations	Saugor	39, 40
19	July 1–Sept. 25	Defence	Residency, Lucknow	159–66
20	July 5	Action	Sassiah	43, 44
21	July 9	Action	Jhelum	29, 30
22	July 12	Action	Futtehpore	169, 170
23	July 15	Action	Aong	171
24	July 15	Action	Pandoo River	171, 172
25	July 16	1st Battle	Cawnpore	172–5
26	July 22–Aug. 2	Defence	Arrah	185, 186
27	July 29	Action	Unao	176
28	July 29	1st Action	Bassiratgunge	177
29	Aug. 2	Action	Bibigunge	189
29a	Aug.–Dec.	Operations	Round Jubbulpore	245, 246
30	Aug. 5	2nd Action	Bassiratgunge	178
31	Aug. 11	Action	Jugdispore	191
32	Aug. 12	Action	Boorya-ki-Chowki	179
33	Aug. 16	Action	Bithur	179, 180
34	Aug. 1–7	Operations	Kohlapore	62
35	Aug. 25	Action	Nagafghur	99, 100
36	Sept. 11	Affair	Koondum Puttee	194
37	Sept. 22	Skirmish	Mangalwar	195
38	Sept. 23	Action	Alam Bagh	195, 196
39	Sept. 24–27	1st Relief	Lucknow	196–205
40	Sept. 28–Nov. 22	Defence	Residency Position	205, 223, 224

TABLE OF BATTLES, SIEGES, ACTIONS, ETC.—*continued*

No.	Date.	Description.	Place.	Page.
41	Sept. 28–Nov. 13	Defence	Alam Bagh	196
42	Sept. 28	Action	Bulandshahr	133–5
43	Oct. 2	Action	Chuttra	192
44	Oct. 3	Attack	Jiran	243
45	Oct. 10	Action	Agra	137–40
46	Oct. 22	Action	Near Dhar	244
47	Oct. 30	Skirmish	Bantara	211
48	Oct. 31	Capture	Dhar	244
49	Nov. 1	Action	Khujwa	208, 209
50	Nov. 9	Capture	Atraulia	246
51	Nov. 14–22	2nd Relief	Lucknow	213–28
52	Nov. 16	Action	Nurnaul	142–44
53	Nov. 20	Action	Rewal	245
54	Nov. 21	Action	Mandesar	245
55	Nov. 22	Action	Goravia	245
56	Nov. 26, 27, 28	Actions	Round Cawnpore	231–5
57	Nov. 25–March 1	Defence	Alam Bagh	228, 270–3
58	Dec. 6	2nd Battle	Cawnpore	237–40
59	Dec. 9	Pursuit	Serai Ghat	240, 241
60	Dec. 15	Action	Gangaree (Khasgunge)	251
61	Dec. 18	Action	Patiala	252
62	Dec. 26	Action	Chota Gundak	247
	1858			
63	Jan. 2	Action	Kali River (Khudagunge)	254–5
64	Jan. 5	Skirmish	Pandora	247
64a	Jan. 19	Capture	Awah	315
65	Jan. 22	Action	Sekundra (Nasratpore)	262, 263
66	Jan. 24–28	Siege	Rahatghur	298, 299
67	Jan. 26	Action	Shamshabad	257, 258
68	Jan. 29	Action	Barodia	299–301
69	Feb. 3	Relief	Saugor	302
70	Feb. 4	Skirmish	Bognipore	259
71	Feb. 8	Operations	Shorapore	64
72	Feb. 12	Capture	Garakota	302
73	Feb. 19	Action	Chanda	263–5
74	Feb. 23	Action	Sultanpore	265–8
75	Feb. 23	Capture	Miangunge	259, 260
76	Mar. 2–22	Siege	Lucknow	273–96
77	March 3	Action	Passes of Bundelkhund	303
78	March 4	Attack	Dhowrara	268–70
79	March 5	Skirmish	Futtehabad	304
80	March 5	Action	Amorha	364
81	March 7–17	Siege	Chanderi	303–5

TABLE OF BATTLES, SIEGES, ACTIONS, ETC.—*continued*

No.	Date.	Description.	Place.	Page.
82	March 22–April 6	Operations	Azimghur	365
83	March 23	Action	Kursi	296
84	March 23–April 3	Siege	Jhansi	305–14
85	March 24–30	Siege	Kotah	316, 317
86	March 31	Battle	Betwa	309–13
87	April 2	Attack	Beyt Island	350
88	April 6	Action	Kankar	355
89	April 11	Action	Amba Pawnee	63
90	April 12	Action	Barri	372, 373
91	April 11–20	Operations	Azimghur and Jugdispore	366–8
92	April 15	Attack	Ruiya	356
93	April 17	Skirmish	Najibabad	359
93a	April 17	Skirmish	Kubrae	324
94	April 18	Battle	Banda	324, 325
95	April 21	Action	Nagina	359
96	April 22	Skirmish	Sirsa and Aligunge	357
97	April 23	Affair	Jugdispore Jungle	368
98	May 2	Capture	Lohari	318
98a	May 5	Skirmish	Mirgunge	359
99	May 5	Battle	Bareilly	359–61
100	May 7–9	Operations	Jugdispore Jungle	369, 370
101	May 7	Battle	Kunch	319–22
102	May 10	Affair	Simri (Sirsi)	373, 374
103	May 11–15	Operations	Shahjehanpore	362, 363
104	May 16–17	Skirmishes	About Kalpi	327, 329
105	May 16	Skirmish	Sher Ghat	337
106	May 22	Battle	Gulauli	330–6
107	May 23	Capture	Kalpi	336
108	May 24	Action	Burnai	363
109	May 25	Capture	Mohamdee	364
109a	May 31	Action	Shahabad	364
110	June 1	Capture	Nargund	63
111	June 12	Action	Newabgunge Barabanki	374, 375
112	June 16	Battle	Morar	339–41
113	June 17	Action	Kota-ki-Serai	341–3
114	June 19	Battle	Gwalior	343–7
115	June 22	Action	Jaora Alipore	347–8
116	July 14	Capture	Dhainawan	379
117	July 16	Capture	Tiroul	379
118	July 29	Skirmish	Ajudia Ghat	376
119	Aug. 7	Skirmish	Sanganeer	403
120	Aug. 13	Skirmish	Nyagaon	420
121	Aug. 13	Action	Banas	404
122	Aug. 19–21	Siege	Paori	413
123	Aug. 25	Passage	Gumti at Sultanpore	377

TABLE OF BATTLES, SIEGES, ACTIONS, ETC.—continued

No.	Date.	Description.	Place.	Page.
124	Sept. 2	Outbreak	Mooltan	27
125	Sept. 5	Action	Sahao	418
125a	Sept. 5	Skirmish	Bijepur	413
126	Sept. 6	Capture	Penghali Pass	421
127	Sept. 15	Action	Beora	406
128	Sept. 22	Skirmish	Nr. Derriabad	378
129	Sept. 22	Skirmish	Salimpore	377, 378
130	Oct. 5	Skirmish	Miangunge	380
131	Oct. 7	Action	Kantur	380
132	Oct. 8	Action	Jamo	383
133	Oct. 8	Action	Bunkagaon	385
134	Oct. 10	Action	Mangaoli	406
135	Oct. 19	Action	Sindwaho	407
136	Oct. 20	Affair	Doadpore	380–3
137	Oct. 21	Capture	Birwah	384–5
138	Oct. 25	Action	Kurai	408
139	Oct. 26	Affair	Bagrode	408
139a	Oct. 27	Skirmish	Kandu River	383
140	Oct. 28	Capture	Ruiya	385
141	Nov. 3	Capture	Rampore Kussia	388
142	Nov. 8	Capture	Mithaulee	386
143	Nov. 10	Capture	Amithi	389
144	Nov. 11–16	Siege	Shunkerpore	389
145	Nov. 14	Action	Kundrie	413
146	Nov. 14–18	Operations	Round Simri	390
146a	Nov. 18	Skirmish	Mehndi	386
147	Nov. 19	Skirmish	Rajpore	409
148	Nov. 23	Capture	Rehora and Keoli	391, 392
149	Nov. 24	Affair	Dundeakhera	390
150	Nov. 23–27	Passage	Gogra at Fyzabad	393, 394
151	Nov. 26	Action	Dumurreagunge	370
152	Nov. 29	Skirmish	Banks of Gumti	392
153	Nov. 30–Dec. 3	Capture	Omeriah	392
154	Dec. 1	Action	Chota Udepur	409, 410
155	Dec. 3	Action	Bururiah	386
155a	Dec. 3	Skirmish	Muchligaon	394
156	Dec. 4	Skirmish	Tola Kanjaran	419
157	Dec. 5	Skirmish	Lucherra	419
158	Dec. 9	Action	Gonda	394
159	Dec. 17	Action	Ranode	414
160	Dec. 22	Surprise	Sarpore	414
160a	Dec. 22	Operations	Mullapore Doab	396
161	Dec. 23	1st Action	Tolsipore	397
162	Dec. 25	Skirmish	Pertabghur	411
163	Dec. 25	Defence	Kirwi Palace	421
164	Dec. 26	Action	Burgeediah	397
165	Dec. 27	Capture	Musjeediah	397, 398
166	Dec. 29	Skirmish	Zirapore	411

TABLE OF BATTLES, SIEGES, ACTIONS, ETC.—continued

No.	Date.	Description.	Place.	Page.
167	Dec. 29	Action	Panwari	423
168	Dec. 31	Action	Banki	398
169	Dec. 31	Affair	Chubbra	411
	1859			
170	Jan. 4	Action	Kunda Koti	400
171	Jan. 14	Action	Dosa	415
172	Jan. 15	Action	Sassaya	398
173	Jan. 15	Action	Chicumba	63
174	Jan. 21	Action	Sikar	415
174a	Feb. 9	Action	Soona Valley	400
175	March 4	Skirmish	Tons River	423
176	March 25 and 28	Actions	Nepaul	400
177	March 31	2nd Action	Tolsipore	400
178	April 1	Capture	Bungaon	400
179	May 21	Action	Jerwa Pass	401
179a	Oct. 9	Capture	Beyt Island	351
180	1857–9	Flying Columns, etc.	Bombay and Madras	
See note (v) below				

Notes to Appendix IV

(a) *Abbreviations:* p. stands for posted, t. for transferred, r. for retirement on half-pay, full pay, or pension (also for resignation), w. for wounded, d. for died, and k. for killed. ⊙ Signifies a step of honorary rank on retirement.

F.F. signifies Field Force. D.P.W. the Department of Public Works. D. of O., C. of O., Director of Ordnance, Commissary of Ordnance. Adj., Adjutant. Qrmr., Quartermaster.

N.B.—Each lieutenant-colonel or colonel in command of troops of horse artillery or companies of artillery had two officers on his staff, 2nd captains or lieutenants, who were designated Adjutant and Quartermaster.

(b) In Tables I, III, IX, and XI officers commanding batteries who occasionally acted as officers commanding artillery are not included.

(c) Lieut. E. C. Cuthbert was C. of O. at the Alam Bagh and at Pandora.

(d) When 3/14 R.A. landed at Calcutta, Lieut. J. R. Oliver was detached to Cawnpore to instruct infantry in artillery work.

(e) Lieut. G. Arbuthnot when he landed in Bombay was attached to the 14th Light Dragoons and served with them in the 2nd Division of the Central India F.F.

(f) Capt. J. H. Smyth acted as Staff Officer at Chuttra and was in command of the guns in the bridge-head entrenchment at Cawnpore on Dec. 6th. He then rejoined his troop, but was in command of a mixed force at Nagina.

ABBREVIATIONS—*continued*.

(*g*) Lieut.-Col. J. Abercrombie was on sick leave when 4/1 Ben.H.A. mutinied.

(*h*) Lieut. J. Perceval acted as C. of O. till after the capture of Lucknow.

(*i*) Capt. A. G. Austen was detached to Cawnpore at the end of November 1857.

(*j*) Lieut. H. V. Timbrell was detached to the Alam Bagh in November 1857.

(*k*) Lieut. A. Pearson served during the siege of Lucknow, but the author is unaware in what capacity.

(*l*) Lieut. E. Simeon remained in Behar with Brigr. Douglas until December 1858.

(*m*) This company had no officers present at Cawnpore (see p. 146).

(*n*) The company was commanded during the pursuit of Tantia Topi by officers of the Bombay Artillery (see pp. 404, 415).

(*o*) Captain C. Cookworthy was detailed to command 1/1 Ben.A. at Bunkagaon, as Captain A. G. Austen was on sick leave.

(*p*) Lieut. A. Light acted as orderly officer to Brigr. Seaton at Gungaree and Patiala.

(*q*) Three companies of Sikh artillery were raised, but the author has been unable to unravel their individual services.

(*r*) Major Cotter went on leave after the operations under Brigr. Lugard in Behar, and Major R. Cadell (accompanied by Lieut. S. H. E. Chamier) took over the command of A/3 Mad.A.

(*s*) C/5 Mad.A. appears to have remained at Futtehpore with Carthew's Brigade during 1858.

(*t*) By a Governor-General's Order of April 11, 1843, this troop, then commanded by Capt. and Brev.-Major J. T. Leslie, was authorised to bear an eagle on its appointments. Though the title was in no way official, it was often spoken of as the Eagle Troop.

(*u*) The author fears that possibly Table XV and probably Table XVI are incomplete. Every effort has been made to trace the services of these officers, but the records are often wanting.

(*v*) The number 180 is introduced to cover the services of officers with the numerous small columns that were sent out from time to time in Madras and Bombay, which, though they had no serious fighting, had plenty of hard work.

APPENDIX V

THE INDIAN MUTINY MEDAL

BY MAJOR J. H. LESLIE, RETIRED LIST R.A.

THIS medal was given to all—military and civilian alike—who were engaged in the suppression of the mutiny and rebellion in 1857–8.

The medal is made of silver, circular, 1·4 inch in diameter.

Obverse.—The head of Queen Victoria, diademed, with the legend VICTORIA REGINA around.

Reverse.—Britannia, standing, holding out a wreath in her right hand ; on her left arm the Union shield, and in her left hand a wreath. Behind her a lion, walking. Above, INDIA, and in the *exergue*, 1857–1858.

Mounting.—A silver cusped bar.

Ribbon.—1¼ inch wide. White, with two red vertical stripes, forming 5 stripes, each ¼ inch wide.

Clasps.—Five :

>DELHI.
>DEFENCE OF LUCKNOW.
>RELIEF OF LUCKNOW.
>LUCKNOW.
>CENTRAL INDIA.

The following General Orders by the Governor-General in India give the authority for the grant of the medal and of the several clasps.

No. 363 of August 18th, 1858. For Medal, and for Clasp "Delhi."

No. 733 of May 19th, 1859. For Clasps "Defence of Lucknow," "Lucknow," and "Central India."

(Unnumbered) January 10th, 1860. For Clasp "Relief of Lucknow."

INDEX

Abu, Mount, Rajputana, attack on, 242
Adye, J. M., Lieut.-Col. R.A., at 2nd battle of Cawnpore, 233 ; recovery of a gun by, *ibid.* note [3]
Agency, Central India, 34 etc.
Agra, mutiny at, 41–4 ; panic at, 136 ; relief of, *ibid.* ; action at, 137–40 ; operations against Tantia Topi from, 402, 415
Aislabie, R., Lieut. Ben.A., at Delhi, 15, 18
Aitken, R. H. M., Capt. 13th Ben. N.I., at defence of Residency, Lucknow, 164 note [1] ; 203
Ajudia Ghat, skirmish at, 376
Alam Bagh, Lucknow, capture of, by Havelock, 195, 196 ; McIntyre's defence of, 196, 205 ; occupation of, by Sir Colin Campbell, 213 ; garrison of, under Outram, 228 ; his defence of, 270–3 ; garrison of, partially withdrawn, 277 ; rebel attack on, 294
Alighur, Doab, mutiny at, 41–2 ; Greathed at, 135 ; Seaton at, 251
Aligunge, Rohilkhund, Walpole at, 357
Allahabad, importance of, 55 ; mutiny at, 58 ; defence of fort at, *ibid.* ; Neill's relief of, 59 ; despatch of small columns from, 247
Amalgamation of the Royal and Indian Artillery, 427 etc.
Amba Pawnee, Satpura, action at, 63
Amithi, near Lucknow, skirmish near, 377, 378
———, Baswarra District, capture of, 388
Amorha, Oudh, action at, 364
Anderson, J. R., Capt. and Brev.-Major R.H.A., at Pandora, 248
Anson, Gen. the Hon. G., Commander-in-Chief in Bengal, 5 ; action by, at outbreak of Mutiny, 19, 20 ; death of, 21
Aong, Doab, action at, 170–1
Arbuthnot, G., Lieut. R.A., at Kunch,

320 ; recommended for V.C., 320 note [1]
Arbuthnot, H. T., Lieut. R.H.A., at Dhowrara, 268–70 ; at Banki, 399
Arrah, siege of, 185–6 ; relief of, 160
Arsenal, Ajmere, 32 ; Delhi, defence of 15–17 ; Phillour, 24 ; Ferozepore, *ibid.* ; Cawnpore, 53 ; defensibility of, *ibid.* ; value of, to Nana Sahib, 150
Artillery, scarcity of, at Delhi, 78, 80 note [1] ; difficulty in relief of, at Delhi, 127 ; good record of Indian, 426 ; composition of, 433 etc.
Ashe, St. G., Lieut. Ben. A., at Cawnpore, 147
Assault, Birwah, Oudh, 384, 385 ; Chanderi, 305 ; Delhi (possibility of), 103 etc. ; of Sept. 14, '57, 118 etc. ; Jhansi, 313 etc. ; Lucknow, Begum's Palace, 285 ; Kaisar Bagh, 290
Atraulia, fort in Oudh, captured by Longden, 246 ; assemblage of rebels at, 365
Auchinleck, J. C., Lieut. R.A., at Kankar, 355
Awah, revolt of Thakhur of, 242, 243 ; in rebel hands, 244 ; capture of, by Roberts, 315
Azimghur, mutiny at, 56, 60–1 ; Gurkha force at, 191 ; held by Franks, 247, 261 ; threatened by Mahndi Husain, 364 ; combats round and relief of, 365 ; Kelly's force at, 387

Badli-ki-Serai, action at, 70 etc.
Badshahgunge, Sultanpore, fighting at, 266, 267
Bagrode, Central India, affair at, 408
Bailey Guard, Lucknow Residency, 161, 203, 227, *ibid.* note [1]
Baird Smith, R., Lieut.-Col. Ben.E., chief engineer at Delhi, 104 ; views as to assault on Delhi, 105 ; his discussions with Wilson, 105 etc. ; his relations with Alexander

Taylor, plans and discussions, 107 etc.
Bala Rao, Nana Sahib's brother, activity of, 352; his defeat at Tolsipore, 397; his defeat in Soona Valley, 401
Banas River, affluent of Chumbal, action on, 404
Banda, mutiny at, 36; disaffection of State troops, 37; rebels from Behar congregate in, 191; 207; battle of, 324
—— Nawab of, asserts his independence, 36; welcomes Behar rebels, 191; strong position of, 207; defeated by Whitlock, 324–5; joins Rao Sahib and Begum of Jhansi at Kalpi, 323, 325; defeated at Gulauli, 330, etc.; joins Tantia Topi, 337; defeated at Morar, Gwalior, and Jaora-Alipore, 339 etc.; his hopes of organising a Mahratta rising, 402; his adventures in company with Tantia Topi, June—Nov. '58, 402–9; submits to Government, 409
Banki, Nepaul, action at, 398, 399
Bank's House, Lucknow, 215, 225, 281–5
—— J. S., Capt. and Brev.-Major 35th Ben.N.I., Asst. Commissioner at Lucknow, 165
Banpore, Rajah of, seizes Chanderi, 37; joins Rajah of Sharghur, occupation of passes into Bundelkhund, 40; seizure of forts of Rahatghur and Sarakota, *ibid.*; attempts to raise siege of Rahatghur, 298; defeated at Barodia, 300–1; defeated at Bundelkhund passes, 302; palace captured, 303; checked by column from Jhansi, 317–18; reasserts his authority in Chanderi, 349; submits, with Rajah of Sharghur, to Government, 423
Bantara, Oudh, skirmish at, 211; camp at, *ibid.*; siege train at, 259
Bareilly, mutiny at, 41; British pretence of attack on, 250; concentration of columns upon, 354–9; battle of, 359–61; garrison of, 363
Bari, Oudh, action at, 372–3
Barker, G. R., Lieut.-Col. and Brev.-Col. R.A., in command of a column from Allahabad, 248; brigadier commanding siege artillery, Lucknow, 276 etc.; in command of a column at Jamo, Oudh, 383; captures fort of Birwah, 384, 385; captures fort of Ruiya, 385
Barnard, Major-Gen. Sir H., commands Sirhind Division, 21; commands Delhi Field Force, *ibid.*; his advance on Delhi, 70; defeats rebels at Badli-ki-Serai, 70–2; occupies Ridge, 72–4; his attitude regarding a *coup de main* on Delhi, 102 etc.; his death, 88
Barnes, C. H., Lieut. Ben.H.A., at Jiran, Rajputana, 243
Barodia, Central India, action at, 300, 301
Barrackpore, outbreak at, 8, 9
Barracks, European, Lucknow, 219, 286
Barry, W. W., Capt. and Brev.-Major R.A., assists to serve guns of 6/13 R.A., 223
Bassiratgunge, Oudh, actions at, 177–9
Basti, Benares District, operations near, 370, 386
Baswarra District, Oudh, trouble in, 374 etc.
Batson, S. H., Surgeon Indian Medical Establishment, attempts to reach Meerut from the Ridge, 19 note [1]
Bayly, N. S. K., Capt. R.A., at Beyt Island, 350
Becher,[1] C. G., Capt. and Lieut.-Col. 5th Ben.L.C., at Bagrode, 408
Begum: *see* Bhopal, Delhi, Jhansi, Lucknow, Oudh
Begum's Palace at Lucknow, capture of, 285–6
Behar, mutiny in, 60, 61; dangerous state of, 185, 192; arrival of Rowcroft's Field Force in, 247; serious trouble in, 353; Lugard's campaign in, 366 etc. *See* Arrah, Dinapore, Kunwar Sing.
Belwa, Benares District, Mahndi Husain's entrenched camp in, 364; its strength, *ibid.*; rebels retire from, 370
Benares City, unprotected state of, 55, 56; the disarming parade at, 56, 57; dispersal of mutineers at, 57; its tranquillity assured, 59
—— District, mutinies in, 60–1; arrival of force of Gurkhas from Nepaul, 191; Longden and Frank's operations in, 247; Rowcroft's force and Gurkhas in, 261; Mahndi Husain and Kunwar Sing's at-

[1] The author is not absolutely sure of this officer's identity.

INDEX   497

tempts in, 364 etc.; expulsion of rebels from, 370
Bengal Army, reorganisation of, 424
Beni Madho, Oudh Talukdar, his action against the British, 389–92
Benson, H. R., Lieut.-Col. 17th Lancers, at Zirapore, 411
Beora, Central India, action at, 406
Berhampore, mutiny at, 7, 8
Berkeley, Brig. C. A. F., Lieut.-Col. and Brevet-Col. 32nd Foot; his operations, 378–9
Betwa River, affluent of Jumna, battle of the, 309–13
Bewsey, J., Farrier Serg. Ben.A., escape from Secrora, 51; gallant conduct after battle of Chinhut, 158 note [1]
Beyt Island, attack on, 350; capture of, 351
Bhopal, loyalty of Begum of, 38; mutiny of contingent, *ibid.*; reinforcements from, to Central India Field Force, 298
Bibigunge, Behar, action at, 189
Bijipur, Central India, affair at, 413
Birwah, Oudh, siege and capture of, 384
Bithur, Doab, residence of deposed Peshwa, residence of Nana Sahib, 51, 52; sack of palace at, 175; action at, 179–81
Blunt, C. H., 1st Lieut. Ben.H.A., commands mortar battery at Delhi, 116; commands 2/3 Ben. H.A., with Greathed's column, 132 etc.; at Bulandshahr, 133–6; at Agra, 137–40; at Bantara with Grant, 211; at 2nd Relief of Lucknow, 213 etc.; at Sekundra Bagh, 217; at 2nd battle of Cawnpore, 236 etc.; at Shamshabad, 257 etc.; at Gulauli and Kalpi, 329 etc.
Bognipore, Doab, skirmish at, 259
Bombay Presidency, energetic action of Governor of, 61, 62; risings in, 62, 63; loyalty of army of, 62
Bone dust said to have been mixed with commissariat flour, 5 note [2]
Bonham, J., 2nd Lieut. Ben.A., escape from Secrora, 49–51; at Chinhut, 154–9; at Lucknow Residency, 161 etc.; his "ship," 162; his lucky mortar shot, 163 note [2]
Bonus, J., Lieut. Bom.E., at Lohari, 319
Boorhya-ki-Chowki, Oudh, action at, 179
Bourchier, G., Capt. Ben.A., pursuit of rebels after 2nd battle of Cawnpore, 238
Boyle, Mr. Vicars, civil engineer at Arrah, 185 etc.
Brahmapootra, s.s., sent up Ganges from Allahabad, 167; arrives at Cawnpore, 176
Brasyer, J., Lieut. Ferozepore Reg., at Allahabad, 59; with Havelock's column, 166 etc.; at 1st Relief of Lucknow, 195 etc.; at Alam Bagh with Outram, 228, 270–3; at Siege of Lucknow, 276 etc.; at assault of Kaisar Bagh, 290
Brennan, P., Bombardier 5/14 R.A., gains V.C. at Jhansi, 309
Brind, Brig. F., Colonel Ben.A., murdered at Sialkote, 30
—— J., Major Ben.A., at Siege of Delhi, 114; with Troup's Column, 386; in command of troops at Mehndi, *ibid.*
Buckley, Conductor, at Magazine Delhi, 16, 17; his escape, 17; is given a commission, *ibid.*, note [2]; gains V.C., *ibid.*
Bundelkhund, passes of, action at, 303
Bungaon, fort in Oudh, capture of, 400
Bunkagaon, Rohilkhund, action at, 385
Bukt Khan, rebel leader, 41
Bulandshahr, Doab, action at, 133–5
Bullocks' blood said to have been used to refine sugar, 5 note [2]
Burgeedia, Northern Oudh, action at, 397
Burgess, F., Corpl. Ben. Sappers and Miners, at Kashmir Gate, Delhi, 121
Burnai, Rohilkhund, action at, 363
Bururiah, on the Rapti, action at, 386
Bygrave, G., Gunner Ben.A., at Arrah, 369

Camel Corps, at Gulauli, 329–35; in pursuit of Tantia Topi, 414, 416 note [1]
Campbell, Brig. G., Lieut.-Col. and Brev.-Col. 52nd Foot, commands No. 3 Column at Siege of Delhi, 119
—— Brig. W., Lieut.-Col. and Brev.-Col. 2nd D.Gs., in command at Allahabad, 247; in command 2nd Cavalry Division at Siege of Lucknow, 275; on western flank of besieging army, 277; mutineers evade his pursuit, 293, 295

Campbell, J. H., Major Ben.A., commands left portion of No. 2 Siege Battery at Delhi, 115; severely wounded, *ibid.*
—— R. D., Lieut.-Col. 71st Foot, in temporary command of 2nd Brig. Central India Field Force, 327; *ibid.* note [1]
Campbell, R. P., Lieut.-Col. 90th Foot, at 1st Relief of Lucknow, 200 etc.
—— Sir Colin, Lieut.-Gen., appointed Commander-in-Chief in India, 206; his preparations, 206-7; goes to the front, 210 etc.; his narrow escape, 211 note [1]; carries out 2nd Relief of Lucknow, 213-27; his plans, 228; his timely return to Cawnpore, 235; defeats rebels at 2nd battle of Cawnpore, 237-41; his anger with 53rd Foot at Khudagunge, 255; his feint on Rohilkhund, 256; in camp at Bantara, 260; assembles Army of Oudh, 275, 276; carries out Siege and Capture of Lucknow, 277-96; allows mutineers to escape, 293; his plans after capture of Lucknow, 353; his advance on Bareilly, 357 etc.; defeats rebels at Bareilly, 359; pacifies Rohilkhund, 363 etc.; his measures for defence of Lucknow, 371; his plans for clearing Oudh, 379, 380; created Baron Clyde, 387 note [1]; his winter campaign in Oudh, 387 etc.; defeats rebels at Burgeediah and Banki, 397-9; returns to Lucknow, 400
Canning, Lord, Governor-General of India, action on outbreak of Mutiny, 6; his Oudh proclamation, 352; his reasons for pacifying Rohilkhund before clearing Oudh, 353, *ibid.* note [2]; meeting with Commander-in-Chief at Allahabad, 379; becomes first Viceroy of India, 387 note [1]
Captan Bazar; Lucknow, plan of, 161; cleared of enemy, 205
Carmichael, J. D., Major and Brev.-Lieut.-Col. 32nd Foot; commands a column, 391 etc.
—— Serg. Ben. Sappers and Miners, at Kashmir Gate, 121
Carpenter, T. D., Lieut.-Col. and Brev.-Col. 1st Mad.N.I., in command at Kirwi, 419 etc.
Carthew, Brig. M., Lieut.-Col. 1st Mad.E.F., commands brigade from Madras, 61; arrives at Calcutta, 192 note [2]; joins Windham at Cawnpore, 229 etc.; in action there, 232-5; at Futtehpore, 248
Cartridges, Greased, 5, 7, 10
Casualties, British and Native, given after descriptions of actions, etc., when recorded; special British lists, 444 etc.
Cawnpore, weak European garrison of, 51; Wheeler's and Hillersdon's mistakes, 52, 53; the entrenchment at, 53; mutiny of native garrison, 53, 54; defence of the entrenchment, 145-51; treachery of the Nana Sahib, 151; massacre at Satti Chaura Ghat, 152; imprisonment of women and children, 153; formation of relief columns for, 167; 1st battle of, 172-5; massacre at the Bibi Ghur, 175; Havelock's occupation of, *ibid.* etc.; his dangerous position at, 181; Outram's arrival at, 194; Wilson's Garrison, 195; Windham's position at, 200; its dangers, 229; combats round, 231-35; relief of, by Sir Colin, 235; 2nd battle of, 235-40
—— Battery, Lucknow Residency, 161, 163
Central India, Outbreaks in, 34 etc.; campaign in, 297 etc.; final pacification of, 401 etc.
—— Field Force, formation of, 249; composition and operations of, 297 etc.
Chamberlain, C., Major 1st Punjaub Cavalry, at Mooltan, 27
—— N., Colonel Commanding Punjaub Irregular Frontier Force, commands Punjaub Movable Column, 30; appointed A.G. Delhi Field Force, *ibid.*; at Subzi Mundi, 92, 93; wounded, 93; his opinion of possibility of *coup de main*, 105; supports Baird Smith as to plan of attack, 111
Chamier, S. H. E., Lieut. Mad.A., at Cawnpore, 230, 232
Chanda, action at, 263-5; rebels defeated by Gurkhas at, 207 note [1]
Chanderi, seized by Rajah of Banpore, 37; Siege of, 303-5; re-occupied by rebels, 349; rebels ejected, *ibid.*; final conquest of, 423
Chandni Chouk, at Siege of Delhi, 121
Char Bridge, Lucknow, capture of, 196-9
Charkari, Rajah of, his loyalty, 309; attacked by Tantia Topi, *ibid.*
Chattar Manzil, Lucknow, 202, 216

INDEX

Chester, Major and Brevet-Col. 25th Ben.N.I., A.G., killed at Badli-ki-Serai, 74
Chicumba, Berar, action at, 64
Chikrata, village near Kirwi, 419
Chinhut, Battle of, 154-9
Chota Gundak, affluent of the Gundak, action at, 247
—— Nagpore, outbreak in, 192, 193
—— Udepur, action at, 409
Christie, S. T., Major and Brev.-Lieut.-Col. 80th Foot, commands a column, 396
Chubbra, on the Dhasan, affair at, 411
Chumbal, affluent of the Jumna, rising of, 339, 405
Chupatti, xxiv.; 4 note ²
Chute T., Lieut.-Col. and Brev.-Col. 70th Foot, on the Khyber, 26
Chuttra, Chota Nagpore, action at, 192
Cis-Sutlej States, loyalty of, 20
Clock Tower, Lucknow, rebel battery at, 164; portion of relieving force advances through, 206
Clyde, Baron, title conferred on Sir Colin Campbell, 387 note ¹
Coke, T., Major 1st Punjaub Infantry, at Siege of Delhi, 87, 98
Colvin, J. R., Ben.S.C., Lieut.-Governor N.W. Provinces, 41; his request to Sindiah, *ibid.*; Wilson's letter to, 106; death of, 140 note ²
Connolly, W., Gunner Ben.H.A., gains Victoria Cross at Jhelum, 29
Contingents, nature of, 32 note ¹. *See* Gwalior, Malwa, Jodhpore, etc.
Cooper, G. L., Major Ben.A., commands artillery of Havelock's force, 179; death of, 204
Corbett, Brig. S., Lieut.-Col. and Brev.-Col. 16th Ben.N.I., commands at Mian Mir, 24; prompt action of, *ibid.*
Corfield, F. B., Lieut.-Col. and Brev.-Col. 6th Ben.N.I., commands column at Jugdispore, 369, 370
Cornish miners in 32nd Foot, 162 note ¹
Cotter, G. S., Capt. and Brev.-Major Mad.A., with Jaunpore Field Force at Nasratpore, 262; Chanda, 263, and Sultanpore, 265; at Siege of Lucknow, 273 etc.; in action in Residency grounds, 293; with Lugard's Column, 366 etc.
Cotton, Brig. S. J., Lieut.-Col. and Brev.-Col. 10th Foot, at Peshawar, 26; effective measures of, *ibid.*
Cotton, H., Major and Brev.-Lieut.-Col. 67th Ben.N.I., at Agra, 140

Coup de main, question of, at Delhi, 103, 105
Cracklow, G., Lieut. Ben.A., at Delhi, 117; at Bulandshahr, 135
Crawford, Brig. W. T., Lieut.-Col. R.A., commands artillery at 2nd Relief of Lucknow, 212 etc.
Crow, Conductor, at Magazine Delhi, 16, 17
Crow's Nest, Delhi Ridge, 86
Crump, C. W., Lieut. Mad.A., attached to 3/8 R.A., death of, 204
Cuthbert, E. C., Lieut. R.A., at Siege of Lucknow, 287 note ².

Daboh, on the Sindh, action at, 417
D'Aguilar, C. L., Capt. and Brev.-Lieut.-Col. R.H.A., at Sekundra, 262, 263
Daly, H., Capt. The Guides, at Delhi, 85 etc.
Dames, W. L., Major and Brev.-Lieut.-Col. 37th Foot at Azimghur, 365
Damoh, Saugor, and Nerbudda Territories, operations round, 245, 246
Delafosse, H. G., Lieut. 53rd Ben.N.I., at Cawnpore, 149 note ¹; his escape, 153 note ¹
Delhi, excited state of, 12; arrival of Meerut mutineers, loss of city, *ibid.* etc.; blowing up of magazine in Arsenal, 15; flight from cantonments, 17; seizure of ridge by British, 72; rebel defences, 74; rebel strength, 76; British strength 78, 79; defence of the ridge, 80 etc.; possibility of *coup de main*, 102 etc.; plan of attack, 107 etc.; siege batteries, 112 etc.; action taken by rebels, 117; the assault, 118 etc.; defeat of No. 4 Column, 123; action of the Cavalry Brigade, 124, 125; capture of the city 128 etc.
Dempsey, P., Private 10th Foot, gains Victoria Cross during retreat from Arrah, 188 note ¹
Dennis, J. L., Major and Brev.-Lieut.-Col. 52nd Foot, at Delhi, 127
—— M. G., Lieut.-Col. and Brev.-Col. King's Royal Rifles, commands column in Oudh, 398
Deopura, near Kalpi, skirmishes round, 327, 328
Derriabad, Oudh, skirmishes round, 378
de Teissier, H. P., Capt. Ben.A., at Delhi, 12 etc.

Dhainawan, Oudh, capture of, 379
Dhar, Rajputana, occupied by Feroz Shah, 243; action at, 244; siege and capture of, 244, 245
Dhasan, affluent of the Betwa, skirmishes on, 417
Dholpore, Central India, assembly of mutineers at, 136
—— Rajah of, his friendly conduct towards British, 37 note [2]
Dhooly Square, Lucknow, 202; death of Neill in, ibid.; massacre of sick and wounded in, 204
Dhowrara, attack on fort of, 268–70
Diamond, A., Serg. Ben.H.A., gains Victoria Cross at Bulandshahr, 135
Dilkhusha, Lucknow, 214; Sir Colin Campbell's operations about (2nd Relief), 214 etc.; (Siege and Capture of Lucknow), 277; establishments of siege batteries at, 278
Dinapore, preponderating native garrison of, 60 note [2]; mutiny at, 182 etc.; weak conduct of Division Commander, 182, 183; escape of mutineers from, 183, 184
—— Brigade (7th, 8th, and 40th Ben.N.I.), escapes to Shahabad and joins Kunwar Sing, 184; lays siege to Arrah, 186; defeats Dunbar's column, 187; defeated at Bibigunge, 189; driven from Jugdispore, 190, 191; flies to Banda, 191; defeated at Khujwah, 208; threatens Damoh, 246; final defeat of, 364 etc.
Dinkar Rao, Sindiah's minister, 37; his loyalty to the British, ibid.
Doadpore, Oudh, affair at, 380 etc.; capture of guns of Jellalabad Battery, 381, 382
Dosa, Rajputana, action at, 415
Dosta, Oudh, occupied by Kelly, 387
Douglas, Brig. J., Lieut.-Col. and Col. 79th Foot, commands 5th Infantry Brigade at Siege of Lucknow, 276 etc.; commands infantry of Lugard's Column, 367; his operations against Kunwar Sing, 367 etc.
—— C. R. G., Lieut. and Brev.-Capt. 32nd Ben.N.I., in command of Palace Guard at Delhi, 13; his murder, 14
Dowell, W., Lieut. Ben.A., at Dacca, 193 note [1]
D'Oyly, E. A. C., Capt. Ben.A., mortally wounded at Sassiah, continues to command his guns, 44 note I

Duffy, T., Private 1st Mad.E.F., gains Victoria Cross at Lucknow, 204
Dumurreagunge, on the Rapti, action at, 370
Dunbar, C., Capt. 10th Foot, marches to relieve Arrah, 186; destruction of his force, 187, 188
Dundeakhera, Fort in Oudh, 374; found evacuated, ibid.; affair at, 390
Dupuis, J. E., Major-Gen. R.A., arrives in India to command the Royal Artillery, 233 note [1]; in actions round Cawnpore (Oct. 26 to Dec. 6, 1857), 231 etc.
Durand, H. M., Lieut.-Col. Ben.E., Acting Agent-General in Central India, 38; besieged at Indore, ibid.; his escape, ibid.

Edwardes, H. B., Major and Brev.-Lieut.-Col. 1st Ben.E.F., Deputy Commissioner of Peshawar, 25; ibid. note [3]
Edwards, J. B., Lieut. R.E., at Gulauli, 331, 336
Edwards, Serg. Ben.A., at Magazine Delhi, 16, 17
Eedghur, Delhi Ridge, 77; combats round, 83, 122
Elephant, killed by round shot under rebel leader at Futtehpore, 169
—— intractability of, with howitzer, at Chinhut, 157
Ellice, C. H., Lieut.-Col. and Brev.-Col. 24th Foot, at Jhelum, 29
Elliot, M., Lieut. Ben.A., at Main Picket, Delhi Ridge, 81, 85 etc.
Elphinstone, Lord, Governor of Bombay, energetic action of, at outbreak of Mutiny, 61, 62
English, F., Major 53rd Foot, at Chuttra, 192
Entrenchment, rebel, at Delhi, 118
—— Wheeler's, at Cawnpore. See Cawnpore
Evelegh, Brig. F. C., Lieut.-Col. 20th Foot, at Newabgunge, 380; at Shunkerpore, 389; at Dundeakhera, 390; at Oomeriah, 392
Explosion, accidental, in Kaisar Bagh (Siege of Lucknow), 292; note [3]; gallant conduct of Shoeing Smith Lever, 3/14 R.A., at, ibid.
—— near Jumma Musjid (Siege of Lucknow), serious loss incurred by the Royal Engineers, 294, 295
Eyre, V., Capt. and Brev.-Major Ben.A., at Dinapore, 188; at

INDEX

Buxar, *ibid.*; his measures for relief of Arrah, 188, 189; defeats rebels at Bibigunge, 189; relieves Arrah, 190; captures Jugdispore, 191; defeats rebels at Koondum Puttee, 194; at 1st Relief of Lucknow, 194 etc.; in garrison at Lucknow, 212 etc.; acts as Brigadier of Artillery, 216 note[1]; with Outram at Alam Bagh, 228, 270-3, 294

Faddy, Brig. P. P., Lieut.-Col. R.A., on the Tons (Kentee),[1] 423

Fagan, R. C. H. B., Capt. Ben.A., helps to repulse Cavalry Raid at Delhi, 91; killed in No. 3 Siege Battery, 118

Farha Baksh, Lucknow, 202, 203

Farquhar, R., Major and Brev.-Lieut.-Col. Beluchi Battalion, at Alighur, 251

Farrell, Gunner, 5/14 R.A., good service with mortars at Gulauli, 336

Feroz Shah, self-styled Prince of the Royal House of Delhi (Shahzada), appears at Mandesar, 243; besieges Neemuch, *ibid.*; defeated at Mandesar and Goravia, 245; goes to Rahatghur, 299 note[1]; in Rohilkhund, 362; defeated at Mithoulee and Mehndi, 386; escapes across Jumna, *ibid.*, 412; defeated at Ranode and surprised at Sarpore, 414; joins Tantia Topi, 415; defeated at Dosa and Sikar, 415; flies with the Rao Sahib to Kushana, 416; escapes disguised as a pilgrim, *ibid.* note[2]

Fischer, T. M., Lieut.-Col. 18th Mad.N.I., marches from Cuttack, 61; approaches Chota Nagpore, 192; his troops, 230 note[1]; arrives at Gorruckpore, 370

Fitzgerald, R., Gunner Ben.H.A., gains Victoria Cross at Bulandshahr, 135

Forbes, J., Capt. 3rd Bom. L.C., at Barodia, 300, 301; at Kalpi, 327

Ford, A., Lieut. R.A., breaches Sekundra Bagh, 218; wounded, 221

Forrest, G., Lieut. Veteran Establishment, at Magazine Delhi, 16, 17; his escape, 17; is promoted Captain, *ibid.* note[2]; gains V.C., *ibid.*

Forts of Oudh, nature of, 46, *ibid.* note[1]; their value to the rebels, 352

Franks, Brig. T. H., Lieut.-Col. and Brev.-Col. 10th Foot, commands Jaunpore Field Force, 247; operations against Mahndi Husain, 262 etc.; defeats him at Nasratpore, Chanda, Hamirpore, and Sultanpore, 263-8; joins Sir Colin Campbell at Lucknow, 270; at Siege and Capture of Lucknow, 275 etc.; placed in command of 1st Division, 290; at capture of Kaisar Bagh, 290, 291

Fraser, Mr. S., Ben.S.C., Commissioner of Delhi, 13; murdered, 14

—— the Hon. D. McD., Capt. and Brev.-Major R.H.A., commands F Troop in Oudh Campaign, 388 etc.

Fulton, G. W. W., Capt. Ben.E., at defence of Lucknow Residency, 154 note[2]; 163 note[2], 165

Furruckabad, Nawab of, proclaimed by rebels, 45; his threatening attitude, 181; his flight to Oudh, 256

Futtehabad, village near Chanderi, skirmish at, 304

Futtehghur, outbreak at, 44; defence of fort, 45; attempt of Europeans to escape from, *ibid.*; concentration on, by Sir Colin Campbell, 250, 256

Futtehpore, Doab, Havelock's advance on, 168; action at, 169, 170; district cleared by Brig. Barker, 248; occupied by Madras troops, *ibid.*

—— Churassie, fort on the Ganges, 259; Nana Sahib reported to be in hiding at, *ibid.*

—— Sikri, village near Agra, 43

Fyzabad, operations round, 393 etc.

—— Moulvie of. *See* Moulvie

Gall, R. H., Major 14th L.Ds., at Jhansi, 308; recommended for Victoria Cross at Kunch, by Sir Hugh Rose, 320 note[1]

Galwey, M., Major and Brev.-Lieut.-Col. 1st Mad.E.F., crosses Gumti, 377; captures Rehora and Koeli, 391, 392

Garakota, fort near Saugor, seized by Rajah of Sharghur, 40; captured by Sir Hugh Rose, 302

Garvey, H. P., Acting Mate R.N., accidentally killed, 286 note[2]

Generalgunge, Cawnpore, 235 etc.

General Service Enlistment Act, 3

[1] This skirmish is sometimes known as "Kentee."

II—34

Gerrard, Brig. J. G., Lieut.-Col. 14th Ben.I., commands column at Nurnaul, 142; defeats rebels, 143; is killed, *ibid.*

Ghazis, attempt to intercept fugitives from Gwalior, 37 note [2]; desperate charge of, at Bareilly, 360

Ghazi-ud-deen-Nuggur (on the Hindun), actions at, 68 etc.

Gogra, affluent of the Ganges, Rowcroft's operations on, 247, 261; Sir Colin Campbell's operations east of, 391 etc.

Gonda, fort in Oudh, captured by Grant, 394

Goodenough, W. H., 2nd Capt. R.A., wounded severely at assault on Birwah, 385

Goravia, Rajputana, action at, 245

Gordon, D. W., Lieut. Ben.A., at Ghazipore with Eyre, 188; at Koondum Puttee, 194
—— S. E., Capt. and Brev.-Lieut.-Col. R.A., at Pandora, 248; commands column in Oudh, 391 etc.

Gorruckpore, Benares District, loyalty of portion of 12th Ben.Irr.C. at, 61; entered by Gurkha force, 191; temporarily abandoned by the British, *ibid.*; occupied by Jung Bahadur, 247, 261; held by Rowcroft, 261; arrival of Madras troops at, 370

Graham, Serg.-Major 2/4 Bom.A., at Gulauli, 332

Grand Trunk Road, from Calcutta to Allahabad, dangers of, 191-3

Grant, Brig. J. H., Lieut.-Col. and Brev.-Col. 9th Lancers, commands brigade at Badli-ki-Serai, 71; at Ochterlony Gardens, Delhi, 84, 85; commands Cavalry Brigade at assault of Delhi, 124 etc.; his tribute to the stoical courage of his command, 125 note [1]; assumes command of Greathed's Column, 140; skirmish at Bantara, 211; is second in command to Sir Colin Campbell at 2nd Relief of Lucknow, 212 etc., and at 2nd Battle of Cawnpore, 236 etc.; at the Kali River, 255; captures Miangunge, 260; commands Cavalry Division at Siege and Capture of Lucknow, 275 etc.; patrols left bank of Gumti, 284; promoted Major-General, *ibid.*, note [1]; at Seetapore, 293 etc.; in command of troops at action of Kursi, 295, 296; in command of garrison of Lucknow, 354; in command at Bari, 372, 373; at Simri, 373, 374; at Newabgunge Barabanki, 374, 375; relieves Maun Sing and enters Fyzabad, 376; at Sultanpore, *ibid.*; marches to Jugdispore (Oudh), 387; made K.C.B., *ibid.* note [1]; in Baswarra district, 388-91; superintends trans-Gogra operations, 393; at Muchligaon, 394; at Secrora, 395; marches towards N.E. frontier of Oudh, 396; orders Rowcroft to Tolsipore, 397; follows up the rebels, 400; defeats them at Kunda Koti, *ibid.*; placed in charge of frontier, *ibid.*; at Jerwa Pass, 401

Grant, General Sir P., Commander-in-Chief in Madras; appointed to succeed Anson, 65

Graves, Brig. H. M., Lieut.-Col. and Brev.-Col. 13th Ben.N.I., in command at Delhi, 12; actions during outbreak at Delhi, 14 etc.; goes to Meerut, 19; at Badli-ki-Serai, 71, 72; at the seizure of the Ridge, 72, 73; action with regard to proposed assault on Delhi, June 12, 103

Greathed, Brig. E. H., Major and Brev.-Lieut.-Col. 8th Foot, at capture of Delhi, 129; commands column from Delhi, 132; at Bulandshahr, 133 etc.; his march to Agra, 135 etc.; action at Agra, 137 etc.; relinquishes his command to Grant, 140; in command of 3rd Infantry Brigade at 2nd Relief of Lucknow, 213 etc.; at Cawnpore, 235 etc.; at the Kali River, 254

Great Imambara, Lucknow, 293

Green, D. S., 2nd Capt. R.A., at Cawnpore, 232, *ibid* note [2]

Gulauli, village on the Jumna, 326; Sir Hugh Rose's position at, 328, 329; battle of, 330-6

Gumti, affluent of the Ganges, passage by Outram, 279; by Galwey at Sultanpore, 377; skirmish on banks of, 392

Gundak, affluent of the Ganges, 207

Gungaree, Doab, action at, 251, 252

Guns, temporary loss of, at Jhelum, 30; at Sassiah, 43; at Cawnpore, 233 note [3]; at Rawal, 245

Gurkhas, enter Gorruckpore, 191; their operations in Benares District, 207; join Rowcroft, 261; march on Lucknow, *ibid.*; occupy Char Bagh and Ganeshgunge, 294;

INDEX

successful action against rebels, *ibid.*
Gurwa, affluent of the Gogra; Christie's operations on, 398
Gwalior, mutiny at, 37, 38; seized by Tantia Topi and the Begum of Jhansi, 337, 338; battle at, 344–6; capture of fort, 347
—— Contingent, confidence reposed in, 37; mutiny of detachments, *ibid.*; main body held in hand by Sindiah and Dinkar Rao, 37, 38; reinforcements from, sent to the Doab, 41; their mutiny, *ibid.*, 42; persuaded to rise by Begum of Jhansi, 210; threaten Cawnpore from Kalpi, 210; attempt on Cawnpore, 231–5; defeat of, in 2nd battle of Cawnpore, 237 etc.; capture of their camp, 238

Haggard, T. T., Lieut. Bom. A., commands siege train at Chanderi, 303 etc., at Gulauli, 331, 332; at battle of Gwalior, 344 note [1]
Hale, E. B., Lieut.-Col. 82nd Foot, commands at Shahjehanpore, 358; besieged for eight days, 362; joins Brig. Barker, 385
Hamirpore District, mutiny at, 36; Whitlock's operations in, 419
——, Oudh, skirmish at, 265
Hansi, Delhi District, mutiny at, 45
Hardy, W. N., Capt. R.A., at 2nd Relief of Lucknow, 210 note [2], 214 etc.; killed, 221
Harward, T. N., Lieut. Ben.A., at Allahabad, 58; his narrow escape, *ibid.*; with Renaud's column, 167; attached to 3/8 R.A., 168 note [1]
Havelock, H., Major-General, arrives at Allahabad, 60; approves Neill's measures for relief of Cawnpore, 167; at Futtehpore, 168; defeats rebels there, 169, 170; at Aong and Pandoo River, 170–2; at 1st Battle of Cawnpore, 172–5; crosses Ganges, 176; defeats rebels at Unao and Bassiratgunge, 176–8; retires to Mangalwar, 178; advances to Bassiratgunge and defeats rebels at Boorhya-ki-Chowki, 179; returns to Mangalwar, *ibid.*; defeats rebels at Bithur, 179–81; his dangerous position, 181; joined by Outram, 194; retains command of Field Force, *ibid.*; captures the Alam Bagh, 194, 195;

relieves Residency (1st Relief of Lucknow), 196–205; continues to command Field Force under Outram, who takes command of Field Force and Garrison, 205; watches Sir Colin Campbell's advance, 223; his meeting with him, 225; made K.C.B., 228 note [2]; his death, 228
Hawthorn, Bugler 52nd Foot, at Kashmir Gate, 121; awarded the Victoria Cross, *ibid.* note [1]
Hazaribagh, Behar, mutiny at, 192
Hewitt, W. H., Major-General in command of Meerut Division, 9; fails to deal with the situation when the mutiny breaks out, 10–12; replaced by Penny, 12
Hicks, T. W., Lieut.-Col. Bom.A., commands artillery at Kotah-ki-Serai and Gwalior, 341 etc.
Hillersdon, C. G., magistrate at Cawnpore, 51; his trust in the Nana Sahib, 52
Hilliard, G. T., Capt. 50th Mad.N.I., defeats rebels at Tola Kanjaran, 419
Hills,[1] J., Lieut. Ben.H.A., on picket at the Mound, Delhi, July 9, 88; is attacked by rebel cavalry, 89; his single-handed resistance, 90; rescued by Tombs, *ibid.*; wounded, 91; gains the Victoria Cross, 91 note [1]
Hindun, affluent of Jumna, 68; actions on, 69 etc.
Hindu Rao's House, Ridge, Delhi, 80; *ibid.* note [1]; main picket established at, 81
Hir, village north of Rapti, Rowcroft at, 386, 397
Hirun Khana, Lucknow, 203; mined for sortie, 216
Hissar, Delhi District, mutiny at, 45
Hodson, W. S. R., Capt. Hodson's Horse, 130; shoots Delhi Princes, *ibid.*; with Greathed's column, 133 etc.; with Sir Colin Campbell at 2nd Relief of Lucknow, 2nd Battle of Cawnpore, etc., 212 etc.; at Gungaree with Seaton, 252; at Shamshabad with Hope, 257; with army of Oudh, 257 [2]
Holkar, Maharajah of Indore, 34; apologises for mutiny of his troops and offers to co-operate in preserving order, 39
Holkar's State troops, mutiny of, 38, 39; disarmed, 245

[1] The present Sir James Hills-Johnes; the second name was assumed in 1883.
[2] Hodson was killed in the attack on the Kaisar Bagh, March 14, 1858.

INDEX

Holmes, J., Major and Brev.-Lieut.-Col. 12th Bom.N.I., commands column in Rajputana, 403; at Sanganeer, *ibid.*; at the Banas, 404
Home, D. C., Lieut. Ben.E., commands Engineer party at assault on Kashmir Gate, Delhi, 121; gains Victoria Cross, *ibid.*, note [1]; accidentally killed at Malaghur, 135
Honner, Brig. R. W., Lieut.-Col. and Brev.-Col. 4th Bom.N.I., in pursuit of Tantia Topi, 416
Hoolahan, J., Corporal Ben.A., commands two howitzers at Arrah, 368; killed, 369
Hope, Brig. the Hon. A., commands 4th Infantry Brigade at 2nd Relief of Lucknow, 213; at the Sekundra Bagh, 218-20; at the Shah Najaf, 222, 223; at 2nd Battle of Cawnpore, 236 etc.; at Kali River, 254 etc.; in command at Shamshabad, 257 etc.; with Army of Oudh, 276; at Siege and Capture of Lucknow, 277 etc.; at the Kaisar Bagh, 286; with Walpole's column, 356; killed at Ruiya, *ibid.* note [1]
Horsford, Brig. A. H., Lieut.-Col. and Brev.-Col. Rifle Brigade, commands 6th Brigade at Lucknow, 276; his operations at Sultanpore, 376, 377; at Doadpore, 380-383; in Baswara District, 391; enters Nepaul, 400
Houghey, A., Gunner Ben.A., at Arrah, 369
Hungerford, T. J. W., Capt. Ben.A., at Mhow, 38; saves the fort, 39; with Malwa Field Force, 244, 245
Hunter, C., Lieut. Ben.H.A., at the Sekundra Bagh, 219, 220; at Bari, 373
Hyderabad, a potential danger, 63; loyalty of Nizam, *ibid.*; contingent sent to reinforce Bombay troops, 63; outbreak in, by Rajah of Shorapore, 64

Indore, Holkar's capital, 38; defence of the Residency at, *ibid.*; Holkar's troops disarmed at, 245; Tantia Topi's attempt to reach, 405 etc.
Inglis, J. E. W., Lieut.-Col. and Brev.-Col. 32nd Foot; marches to Kokral Bridge, 154; at Battle of Chinhut, *ibid.* etc.; appointed Brigadier to command Lucknow Residency, 165; *ibid.* note [1]; relieved by Havelock, 204; retains command of garrison, 205; the last to leave the Residency, 227, note [1]; appointed to command 5th Brigade, 235, *ibid.* note [1]; at 2nd Battle of Cawnpore, 236 etc.
Innes, J. J. McL., Lieut. Ben.E., gains Victoria Cross at Sultanpore, 267; at Dhowrara, 269
Iron Bridge, Lucknow, 159; Outram's operations about, 287 etc.
Ismailgunge, village near Lucknow, 155, 280 etc.
Issuri Pandy, Jemadar, mutinies at Berhampore, 8, *ibid.* note [2]; court martial on, 13

Jacob, G. Le G., Col. 27th Bom.N.I., at Kohlapore, 62
—— G. O., Major 1st Ben.E.F., at Metcalfe's House, 98
Jajjhar, Delhi District, disturbance in, 142
Jalaon and Hamirpore District, mutiny at, 36; operations about, 418
Jamo, Oudh, affair at, 383, 384
Jaora-Alipore, action at, 347, 348
Jaunpore, Benares District, Longden at, 246; made a brigade commander under Franks, 247, 261
Jellalabad, mural crown for, on captured rebel guns, 383
Jerwa Pass, action at, 401
Jeypore, Tantia Topi's attempts to reach, 402, 415; troops of, defeated by Jodhpore Legion, 141
Jhalawar, Rajah of, his loyalty, 405; deserted by his troops and robbed by Tantia Topi, 405
Jhansi, sacred Hindu city, 35, *ibid.* note [2]; outbreak at, 36; defence of Star Fort, *ibid.*; massacre of Europeans, *ibid.*; defences of, 305, 306; *ibid.* note [1]; siege of, 307 etc.; Tantia Topi's attempt to relieve siege of, 309 etc.; storming of, 313 etc.; garrison left by Sir Hugh Rose, 318, *ibid.* note [1]; its weakness, 349; reinforced, 401; operations in the district, 417 etc.
—— Begum of, her hatred of the British, 35; her treachery, 36; persuades Gwalior Contingent to declare for the Peshwa, 210; her escape from Jhansi, 314; at Kalpi with the Rao Sahib, 319; begs for an army, *ibid.*; her plan to seize Gwalior, 337; her temporary success, 338; her death at Kotahki-Serai, 343 note [1]
Jhelum, action at, 29, 30

INDEX 505

Jind, Cis-Sutlej State, loyalty, 20; contingent supplied by, 70
Jiran, Malwa, attack on, 243
Jodhpore, State troops of, sent to garrison Ajmere, 32
—— Legion, composition of, 242 note [1]; portion of, attacks Mount Abu, 242; joins in revolt at Awah, 242-3; defeats Jeypore troops, 141; defeated and scattered at Nurnaul, 142 etc.
Johannes House, Lucknow, 161; capture of, 163
Johnson, A. W., 2nd Capt. R.A., at Beyt Island, 350
Jokun Bagh, Jhansi, 308 note [2], 313
Jones, Brig. H. R., Lieut.-Col. and Brev.-Col. 6th D.Gs.; succeeds Penny in command of Meerut column, 358; at Battle of Bareilly, 359 etc.; with Sir Colin Campbell in Oudh, 390 etc.
—— Brig. J., Lieut.-Col. and Brev.-Col. 60th Rifles, commands Rurkhi column, 354; at Battle of Bareilly, 359 etc.; sent to relieve Shahjehanpore, 362 etc.; pacifies Rohilkhund, 363
—— Brig. W., Lieut.-Col. and Brev.-Col., 61st Foot; commands No. 2 Column in assault of Delhi, 119, 120
Jubbulpore, good service of 52nd Ben.N.I. at, 40; arrival of Nagpore Movable Column at, 245; mutiny of 52nd Ben.N.I. at, 246; saved by Madras troops, *ibid.*; Whitlock at, 323
Jugdispore, Behar, Kunwar Sing retreats to, 190; is defeated at, by Eyre, 191; Kunwar Sing again retreats to, 367: jungle round, cleared, 369 etc.
—— Oudh, operations near, 388, *ibid.* note [1]
Jullundur, disarming fiasco at, 28; escape of rebels, *ibid.* 29
Jumma Musjid, Delhi, 121, 129
—— —— Lucknow, 294
Jumni, affluent of the Betwa, 407; the Rao Sahib's attempt to turn the protected fords of, *ibid.*
Jung Bahadur, Chief Minister of Nepaul, 191 note [1]; tenders help to Government, *ibid.*; enters British territory, 247; on the Gogra, 261; at Lucknow, 294

Kabul Gate of Delhi, 120, 123, 125
Kadam Rasul, Lucknow, 200, 220, 221

Kaisar Bagh, Lucknow, occupied by rebels during 1st Relief, 200; their citadel during siege, 273; storming of, 290, 291
Kali River, affluent of Ganges, action on, 254, 255
Kalpi on the Jumna, 181; importance of bridge at, 210; assembly of mutineers at, 319; its strong position, 326; skirmishes round, 327; capture of, 336, 337
Kampti, movable column assembles at, 61; Saugor Field Force assembles at, 250; departure of Whitlock from, 323
Kandu River, affluent of Gumti, Horsford's operations at, 380, 381; cavalry skirmish at, 383
Kankar, Rohilkhund, action at, 355, 356
Kano j, Doab, skirmish near, 140
Kantur, Oudh, skirmish at, 380
Karabala, Lucknow, capture of, 285
Karee River, affluent of Jumna, 137, 140
Karpurthala, Punjaub, contingent from, 378, *ibid.* note [1]
Kashmir Bastion, Delhi, 107; silenced and breached, 115; breach reported practicable and reconnoitred by daylight, 118; *ibid.* note [2]; assault on, 120
—— Gate, Delhi, 14; main guard in proximity of, *ibid.*; events of May 10 at, 14, 15; blown in by Home and his party, 121
Kavanagh, Mr. T. H., escapes from Lucknow and reports to Sir Colin Campbell, 213
Kaye, E., Capt. and Brev.-Major Ben.A., at Badli-ki-Serai, 71, 72; at Delhi, 114
Kelly, R. D., Lieut.-Col. 34th Foot, at Azimghur, 369; occupies Dosta and Tunda, 387; enters Nepaul and defeats [5] rebels near Butwul, 400
Ken River, affluent of Jumna, 324
Kerr, Lord Mark, Lieut.-Col. and Brev.-Col. 13th Foot, relieves Azimghur, 365
Khan, Bahadur Khan, self-styled "Regent" in Bareilly, 41; defeated by Troup, 386
Khas Bazar, Lucknow, 202 etc.
Khujwa, Doab, action at, 208, 209
Kirwi, hostility of Raos of, 348; capture of, 349
—— Palace, attack on, 421 etc.
Kishangunge, Delhi, actions in, 82,

83 ; rebel guns placed in, 118 ; strongly held, 124 ; evacuated, 128
Kishenghur, Rajputana, 415
Koeli, fort in Oudh, capture of, 392
Kohlapore, South Mahratta Country, outbreak in, 62
Kolwan on the Betwa, 310, 311
Koondum Puttee on the Ganges, affair at, 194
Kotah, Rajputana, mutiny at, 243 ; siege and capture of, 316 etc.
—— Contingent, trusted and sent to N. W. Provinces, 33; temporary good behaviour of, 42 ; joins rebel Neemuch Brigade, *ibid*. ; in action at Sassiah, *ibid*.
—— State troops ; used as temporary garrison at Neemuch, 34 ; mutiny at Kotah and murder Resident, 243 ; besiege Maharao in his palace, *ibid*.
Kotah-ki-Serai, Gwalior District, arrival of Smith at, 341 ; action at, 342, etc.
Kothi, Bundelkhund, Carpenter's march on, 421
Kubrai, near Banda, skirmish at, 324
Kudsia Bagh, Delhi, 109 ; occupation of, proposed by Taylor, 109 ; and carried out, 112
Kunch, Hamirpore district, battle of, 319–22
Kunda Koti, Nepaul Hills, action at, 400
Kundri, Central India, affair at, 413
Kunwar Sing, Rajput, 184 ; joined by mutineers from Dinapore, *ibid*. ; attacks Arrah, 184 etc. ; defeats Dunbar, 186, 187 ; driven back to and defeated at Jugdispore, 190, 191 ; flies to Banda, 191, 207 ; marches to join Mahndi Husain, 364, 365 ; at Atraulia, 365 ; defends bridge over Tons, 366 ; retreats to Jugdispore, 367 ; wounded at Sheopur Ghat, 368 ; defeats Le Grand, 368, 369 ; dies of wound, 369
Kurai on the Bina, action at, 408
Kurshid Manzil (Mess House), at Lucknow, 199, *ibid*. note [2] ; occupied by rebels during 2nd Relief, 223 ; captured, 225
Kursi, action at, 295, 296
Kyle, S., Lieut. R.A., commands mortars in operations against Kunwar Sing, 366–70

Lahore, steps taken at, when Mutiny breaks out, 23, 24
—— Gate, Delhi, 4th Column enters Delhi by, 119 ; fire opened from, on Cavalry Brigade, 125 ; capture of, 129
Lall Madho, Talukdar of Amithi, 388 ; his surrender, 389
Lang, A. M., Lieut. Ben. E., reconnoitres breach in Kashmir Bastion by daylight, 118 note [2]
Lashkar, New City of Gwalior, 343, *ibid*. note [2] ; capture of, 346
Lavie, T., Major Mad.A., at Banda, 324
Lawrence, G. St. P., Colonel and Agent, General, Rajputana, actions on outbreak of Mutiny, 32, 33 ; his critical position, 34 ; appointed Brigadier, *ibid*. ; his position at end of 1857, 243 ; relieved in military command by Roberts, 315
—— Sir H. M., Lieut.-Col. and Brev.-Col. Ben.A., Chief Commissioner of Oudh, 46 ; his wise measures, *ibid*. ; his prescience, 47 ; his actions after outbreak of Mutiny, 48 ; his anxiety for Cawnpore, 145 ; his inability to help, 151 ; at Battle of Chinhut, 154 etc. ; his defence of the Residency, 160 etc. ; mortally wounded, 164 ; promoted after his death, 463 note [1]
—— Sir John, Chief Commissioner of the Punjaub, 23 ; his views as to Delhi and his strong measures, 25 etc. ; his pressure on Wilson, 102 etc.
—— R. C., Colonel in political charge of Kashmir Contingent, 122 note [1] ; at the assault of Delhi, 123
L'Estrange, F. W., Capt. 5th Foot, at Bibigunge, 189
Le Grand, A. J., Capt. 35th Foot, his disastrous attempt at Arrah, 368, 369
Lever, Shoeing Smith, R.A., gallant conduct at capture of Kaisar Bagh, 292, note [3]
Liddell, J., Lieut.-Col. 3rd Bom. E.R., at storming of Jhansi, 313, 314 ; commands Jhansi garrison, 318 note [1] ; intercepts Tantia Topi on Betwa, 408 ; pacifies country between Betwa and Dhasan, 417
Lightfoot, J. G., Capt. Bom.A., with 2nd Brig. Central India Field Force, 297 ; commands 1st Troop Bom.H.A. after Turnbull is killed,

INDEX 507

314; at Kunch, 320 etc.; at Gulauli, 331 etc.; at Morar, 340; at Jaora-Alipore, 347, 348

Lindsay, A. H., Lieut. Ben.H.A., with Cavalry Brigade at assault of Delhi, 125; at Shamshabad, 259

Lines of communication, Calcutta to Allahabad, 65, 166, 167; dangers to, 181 etc.; 191 etc.; 207 etc.

—— ——— Punjaub to Delhi, protected by local chiefs, 20; critical position regarding, 65, 132

Little, Brig. A., Lieut.-Col. and Brev.-Col. 9th Lancers, commands Cavalry Brigade in 2nd Relief of Lucknow, 212 etc.

Little Imambara, Lucknow, 289; breached and captured, 290

Lloyd, G. W. A., Maj.-Gen. commanding Dinapore Div., 182; his inaction at Dinapore, *ibid.*; allows mutineers to escape, 183; removed from his command, 193

Lohari, village near Poonch, affair of, 318, 319

Longden, C. S., Capt. R.A., at 2nd Relief of Lucknow, 212, 217, 221; at 2nd Battle of Cawnpore, 237, 239

—— H. E., Major and Brev.-Lieut.-Col. 10th Foot at Atraulia, 246

Longfield, J., Lieut.-Col. and Brev.-Col. 8th Foot, commands Reserve at Assault of Delhi, 126

Lowry, J. C. J., Lieut. R.A., in skirmishes outside Kalpi, 327, 328

Lowth, R. H., Lieut.-Col. 86th Foot, at storming of Jhansi, 313; at Gulauli, 330, etc.

Loyal Sepoys. *See* Services and Chap. II., App. 66, 67

Lucherra, on the Dhasan, skirmish at, 419

Lucknow, attracts many rebels, 6; Lawrence's precautions at, 48; outbreak in, 49; Lawrence's hold on, 51; loss of city except Residency after Chinhut, 159; defence of Residency at, 159-66; 1st Relief of, 164-6; 196-205; 2nd Relief of, 215-26; evacuation of, by British, 226-8; invested by Sir Colin Campbell, 277; rebel defences in, 273-5; Siege and Capture of, 278-95; escape of mutineers from, 295; prepared for defence by British, 371; garrison of, 372

Ludlow Castle, Delhi, Commissioner of Delhi's residence, 13; slackness of rebel pickets at, noted by Taylor, 109; Nicholson's nocturnal visit to, 111; occupied by British, 112

Lugard, Brig. Sir E., Major and Brev.-Col. 29th Foot; commands 2nd Div. Army of Oudh, 276; at Siege and Capture of Lucknow, *ibid.* etc.; marches with a column to Benares District and Behar, 353, 366; his pacification of those districts, 366-70

Lumsden, W. H., Lieut. 1st P.I., at Najafghur, 100; killed, *ibid.*

Lyon, F., Lieut. R.H.A., at Doadpore, 380-2

Maberly, E., Lieut.-Col. R.A., commands artillery of Jaunpore Field Force, 261; at Chanda, 264; with siege artillery at Lucknow, 276; commands artillery at Kursi, 296; and Bari, 373

MacDuff, Brig. J., Lieut.-Col. and Brev.-Col. 74th Foot, commands 2nd Brig. Saugor Field Force, 348; at Banda, *ibid.*; at Jalaon, 418; at Sahao, *ibid.*; clears district, *ibid.*

MacFarlan, D., 2nd Lieut. Ben.A., at Chinhut, 157; severely wounded, 165 note [4]

Machi Bhawan, Lucknow, prepared for defence, 48; magazine is blown up, 160; occupied by British during Siege of Lucknow, 293

Macintyre, A. W., Capt. Mad.A., at Banda, 324

Macneill, J. M., Lieut. 53rd Foot, at Khujwah, 209, note [1]

Madanpore, Pass of, Bundelkhund, 303

Madhoo, Havildar Ben. Sappers and Miners, at Kashmir Gate, Delhi, 121

Madras, little affected by mutiny, 61. *See* Hyderabad

Magazine, Allahabad, prepared for destruction, 58, 59; Main Cartridge, at Delhi, blown up, 15-17; outside Delhi, falls into rebels' hands, 19

Mahrattas, the hopes and aims of the, 54, 209, 210, 402

Mahdi Husain, self-styled "Nazim," appears at Sultanpore, 262; defeated at Chanda, 263; and at Sultanpore, 265 etc.; disappears, 268; reappears at Belwa, Benares District, 364; defeated by Rowcroft, *ibid.*; assists Kunwar Sing, 365;

driven across Rapti at Dumurreagunge, 370 ; defeated at Bururiah, 386 ; surrenders in Nepaul, 400
Main Guard, Delhi, May 10, events at, 14, 15, 17, 18
Main Picket, Delhi Ridge, at Hindu Rao's House, 80 etc.
Mainpuri, Doab, skirmish at, 253
Malaghur, fort near Delhi, blown up, 135 ; accidental death of Lieut. D. C. Home, Ben.E., at, *ibid.* note [2]
Malcolm, G., Colonel South Mahratta Horse, at Nargund, 63
Malthura, Golandauz, Kotah Contingent, loyal behaviour of, 43 note [1]
Malwa, geographical area in Central India, 34
—— Contingent, composition of, 35 note [1] ; attacked by Feroz Shah, 245; desertions from, *ibid.*; Field Force, operations of, 244-6
——Division, formation of, 349 ; at Mhow, 402
Mandesar, Feroz Shah appears at, 245 ; action at, 245
Mangalwar, Oudh, Havelock at, 176, 178 ; skirmish at, 195
Mangaoli, on the Betwa, action at, 406
Mangles, R. L. R., Bengal Civil Service, gains V.C. for his conduct during retreat from Arrah, 188 note [1]
Mannahar, Benares district, skirmish at, 367
Mansfield, Major.-Gen.[1] W. R., Lieut.-Col. unattached, Chief of the Staff, 236 ; at the Soubadar's Tank, Cawnpore, 239, 240 ; at Siege of Lucknow, 292, note [2]
Marches, Rapid, Guides Corps to Delhi, 25, 81 ; Greathed's to Agra, 136; Mhow columns in pursuit of Tantia Topi, 412
Mario Cantonments, Lucknow, 48 etc.
Martinière, Lucknow, captured by Sir Colin Campbell, 214 ; occupied by rebels during Siege of Lucknow, 278-81 ; recapture of, 282, 283
Massacre, at Satti Chaura Ghat, Cawnpore, 152, 153 ; at Bibighur, Cawnpore, 153, 175 ; at Jhansi, 36, 308, note [2]
Mathias, C., Assistant Surgeon Ind. Med. Est., destroys ammunition of Kotah Contingent Artillery, 43 note [1]

Matra, village near Kalpi, skirmish at, 328
Maude, F. C., 2nd Capt. R.A., arrives with 3/8 R.A. from Ceylon, 9 ; with Havelock's advance on Cawnpore, 167, etc. ; at Futtehpore, 169, 170 ; at Aong and Pandoo River, 172 ; at 1st Battle of Cawnpore, 173-5 ; at Unao, 176 ; at Bithur, 180 ; at Alam Bagh, 196 ; at Char Bridge, 196-8 ; gains Victoria Cross, 198 note [1] ; in Lucknow Residency, 216, 224 ; with Outram in Alam Bagh, 270 ; at Siege of Lucknow, 276
Maunsell, F. R., Lieut. Ben.E., holds together two companies of Ben. Sappers and Miners at Meerut, 11 ; at Siege of Delhi, 112, 113
Maun Sing, feudatory of Sindiah, 410 ; in arms against Government, *ibid.* ; seizes Paori, 412 ; his escape thence, 413 ; defeated at Bijepur and disappears, 413 ; reappears at Mangaoli, *ibid.* ; defeated at Kundri, 414 ; his association with Tantia Topi, 412-16
—— Oudh Talukdar, 375 ; his duplicity, 376 note [1]
Maynard, C. W., Lieut. Ben.H.A., at Jullundur, 28
—— E. G., Major 88th Foot at Sundeela, 383
McDonnell, W. F., Bengal Civil Service, gains Victoria Cross for his conduct during retreat from Arrah, 188 note [1]
McIntyre, C. C., Major and Brev.-Lieut.-Col. 78th Highlanders in command at Alam Bagh, 196 ; 205
McKillop, J., Bengal Civil Service, takes charge of the well at Cawnpore until he is killed, 151 note [1]
Meerut, outbreak at, 9-11 ; escape of mutineers from, 11 ; Field Force at Hindun, 68 etc. ; column from, for Rohilkhund, 354, 358
Mehidpore, Malwa, attacked by Feroz Shah, 245
Mehndi, village near Mithaulee in Oudh, skirmish at, 386
Metcalfe's House at Delhi, 81 etc.; skirmishes round, 95-8
Mhow, outbreak at, 38, 39 ; defence of, 39 ; relief of, 244
Miangunge, Oudh, capture of, 259, 260 ; skirmish at, 380
Michel, J., Major-General commanding Malwa Division, 402 ; succeeds

[1] Local rank.

INDEX 509

Roberts in command of Rajputana Field Force, 405 note ²; operations against Tantia Topi, 406 etc.
Michell, J. E., 2nd Capt. and Brev.-Major R.H.A., commands half E Troop under Lugard, 366; commands cavalry and horse artillery in pursuit of Kunwar Sing, 366-7
Middleton, W. A., Capt. R.A., at 2nd Relief of Lucknow, 212 etc.; in action at Shah Najaf, 221, 222; at Khurshid Manzil, 223; in charge of convoy, 227; at 2nd Battle of Cawnpore, 237 etc.; at Sher Ghat, 240; with Jaunpore Field Force at Chanda, 263; at Sultanpore, 265; at Siege of Lucknow, 277 etc.; in winter campaign in Oudh, 387 etc.; at the Gogra, etc., 393, 394, *ibid.* note ¹
Millar, Brig. J., Lieut.-Col. 33rd Mad.N.I., commands Nagpore Movable Col., 245; his operations round Jubbulpore, 245, 246
Miller, Brig. W. H., Lieut.-Col. Mad.A., commands artillery of Saugor Field Force, 323; wounded at Banda, 324
Milman, E. C. W. M., Lieut.-Col. and Brev.-Col. 37th Foot at Azimghur, 365
—— W. D., Lieut. R.A., at Shah Najaf, 223; at Serai Ghat, 241
Mines, at Lucknow Residency, 163; at Arrah, 186; at Chattar Manzil, Hirun Khana, and Engine House, Lucknow, 216, 223, 224
Minié, whiz of bullet recognised by defenders of Residency, 166
Miranpore Kutra, Rohilkhund, rendezvous of columns, 358
Mirgunge, village near Bareilly, skirmish at, 359
Mirza Guffur Bey, general in the late King of Oudh's Army, 265; commands rebels at Sultanpore, *ibid.*
Mitchell, W. St. L., Lieut.-Col. 19th Ben.N.I., at Berhampore, 8
Mithaulee, fort in Oudh, captured, 386
Moghuls, hopes of the, 12; empire of, proclaimed at Delhi, 19; the other rebels suspicious of, 64, 65; fall of power of, with loss of Delhi, 210
Mohamdee in Oudh, abandoned by rebels, 363, 364
Mongol Pandy, rebel sepoy at Barrackpore, 8, *ibid.* note ²
Mooltan, importance of, 26; sepoys disarmed at, 27; subsequent outbreak, 27, 28
Moore, J., Capt. 32nd Foot, commands the entrenchments at Cawnpore, 146 etc.; his successful sortie, 150
Moorsom, W. R., Lieut. 52nd Foot, on Havelock's Staff, 202; guides portion of force into Residency, 202, 203
Morar, Battle of, 339-41
Moresby, P., Lieut. R.A., killed at Chanderi, 305
Mori Bastion, Delhi, temporarily silenced, 82; opposed by No. 1 Siege Battery, 109, 113, 114; on day of assault, 120
Mortar Battery at Gulauli, 329, 332; danger of capture of, 335; saved by Camel Corps, *ibid.*
Mosque Picket, Delhi, 80, *ibid.* note ¹
Moti Mahal, Lucknow, Havelock at, 199; capture of, during 2nd Relief of Lucknow, 225
Moulvie of Fyzabad, religious fanatic, 47 note ¹; with Begum of Oudh in Lucknow, 248, 272; his followers, 353; defeated at Bari, 372; flies to Rohilkhund, 373; at Shahjehanpore, 362; death of, 364 note ¹
Mound, General's, Delhi, 80; cavalry raid at, July 9, 88 etc.
Mound Picket at Metcalfe's House, 81 note ²
Movable column, Nagpore, assembles at Kampti, 61; operations round Jubbulpore, 245, 246; joined by Whitlock's 1st Brigade, 323
—— —— Punjaub, organised by Lawrence, 25; operations in Punjaub under Nicholson, 30, 31; arrives at Delhi, 98
Muchligaon, skirmish at, 394
Mullapore, Doab, operations in, 396
Musjeediah, capture of, 397, 398
Muttra, outbreak at, 42; good service of Kotah Contingent at, *ibid.*

Nagina, action at, 359
Naharghur, Central India, operations about, 412, *ibid.* note ¹
Najafghur, Battle of, 99, 100
Najibabad, near Rurkhi, skirmish at, 359
Nana Sahib, his parentage and disposition, 51, 52; his hopes and open hostility, 54; his siege of the Cawnpore entrenchment, 145 etc.; his treachery, 151; his orders regarding the women and children, 153; declared Peshwa, *ibid.*;

despatches a force of sepoys towards Allahabad, *ibid.*; his defeat at 1st battle of Cawnpore, 175; orders massacre of women and children, *ibid.*; his flight, *ibid.*; palace sacked, 179 etc.; his movements in Oudh, 259, 260; his brother's activity, 352; Mahratta hopes about, 402; reported death of, 401, *ibid.* note [1]
Napier, R., Lieut.-Col. Ben.E., with Havelock at Lucknow, 204; chief engineer (Brigadier) at Siege and Capture of Lucknow, 276 etc.; at assault on Begum's Palace, 286; reports on Lucknow defences, 371, 372; commands Gwalior Division, 401; at Siege of Paori, 412, 413; operates against Feroz Shah, 414
Nargund, South Mahratta Country, outbreak in, 63
Naval Brigade, H.M.S. *Pearl* lands at Calcutta, 207; its guns, 247 note [3]
―― H.M.S. *Shannon* lands at Calcutta, 207; its guns, 208. *See* Services
"Nazim, The." *See* Mahndi Husain
Neemuch, Rajputana, outbreak at, 33; besieged by Feroz Shah, 243; relieved, 245
―― Brigade, mutinies and marches off in military order, 33; reaches neighbourhood of Agra and is joined by Kotah Contingent, 42; defeats Polwhele at Sassiah, 43; marches to Delhi, 44; enters city, 96
Neill, J. G. S., Major and Brev.-Lieut. Col. 1st Mad.E.F., at Benares, 56; safeguards that city, 59; relieves Allahabad, *ibid.*; arranges for advance on Cawnpore, *ibid.*; despatches Renaud and Spurgin, 167; appointed Brigadier under Havelock, 195; leads advance on Lucknow, 196 etc.; killed, 202
Nepaul, on Oudh frontier, 49; help from, offered and accepted, 191. *See* Gurkhas
Neville, G., Capt. R.E., killed at Barodia, 301
Newabgunge-Barabanki, rebels concentrate after fall of Cawnpore, 153; action at, 374, 375
Nicholson, J., Capt. and Brev.-Lieut.-Col. 27th Ben.N.I., Deputy Commissioner Punjaub, 25, *ibid.* note [3]; appointed Brigadier to command Punjaub Movable Column, 30, *ibid.* note [2]; at Trimmoo Ghat,

31; marches to Delhi and arrives on Ridge, 98; defeats rebels at Najafghur, 99-101; his opinion of Wilson's position, 106, *ibid.* note [2]; visits Ludlow Castle, 111; commands No. 1 Column at the Assault of Delhi, 119; mortally wounded, 120
Nizam of Hyderabad, loyalty of, 63
Now Gong, Central India, mutiny at, 36
Nowshera, Punjaub, mutiny at, 26
Nubha, loyal Cis-Sutlej State, 20
Nuggur, Baswarra District, operations round, 390
Nunparah, Northern Oudh, Lord Clyde at, 398
Nurnaul, Delhi District, action at, 142
Nusseerabad, Rajputana, outbreak at, 33; escapes from, 66, 67; rebels from, reach Delhi, 83
Nyagaon, village near Kirwi, 419; affair at, 420

Oakes, T. G. A., Major and Brev.-Lieut.-Col. 12th Lancers, at Lucherra, 419
Observatory, Delhi Ridge, 80, *ibid.* note [1]
Ochterlony Gardens, near Delhi, skirmish at, 83, 84
Oliver, J. R., Lieut. R.A., commands *ex-tempore* battery at Cawnpore, 231, *ibid.* note [1]; superseded on account of his youth, 232 note [2]; at the Gogra, 393, 394
Olpherts, H. A., Capt. and Brev.-Maj. Ben.H.A., at Jullundur, 28, 29; at Delhi, 85, 86
―― W., Capt. Ben.A., at disarming parade, Benares, 57, *ibid.* note [1]; at 1st Relief of Lucknow, 195 etc.; at Alam Bagh, 196; at Char Bridge, 199; gains Victoria Cross, *ibid.* note [1]; at Dhooly Square, 202; his plan to recover a gun, 204; with Lucknow Garrison, 212, 216, 223; with Outram at Alam Bagh, 270-3; at Siege and Capture of Lucknow, 275 etc.
Ommaney, F. M. M., Capt. R.A., commands artillery at Siege of Chanderi, 304, 305; at Siege of Jhansi, 307 etc.; at the Battle of the Betwa, 311 etc.; becomes O.C.A. Central India Field Force, 314; at battle of Kunch and Gulauli, 319 etc., and 330 etc.; his

INDEX 511

somewhat useless brevet, 331 note [1]; at Kalpi, 336
Oodeypore, Rajputana, Tantia Topi at, 403
Oomeriah, Oudh, fort, capture of, 392
Orcha Gate, Jhansi, 313
Ordnance, mounted in Delhi, 75; captured in Lucknow, 275
Orr, W. A., Capt. and Brev.-Major Mad.A., commands Hyderabad Contingent during operations in Central India, Chap. VIII. *passim*
Oudh, principal recruiting ground, 3; trouble after annexation of, *ibid.* 45; Henry Lawrence's wise administration of, 46, 47; outbreaks in, 48, 49; all lost except capital, 51; activity of rebels in, after Cawnpore Massacre, 153; Havelock's advance through, 176 etc. (1st Relief); Sir Colin Campbell's advance through, 213 etc. (2nd Relief); Alam Bagh alone held in, 228; difficult position of Talukdars of, 249, 352; Sir Colin's second advance through, 277 etc. (Siege and Capture of Lucknow); mutineers escape from Lucknow and spread through, 292, 296; different views as to subjugation of, 353; summer campaign in, 371-9; operations in, during October 1858, 379-87; winter campaign in, 387-401
Outram, Sir James, Major-Gen., appointed to command the combined Dinapore and Cawnpore Divisions, and Chief Commissioner of Oudh, 193; generously allows Havelock to retain command of Field Force, 194; at 1st Relief of Lucknow, 195-203; resumes his military position, 204; watches advance of and meets Sir Colin Campbell during 2nd Relief, 223, 225; left in command at Alam Bagh, 228; his defence of Alam Bagh, 270-3; commands Right Attack at Siege of Lucknow, 278; crosses Gumti, 279; his conduct of the attack, 280 etc.; his position at the Stone Bridge, 291; forbidden to cross, *ibid.*; crosses by bridge of boats into Lucknow, 293; at the Musa Bagh, 295

Paeen Bagh, Lucknow, 203
Pahuj, affluent of the Sindh, skirmish on, 417
Palmer, E., Capt. R.A., at Banda,
325; at Nyagaon, 420; at Kothi, *ibid.*; at Panwari, 423
Pandoo River, affluent of the Ganges, Havelock defeats rebels at, 171; Windham's successful skirmish at, 231
Pandora, near Allahabad, skirmish at, 247, 248
Panghali Pass, between Kirwi and Kothi, 421
Panna, Bundelkhund, Whitlock at, 324
Panwari, Bundelkhund, action at, 423
Paori, Central India, seized by Maun Sing, 412; siege and capture of, 413
Parbati, affluent of the Chumbal, flight of Kotah rebels to, 316; Maun Sing on, 412
Parke, Brig. W., Lieut.-Col. 72nd Highlanders, Rajputana Field Force, 401; pursues Tantia Topi to the Chumbal, 405; at Chota Udepur, 410
Parwan, affluent of the Chumbal, 405
Passees, indigenous race in Oudh, 162 note [1]
Patiala, loyal Cis-Sutlej State, 20; action at, 252
Patna, Hindu City on Ganges, 60 note [1]; firmness of Commissioner of, 182, *ibid.* note [2], 185
Paton, J., Sergeant 93rd Highlanders, at Shah Najaf, 222
Pearson, A., Lieut. Ben.A., his narrow escape, 42 note [1]
Peel, W., Capt. R.N., commands Naval Brigade, H.M.S. *Shannon*, at Khujwah, 208; at 2nd Relief of Lucknow, 212 etc.; at Shah Najaf, 221, 222; at Martin's House, 226; at Cawnpore, 235; at 2nd Battle of Cawnpore, 237 etc.; at Kali River, 254; at Siege of Lucknow, 275 etc.; wounded, 283; promoted K.C.B., 283 note [2]; death of, *ibid.*
Penny, N., Major-Gen., succeeds Hewitt at Meerut, 12; and Wilson at Delhi, 132; commands Meerut Column, 354; marches to Miranpore Kutra, 358; killed at Kakraula, Rohilkhund, 358
Perkins, H. G., Lieut. Ben.H.A., killed at the Hindun, 69, *ibid.* note [2]
Peroo, Behar, operations near, 370
—— Oudh, operations near, 389, *ibid.* note [1]
Pertabghur, Rajputana, affair at, 411

Peshawar, critical position at, 24 etc.;
disarming parade at, 26; punishment parade at, *ibid.*
Peshwa, title of Chief of the Mahrattas, 52, *ibid.* note [2]; relics of the tradition of, 54, 210, 337, 402
Phul Bagh, Gwalior, 343, 346
Phulchadar Aqueduct, Delhi, 94
Picket, Rebel, at Ludlow Castle, slackness of, 109
Pickets, British, at Delhi, 80, 81, *ibid.* note [2]; 85, 86
Pinckney, Brig. F. G. A., Lieut.-Col. and Brev.-Col. 73rd Foot, in Winter Campaign in Oudh, 387 etc.
Plans of Attack, Delhi, discussion of, 102-12
Plassey, Battle of, 1757, prophecy concerning, 4; rebel activity on Centenary of, at Delhi, 85; at Cawnpore, 151, 152.
Platt, J., Lieut.-Col. and Brev.-Col. 32nd Ben.N.I., commands at Mhow, 38; his belief in the sepoys, *ibid.*; his murder, 39
Polwhele, Brig. T., Lieut.-Col. and Brev.-Col. 36th Ben.N.I. in command at Agra, 41; defeated at Sassiah, 43
Poona, column of Bombay troops assembled at, 62
Poonch, Bundelkhund, operations round, 318
Poorwah, Banswarra District, operations round, 390
Port-fire, use of, in India, 17 note [1]; use of, to deceive enemy at Kokral, 158; unsuitable for night work at Lucknow, 286 note [2]
Post guns, use of, 8 note [1]
Powain, Oudh, death of Moulvie of Fyzabad at, 364 note [1]; loyalty of Rajah of, 385
Powell, T. S., Lieut.-Col. 53rd Foot, at Khujwah, 208; killed, 209
Price, E., Lieut.-Col. R.A., commands artillery at Kolah, 315 etc.; and on the Banas, 404
Priming pockets, hidden by rebels in Delhi Arsenal, 16; use of, 17 note [1]
Punjaub, our advanced post, 24, 25; John Lawrence's bold policy, 25 etc.; our base of supply for Delhi Field Force, 65; critical position of, during Siege of Delhi, 102, *ibid.* note [1]; anxiety for, removed by capture of Delhi, 249; Sikh prophesy, *ibid.* note [1]

Racecourse, Delhi, picket at, 85; demonstration against, 92
Race Stand, Lucknow, capture of, 281; colours of, 1st Ben.E.F. mounted on, as a signal, 282
Radcliffe, C. W., Capt. 7th Ben.L.C., commands Volunteer Cavalry at Chinhut, 154; his successful charge, 158, 159
Rahatghur, Fort of Bhopal, seized by rebels, 40; siege and capture of, 298, 299
Raid, rebel cavalry, at Delhi, 88-92
Raines, J. A. R., Lieut.-Col. 95th Foot, at Kotah-ki-Serai, 342, 343; at Gwalior, 345
Raj Ghat Gate, Delhi, 13
Rajpur, ford of the Betwa, 310, 312
Rajpore,'Satpura district skirmish at, 409
Rajputana, position in, at outbreak of Mutiny, 31-33; George Lawrence's difficult task, 242, 243; Feroz Shah's operations in, 243; operation of Malwa Field Force in, 244, 245
—— Field Force, formation of, 250; operations against Awah and Kotah, 315-17; against Tantia Topi, 401 etc.
Rajputs, military caste in Oudh, 3
Ramgunga, affluent of the Ganges, Walpole's operations on, 256; crossed by rebels from Rohilkhund, *ibid.* 257
Ramgurh, Chota Behar, loyalty of Rajah, 192
Ram Het, sepoy Ben. Sappers and Miners, at Kashmir Gate, Delhi, 121
Ramnagar, village near Chanderi, 304
Ramnuggur, Oudh, 373
Rampore Kussia, Oudh, capture of, 388
Ranode, on Sindh River, defeat of Feroz Shah at, 414
Rao Sahib, nephew of Nana Sahib; collects forces at Kalpi, 210; supplies Begum of Jhansi with troops, 319; joined by Nawab of Banda, 323; defeated at Gulauli, 332 etc.; in company with Begum of Jhansi seizes Gwalior, 337; his efforts in company with Tantia Topi, 402-16; arrested and hanged, 416 note [2]
Rapti, affluent of the Gogra, Rowcroft's operations on, 370, 386, 387; cavalry charge on banks of, 399
Rattray's Sikhs, raising of, 185 note [2]. *See* Services

INDEX

Rawal, Central India, skirmish at, 245

Raynor, W., Lieut. Veteran Establishment at Magazine Delhi, 16, 17; his escape, 17; is promoted Captain, *ibid.* note [1]; gains V.C., *ibid.*

Reed, T., Maj.-General, succeeds Anson, 21 note [2]; his illness, *ibid.*; accompanies Delhi Field Force, *ibid.*; assumes nominal command, 88; relinquishes it to Wilson, *ibid.*; discusses question of assault, 103, 104

Rehora, Oudh fort, capture of, 391

Reid, C., Major Sirmur Battalion, at the Hindun, 70; commands Main Picket on Ridge, 81; resists rebel attacks, 81 etc.; commands column in Kishangunge, 82, 83; commands No. 4 Column at Assault of Delhi, 122 etc.; severely wounded, 124

Reinforcements, 439 etc.

Relief of Allahabad by Neill, 59; of Lucknow by Havelock, 194 etc.; of Lucknow by Sir Colin Campbell, 213 etc.; of Mhow by Stuart, 244; of Neemuch by Stuart, 245; of Saugor by Rose, 302; of Shahjehanpore by Sir Colin Campbell, 362

Remmington, F. F., Capt. Ben.H.A., at Samee House Battery, 112; *ibid.* note [2], 118; with Greathed's column, 133 etc.; at Bulandshahr, 134; at Agra, 139; in 2nd Relief of Lucknow, 212 etc.; episode of Nov. 18, 226; at 2nd Battle of Cawnpore and Serai Ghat, 236 etc.; at Shamshabad, 257, 258; reinforces Outram at Alam Bagh, 273; at Siege and Capture of Lucknow, 275 etc.; at Kursi, 296

Renaud, S. G. C., Major 1st Mad.E.F., marches towards Cawnpore, 167; his risk, 168; joins Havelock, *ibid.*; at Futtehpore and Aong, 168 etc.; mortally wounded at Pandoo River, 172

Renny, G. A., Lieut. Ben.H.A., in cavalry raid at Delhi, 91; at capture of Delhi Arsenal, 128; gains Victoria Cross, *ibid.* note [1]

Reorganisation of Indian Army, 424 etc.

Residency, Indore, defence of, 38; —— Lucknow, prepared for defence, 48, 49; investment of, 159; garrison of, 160; defence of, 160-6; relief of, 204; defences of, improved, 205; action of garrison of, during Sir Colin Campbell's advance on Lucknow, 216, 223, 224; evacuation of, 227; *ibid.* note [1]

Rewal, Malwa, action at, 245

Rice, W., Capt. 25th Bom.N.I., at Sarpore, 414

Richard, H. E., Ensign Rifle Brigade, at Birwah, 384

Riddell, C. J. B., Lieut.-Col. R.A., commands artillery of Right Attack at Siege of Lucknow, 279 etc.; commands artillery of Lugard's column, 366 etc.

Rideout, A. K., Lieut. R.H.A., at the Gumti, 392 note [1]

Ridge, Delhi, seizure of, by British, 72 etc.; position on, 77-80; defence of, 81-102

Rifled gun, first used by us in war, 147, *ibid.* note [1]

Roberts, F. S., Lieut. Ben.A., at Lahore Gate of Delhi, 129; at Kali River, 256; gains Victoria Cross, *ibid.* note [1]

—— H. G., Major.-Gen., commands Rajputana Field Force, 250; at Nusseerabad, 315; captures Kotah, 316, 317; in pursuit of Tantia Topi, 401 etc.

Robertson, A., Capt. and Brev.-Major Ben.A., defends fort at Futtehghur, 44, 45; dies of his wounds, *ibid.* note [1]

—— G. H., Major 25th Bom. N.I., at Gulauli, 330, 336; commands column in pursuit, 337; at Bijepur 413

—— J., Lieut. R.A., at Azimghur, 365, 368, *ibid.* note [1]

Robison, H. G., Capt. 3rd Bom.E.R., at Jhansi, 313, 314

Rocke, R., Major 72nd Highlanders at Pertabghur, 411

Rockets, used by Naval Brigade at Shah Najaf, 222; by rebels at Delhi, 117; and at Barodia, 301

Rohilkhund, outbreak in, 40, 41; British feint on, 250; operations on Ramgunga, 256; rebels from, cross Ganges, 256; Governor-General's policy regarding, 353, *ibid.* note [2]; rebels advance on Futtehghur from, 355; are defeated at Kankar and fall back to, 356; subjugation of, 357-64

Rose, General, Sir Hugh, commands Central India Field Force, 249; captures Rahatghur, 298; defeats

514 INDEX

Rajah of Banpore at Barodia, 299–301; relieves Saugor, 302; Captures Garakota, *ibid*.; forces Bundelkhund passes, 302, 303; captures Jhansi, 307 etc.; defeats Tantia Topi on the Betwa, 309 etc.; wins battle of Kunch, 320 etc.; advances to the Jumna, 326; wins battle of Gulauli, 330 etc.; personally leads the camel corps, 334 etc.; captures Kalpi, 336; marches to Morar, 338; defeats rebels there, 339–40; supports Smith at Kotah-ki-Serai, 344; defeats rebels at Gwalior, 344, 345; his splendid services, 351

Rose, W., Lieut. 25th Bom.N.I., killed at capture of Gwalior Fort, 347

Ross, J., Major Rifle Brig., in command of camel corps, 331; at Battle of Gulauli, *ibid*., 334, 335

Rowcroft, Brig. F., Lieut.-Col. and Brev.-Col. 62nd Ben.N.I., commands field force at Chuprah, 247; defeats rebels on Chota Gundak, *ibid*.; joins Jung Bahadur, 261; defeats Mahndi ;Husain at Amorha, 364; observes Belwa, 364, 365; advances to Basti, 370; drives rebels across Rapti at Dumurreagunge, *ibid*.; pursues them to Bururiah, 386; defeats Bala Rao at Tolsipore, 397

Ruiya, unsuccessful attack on, 356; evacuation of, 357; captured by Barker, 385

Rurkhi Column, for Rohilkhund, formation of, 354; operations of, 358 etc.

Russell, Sir W., Bt., Lieut.-Col. 7th Hussars, defeats Beni Madho, 392
—— W. C., Lieut. Ben.A., Commissary of Ordnance, 58; prepares to blow up magazine in Allahabad Fort, 59

Sadat Ali's Mosque, Lucknow, 199; occupied by rebel artillery, *ibid*.
Sadatgunge, Lucknow, 295
Sahao, Jalaon district, action at, 418
St. James's Church, Delhi, 122
St. John's Church, Cawnpore, 150
Salimpore, Oudh, affairs at, 377, 378
Salt, T. H., Lieut. Ben.A., turns guns in Water Bastion, Delhi, upon Selimghur, 128
Salkeld, P., Lieut. Ben.E., killed at Kashmir Gate, Delhi, 121

Samee House, Delhi Ridge, occupation of, 86, *ibid*., note [2]; strengthened, 94; attacks on, 96; battery at, 112, 113
Sanganeer, Rajputana, skirmish at, 403
Sarpore, Central India, Feroz Shah surprised at, 414
Sassaya, Northern Oudh, action at, 398
Sassiah, near Agra, action at, 43, 44
Satpura, mountainous district between the Nurbudda and the Tapti, 63
Saugor, outbreak at, 40; relief of, 302
—— Field Force to assemble at Kampti under Whitlock, 250, 297; its slow formation, 323; its northern march, 323; at Kubrai and Banda, 324–5; at Kirwi, 349; final operations of, 417 etc.
Scott, E. W. S., Capt. and Brev.-Major Ben.A., at the Hindun, 69; at Badli-ki-Serai, 71; at the Subzi Mundi, 92, *ibid*. note [3]; commands No. 3 Siege Battery, 115, 116, *ibid*. note [1]
Scully, Conductor, at Magazine, Delhi, 16, 17; volunteers to fire charge of mine, 17
Seaton, Brig. T., Lieut.-Col. and Brev.-Col. 60th Ben.N.I., at Gungaree, 251; at Patiala, 252; joins Walpole, 253; marches to Futtehghur, 256; in command of garrison there, 355; at Kankar, 355, 356; at Shahjehanpore and Powain, 385; at Bunkagaon, *ibid*.
Secrora, Oudh, Bonham's escape from, 49 etc.; Lord Clyde's operations round, 395 etc.
Segowlie, Behar, mutiny at, 184
Sekundra, village near Allahabad, action at, 262, 263
—— Bagh, Lucknow, 199; occupied by rebels, 217; attack and capture of, 218–20; occupied without opposition at Siege of Lucknow, 286
Selimghur, Delhi, 13, *ibid*. note [1], 75; 117
Serai Ghat, Doab, affair at, 240, 241
Services : [1]
   Naval Brigade :
      H.M.S. *Pearl*, 80, 97, *151*, *155*, *161*

[1] The figures opposite the Naval Brigades, Regiments, etc., refer to the Table XVII, Battles, Sieges, etc., pp. 487–91. The services of the Royal, Bengal, Madras, and Bombay Artillery are given in the Special Lists I to XVI, pp. 450–86.

INDEX

Services (*continued*):
H.M.S. *Shannon*, *49, 51, 56, 58, 63, 76*
Regiments:
2nd D.Gs., *65, 76, 83, 90, 102, 111, 132, 137, 140, 150, 155a, 158, 170, 179*
6th D.Gs., *2, 8–13, 52, 60, 61, 93, 95, 98a, 103, 108, 109, 133, 142–4, 146a, 160a, 164, 168*
7th Hussars, *75, 76, 90, 102, 111, 118, 136, 143, 144, 152, 164, 168, 174a*
8th Hussars, *85, 113, 114, 119, 121, 122, 125a, 135, 138, 145, 154*
9th Lancers, *8–13, 35, 42, 45, 51, 58, 60, 61, 63, 67, 75, 76, 78, 92, 96, 99, 150, 155a, 158, 170*
12th Lancers, *93a, 94, 157, 167, 173*
14th L.Ds., *46, 48, 54, 55, 66, 68, 69, 77, 79, 81, 84, 86, 112, 115*
17th Lancers, *127, 134, 135, 138, 166, 169*
Military Train, *51, 57, 76, 91, 100*
Royal Engineers, *49, 51, 76, 79, 81, 84, 85, 91, 92, 96, 100, 104, 106, 107, 119, 121, 122, 143, 144, 154, 156, 170, 174*
4th Foot, *87*
5th Foot, *29, 31, 37–40, 51, 57, 76, 143, 144, 153*
6th Foot, *100*
8th Foot, *1, 6, 10–13, 42, 45, 51, 58, 63, 140*
10th Foot, *3, 31, 50, 65, 73, 74, 76, 91, 100*
13th Foot, *82, 151, 155, 161*
20th Foot, *65, 73, 74, 76, 130, 146, 160a, 164, 168*
23rd Foot, *51, 58, 76, 99, 129, 146*
24th Foot, *1, 21*
28th Foot, *179a*
32nd Foot, *7, 16, 19, 40, 58, 116, 117, 136, 143, 144*
34th Foot, *56, 75, 76, 91, 100, 176*
35th Foot, *97*
37th Foot, *82*
38th Foot, *58, 75, 76, 90, 102*
42nd Foot, *58, 59, 67, 76, 92, 96, 99, 172*
43rd Foot, *120, 125, 126, 152, 163, 167*
52nd Foot, *1, 12–14*
53rd Foot, *43, 49, 51, 58, 63, 67, 75, 76, 83, 150, 156, 158, 161, 170, 177*
54th Foot, *116, 117, 143, 144*
60th K.R.R., *2, 8–13, 95, 103, 108, 109, 133, 142*

61st Foot, *1, 10–13, 35*
64th Foot, *22–5, 28–30, 32, 33, 37, 40, 49, 51, 56, 58, 63, 93, 95, 103*
70th Foot, *1*
71st Foot, *101, 104, 106, 107, 112, 127, 134, 135, 138, 159*
72nd Foot, *85, 119, 121, 154, 162*
75th Foot, *8–13, 42, 45, 57*
78th Foot, *22–5, 28–30, 32, 33, 37, 39, 41, 51, 57, 76*
79th Foot, *64, 76, 92, 96, 99, 103, 108, 109, 141, 143, 144, 150, 156, 170*
80th Foot, *146, 160a*
81st Foot, *1*
82nd Foot, *51, 56, 58, 88, 103, 108, 133*
83rd Foot, *44, 85, 119, 121, 174*
84th Foot, *7, 22–5, 28–30, 32, 37–40, 39, 57, 76, 91, 100*
86th Foot, *46, 48, 54, 55, 79, 81, 101, 104, 106, 107, 112, 114, 122, 125a*
87th Foot, *1*
88th Foot, *56, 58, 70, 106, 107, 132, 137, 140*
90th Foot, *38, 39, 51, 57, 76, 102*
92nd Foot, *127, 134, 135, 138, 147, 169*
93rd Foot, *49, 51, 58, 59, 63, 76, 92, 96, 99, 142*
95th Foot, *85, 113, 114, 122, 125a, 135, 138, 145, 162*
97th Foot, *65, 73, 74, 76, 78*
Rifle Brigade, *56, 58, 64, 76, 83, 90, 102, 111, 132, 137, 140, 143, 144, 153, 164, 168, 174a*
1st Ben.E.F., *8–13, 35, 52, 60, 61, 76, 90, 128, 131*
2nd Ben.E.F., *8–13*
3rd Ben.E.R., *20, 105, 171*
1st Mad.E.F., *3, 7, 22–5, 28–30, 32, 33, 37–40, 57, 76, 123, 136, 143, 144, 148, 167*
3rd Mad.E.R., *93a, 94*
1st Bom.E.F., *124*
3rd Bom.E.R., *66, 68, 69, 71, 77, 84, 86, 101, 104, 106, 107, 112, 114*
3rd Ben.L.C. (Craigie's Squadron), page 16 note [1]
4th Ben.Irr.C., *142, 146a*
12th Ben.Irr.C., *38, 39, 91, 100*
17th Ben.Irr.C., *99*
1st Punjaub C., *11–13, 42, 45, 51, 58, 63, 76, 83, 95, 99, 141, 143, 144, 155a, 168, 174a*
2nd Punjaub C., *35, 42, 45, 51, 58, 63, 76, 83, 92, 96, 99, 172*

Services (*continued*):
Regiments (*continued*):
5th Punjaub C., 10–13, 42, 45, 51, 58, 63, 76, 99
Guides Corps, 1, 9–13, 35, 52
Sikh C., 88, 91, 130, 146, 153, 175
Hodson's Horse, 11–13, 42, 45, 51, 58, 60, 61, 63, 67, 76, 90, 102, 111, 118, 136, 143, 144, 150, 155a, 158, 160a, 170
Wale's Horse, 76, 90, 102, 111
Mooltani Horse, 12, 13, 21, 35, 52, 93, 95, 98a, 103, 108, 109, 142, 146a
Lahore Horse, 99, 116, 117
Pathan Horse, 142, 144
Ben. Sappers and Miners, 2, 8–13, 37, 42, 45, 51, 60, 61, 76, 83, 92, 96, 99, 132, 137, 140
13th Ben.N.I., 16, 19, 40
11th Ben.N.I. (det.), page 10 note [1]
21st Ben.N.I., page 26 note [2]
31st Ben.N.I., 18
42nd Ben.N.I., (det.), 18
48th Ben.N.I. (det.), 16, 19, 40
53rd Ben.N.I. (det.) 7,
71st Ben.N.I. (det.), 16, 19, 40
Beluchi Battalion, 10–13, 93, 95, 103, 141, 143, 144, 174a
Kumaon Battalion, 11–13
Sirmur Battalion, 8–13
Ferozepore Regiment, 5, 22–5, 28–30, 32, 33, 37–40, 57, 76, 151, 155, 161, 179
1st Punjaub Inf., 10–13, 35, 95
2nd Punjaub Inf., 12–13, 35, 42, 45, 51, 58, 76, 99
4th Punjaub Inf., 13, 42, 45, 51, 58, 67, 76, 92, 96, 99
5th Punjaub Inf., 90, 102, 111, 136, 143, 144, 174a
7th Punjaub Inf., 52, 60, 61, 88, 116, 117, 179
9th Punjaub Inf., 141, 143, 144
17th Punjaub Inf., 95
22nd Punjaub Inf., 93, 95, 103, 108, 109
Sikh Inf., 10–13, 95, 143, 144, 150, 156, 158, 170, 177
Rattray's Sikhs, 26, 31, 43
Harriana F. Force, 52
Shekwattee Battalion. *See* page 192 note [4]
4th Mad.L.C., 29a, 176
6th Mad.L.C., 93a, 94, 116, 117, 143, 144, 164
7th Mad.L.C., 93a, 94

Mad. Sappers and Miners, 51, 66, 69, 73, 84, 93a, 94, 101, 104, 106, 107, 112, 114, 160a, 179a
1st Mad.N.I., 93a, 94, 120, 126, 163, 167
17th Mad.N.I., 50
27th Mad.N.I., 56, 57, 151, 155, 161
33rd Mad.N.I., 29a
50th Mad.N.I., 152, 167
Nagpore Rifles, 29a, 93a, 94
1st Bom.L.C., 85, 113, 114, 122, 135, 138, 145
2nd Bom.L.C., 44, 85, 154
3rd Bom.L.C., 66, 68, 69, 71, 77, 79, 81, 84, 86, 101, 104, 106, 107, 112, 115, 127, 134, 135, 138
Sind Horse, 85, 154
Gujerat Horse, 119, 121, 154
Beluchi Horse, 85, 119, 121
Mahratta Horse, 154
Meade's Horse, 122, 125, 145
Mayne's Horse, 135, 138
Bom. Sappers and Miners, 78, 81, 84, 98, 101, 104, 106, 107, 112, 114
4th Bom.N.I., 127, 134
6th Bom.N.I., 179a
10th Bom.N.I., 85, 87, 113, 114, 122, 125a, 145
11th Bom.N.I., 87
12th Bom.N.I., 44, 85, 119, 121, 174
13th Bom.N.I., 85, 119, 121, 162
19th Bom.N.I., 127, 134, 135, 138
24th Bom.N.I., 66, 69, 71, 77, 84, 86, 101, 104, 106, 107, 112, 114
25th Bom.N.I., 46, 48, 54, 55, 79, 81, 84, 86, 101, 104, 106, 107, 112, 114, 122, 125a, 160
Gurkhas, 62, 73, 74, 80
Jind, Contingent, 8–13
Kashmir, Contingent, 13
Hyderabad Contingent, 44, 48, 53, 54, 55, 66, 68, 69, 71, 77, 84, 86, 93a, 94, 101, 104, 106, 107, 112, 167, 173
Shahabad, Western Oudh, action at, 364
Shahjehanpore, Rohilkhund, Sir Colin Campbell at, 357; Hale left to garrison, 358; attacked by Fyzabad Moulvie, 362; relief of, etc., 363
Shah Najaf, Lucknow, 220; attack and capture of, 221–3
Shamshabad, Doab, action at, 257 etc.
Sharghur, Rajah of, hostile attitude

INDEX 517

of, 40 ; driven from Garakota, 302 ; territory annexed, 303, *ibid.* note [1].
See Banpore
Shaw, Conductor, at Magazine Delhi, 16, 17
Sheopur Ghat, on Ganges, Kunwar Sing crosses at, 368
Shergati, Behar, narrow escape of Sir Colin at, 211 note [1]
Sher Ghat on the Jumna, affair at, 337
"Ship," Bonham's, Lucknow Residency, 162
Shorapore, Deccan, Rajah of, defeated and taken prisoner, 64
Showers, Brig. St. G. D., Lieut.-Col. 2nd Ben.E.F., at Badli-ki-Serai, 71 ; at seizure of Ridge, 72, 73 ; in combats round Delhi, 86, 93, 95, 98 ; in Rohtak district, 141 ; at Agra, 401, 402 ; at Dosa, 415
Shunkerpore, fort in Banswarra district, capture of 388, 389
Sialkote, outbreak at, 30
Sikar, Rajputana, Tantia Topi, surprised at, 415, 416
Sikhs, demeanour of, at outbreak of Mutiny, 24, 25 ; during siege of Delhi, 102, note [1], 132 ; after capture of Delhi, 249 ; well-known prophecy among, *ibid.* note [1]
Simri, Banswarra district, affair at, 374
Sind, Punjaub reinforced from, 27 ; loyalty of, *ibid.* note [2]
Sindh, affluent of Jumna, 414
Sindiah, Maharajah of Gwalior, 34 ; importance of his attitude on outbreak of Mutiny, 37 ; his efforts to support the British aided by his minister Dinkar Rao, *ibid.* ; difficulties of his position, 38 ; his troops declare for the Peshwa, 210 ; his capital seized by Begum of Jhansi, 338 ; his restoration, 347
Sindwaho, Bundelkhund, action at, 407
Singleton, J., Capt. and Brev.-Major R.A., at Mooltan, 27, 28
Sirsa, near Aligunge, skirmish at, 357
Skinner's House, Delhi, Wilson's headquarters established at, 126
Slow match, how carried in action, 17 note [1]
Smith, Brig. M. W., Lieut.-Col. and Col. 3rd D.Gs., at Sipri, 338 ; *ibid.* note [2] ; at Kotah-ki-Serai, 341–4 ; at Gwalior, 345 etc. ; ejects rebels from Chanderi, 349 ; returns to Sipri, 401 ; operates against Tantia Topi, 406
Smith, C. H., Capt. and Brev.-Major R.A., at Cawnpore, 236, *ibid.* note [2] ; at Kali River, 254 ; at Kankar, 355 ; at Ruiya, 385
Smith, J., Sergeant Ben.E., at Kashmir Gate, Delhi, 121
—— W., Lieut. R.A., at Sekundra Bagh, 220 ; at Shah Najaf and Khurshid Manzil, 223 ; at Jerwa Pass, 401
Smyth, J. H., Capt. Ben.H.A., at Chuttra, 192 note [3] ; at 2nd Battle of Cawnpore, 236 ; at Manglaur, 358 ; at Nagina, 359 ; at Sassaya, 398
Somerset, Brig. C. H., Lieut.-Col. and Brev.-Col. 72nd Highlanders at Chubbra, 411, 412
Sone, affluent of Ganges, British disaster on, 187
Soona Valley, Nepaul, Horsford's operation in, 400
Sotheby, E. S., Capt. R.N., commands Naval Brigade, H.M.S. *Pearl*, at Chuprah, 247, *ibid.* note [3]
Soubadar's Tank, Cawnpore, 240
Spurgin, J. B., Capt. 1st Mad.E.F., sent in *Brahmapootra* to Cawnpore, 176
Steuart, Brig. C., Lieut.-Col. and Brev.-Col. 14 L.Ds., commands 2nd Brig. Central India Field Force, 297 ; with Sir Hugh Rose at Rahatghur, Saugor, Jhansi, Kunch, etc., 298–314, 319–23 ; prostrated by the heat, 327 note [1]
Stewart, Serg. Ben.A., at Magazine Delhi, 16, 17
Stillman, J., Lieut. 6th D.Gs., at rebel cavalry raid, July 9, 88, etc.
Stockings, women's use of, to improvise case shot at Cawnpore, 149 note [1]
Stone Bridge, Lucknow, left open for escape of mutineers, 292
Strange, T. B., Lieut. R.A., at Chanda, 264 ; at Dhowrara, 269 ; at Doadpore, 380-2
Strutt, C. H., Lieut. Bom.A., in command of siege guns at Rahatghur, 298 ; at Kunch, 320 ; at Gulauli, 331 etc. ; at Morar, 339, 340 ; at Gwalior, 345, 346
Stuart, Brig., C. S., Lieut.-Col. and Brev.-Col. 1st Bom.E.F., commands Malwa Field Force, 244 ; at Dhar, *ibid.* ; at Mandesar and Goravia, 245 ; disarms Holkar's

II—35

518 INDEX

troops, *ibid.*; commands 1st Brig. Central India Field Force at Chanderi, 303 etc.; at Jhansi and the Betwa, 307 etc.; at Kunch, 319 etc.; in command of right wing at Gulauli, 330; his fight on the hillock, 335; at Morar and Gwalior, 339 etc.

Stuart, W. K., Major 86th Foot, at Jhansi, 313

Subzi Mundi, Delhi, 72; a danger to the British position, 77; combats in, 85, 92, 93, 124

Sultanpore, Oudh, action at, 265–8; operations about, 376, 377

Sundeela, Oudh, rebel attack on, 383

Sutherland, R. M., Major 92nd Highlanders, at Rajpore, 409, *ibid.* note [2]

Talukdars of Oudh, Lawrence's wise treatment of, 46, 47; difficult position of, 352

Tantia Topi, commands troops sent from Bithur to Cawnpore by the Nana Sahib, 52; superintends the massacre at Satti Chaura Ghat, 152, 153; accepted as their military leader by rebel Mahrattas, 210; his attacks on Windham, 231–5; his defeat by Sir Colin Campbell at 2nd Battle of Cawnpore, 237–40; with the Rao Sahib at Kalpi, 309; attacks Rajah of Charkhari, *ibid.*; marches to relieve Jhansi, *ibid.*; his defeat on the Betwa, 310–13; defeated at Kunch, 320–2; at Morar with Begum of Jhansi, 338 etc.; heads the flight from Gwalior, 347; joins the Rao Sahib and Nawab of Banda, 402; his flight through Rajputana and Central India, 402 etc.; his operations with Maun Sing and Feroz Shah, 412 etc.; is captured and hanged, 416

Taylor, A., Capt. Ben.E., second in command of Engineers at Delhi, 94; his energy, *ibid.*; his work with Baird-Smith, 107; his complete reconnaissance of north front of Delhi, *ibid.*; his plans, etc., 108 etc.; useful work during capture of city, 129

—— Brig. R. C. H., Lieut.-Col. and Brev.-Col. 79th Highlanders, in command at Fyzabad, 389; operations on the Gogra, 393

—— Mr. W., Ben. Civil Service Commissioner of Patna, his strong measures, 182; his acts vindicated, *ibid.* note [2], sends detachment of Sikhs to Arrah, 185

Tehri, village near Kalpi, 327, 334

Telawarra, suburb of Delhi, 77; a danger to British, 78; guns placed in, by rebels, 118, 124

Thackeray, E. T., Lieut. Ben.E., at Siege of Delhi, 115; gains Victoria Cross in attack on Arsenal, 128

Thomas, L. F. C., Lieut. Mad.A., blows up magazine in Machi Bhawan, 160

Thomason, C. S., Lieut. Ben.E., at Delhi, 109; *ibid.* note [1]

Thring, J. E., Capt. R.A., with Jaunpore Field Force, 261 etc.; at Siege of Lucknow, 270 etc.; with Outram in Right Attack, 281 etc.; with Lugard's Column, 366 etc.

Tigra, Benares district, capture of, 366

Tiluk, Havildar Ben. Sappers and Miners, at Kashmir Gate, Delhi, 121

Tiroul, fort in Oudh, capture of, 379

Tola Kanjaran (on the Dhasan), skirmish at, 419

Tolsipore, Oudh, rebels collect at, 396; Bala Rao's defeat at (1st Action), 397; Mahndi Husain's defeat at (2nd Action), 400

Tombs, H., Capt. and Brev.-Major Ben.H.A., at the Hindun, 68, 71; at Badli-ki-Serai, 72 *ibid.* note [1]; commands column at the Eedghur, Delhi, 82, 83; in Ochterlony Gardens, 84; saves Hill's life during rebel cavalry raid, 90, 91; his numerous acts of gallantry, gains the Victoria Cross on this occasion, *ibid.* note [1]; commands artillery at Najafghur, 99 etc.; commands No. 4 Siege Battery, 117; with Cavalry Brigade on day of assault, 124 etc.; at siege of Lucknow, 275 etc.; with Walpole's column at Ruiya and Sirsa, 356, 357; with Sir Colin Campbell at Bareilly, 360; at Mohamdee and Shahabad, 364

Tons, northern affluent of Ganges, Kunwar Sing defends bridge over, 366

—— southern affluent of Ganges, affair on, 423

Traill, G. B., Lieut. Ben.H.A., spikes rebel guns, 125, 126

Travers, F. J., Capt. R.A., commands

INDEX

R.A. at 2nd Relief of Lucknow, 212 etc.
Trimmoo Ghat, Punjab, affair at, 31
Troup, Brig. C., Lieut.-Col. and Brev.-Col. 68th Ben.N.I., commands column at Shahjehanpore, 386; at Mithaulee, *ibid.*, sends small force towards Kyseeghur Jungle, 398
Turnbull, S., Capt. and Brev.-Lieut.-Col. Bom.H.A., commands advanced guard at Barodia, 300 etc.; commands artillery of Central India Field Force at Jhansi and the Betwa, 307 etc.; killed at Jhansi, 314
Turner, F., Capt. and Brev.-Major Ben.H.A., at Badli-ki-Serai, 70; at Siege of Delhi, 80 etc.; in Octerlony Gardens, 84; in the Subzi Mundi, 93; commands left half of No. 1 Siege Battery, 114; commands artillery, Greathed's column, 133; at Bulandshahr and Agra, 134 etc.; commands Ben. and Mad. Artillery at 2nd Relief of Lucknow, 212 etc.; provides E Troop R.H.A. with ammunition, 260 note [1]; at Siege and Capture of Lucknow, 275 etc.

Unao, Oudh, action at, 176

Van Courtland, W., General, 141; *ibid.* note [1]
Volunteer Cavalry at Chinhut, 156, 158, 159; with Havelock's column, 168 etc.; 195 etc.; with the Jaunpore Field Force, 261 etc.
Walker, C. P. B., Lieut.-Col. 2nd D.Gs. at Bungaon, 400, 401
Walpole, Brig. R., Lieut.-Col. Rifle Brigade, 232 note [1]; in combats round Cawnpore, 233-4; in 2nd Battle of Cawnpore, 236-8; his feint on the Ramgunga, 256; commands a division at Siege of Lucknow, 276; his march up the Ganges to Rohilkhund, 354, 356; his failure at Ruiya, 356; joins Sir Colin Campbell at Aligunge, 357; at Battle of Bareilly, 359 etc.; commands Bareilly garrison, 363; defeats rebels at Sassaya, 398
Warren, F. G. E., Lieut. R.A., at Miangunge, 260; in the Kaisar Bagh, Lucknow, 291; at the Gumti, 377; at Rehora, 391
Water Bastion, Delhi, 108; breached, 117; assaulted, 120; its guns turned on Selimghur, 128
Water, want of, at Chinhut, 158; at Cawnpore, 151, *ibid.* note [1]
Watson, S., Gunner Ben.A., at Arrah, 369
Wetherall, Brig. E. R., Colonel unattached, in command at Saraon, 387; captures Rampore Kussia, 388
Wheeler, Sir Hugh, Major-Gen., commands Cawnpore Division, 51; his two fatal steps, 51-3
Whinyates, F., Lieut. R.A., at Kankar, 355
Whitlock, Brig. G., Lieut.-Col. and Brev.-Col. 3rd Mad.E.R., commands Saugor Field Force, 250; at Kampti; his advance on Banda, 323 etc.; defeats rebels at Kubrai and Banda, 324 etc.; marches on Kirwi, 348, 349; his operations in Kirwi district, 421-3
Willoughby, G. D., Capt. R.A., in charge of Delhi Arsenal, 15; his defence of the arsenal, 16, 17; his escape, *ibid.*; nominated for V.C., *ibid.* note [1]; murdered on road to Meerut, *ibid.*
Wilson, Brig.-Commandant A., Lieut. Col. and Brev.-Col. Ben.A., at Meerut, 9; *ibid.* note [1], 10; in command of Field Force, 12; on the Hindun, 68, 69; joins Barnard, 70; commands artillery at Badli-ki-Serai and on the Ridge, 70 etc.; succeeds Barnard, 88, *ibid.* note [1]; discusses possibility of *coup de main* at Delhi, 103, 104; his views when in command, 105-7; accepts the plan of attack prepared by the Engineers, 111; at assault of the city, 126; goes on sick leave, 132; commands artillery at Siege of Lucknow, 275
—— N., Lieut. -Col. and Brev.-Col. 64th Foot, at Cawnpore, 195; holds bridge head on Ganges, 232; killed, 234
—— W., Lieut. Ben.A., in Delhi on day of outbreak, 15; mutiny of his men, 18; his escape, *ibid.*; commands 2/1 Ben.H.A. with Cavalry Brigade on day of assault of Delhi, 124, 125
Windham, C. A., Major-Gen., commands at Cawnpore, 210; troops at his disposal, *ibid.* 211; his orders from Sir Colin Campbell, 229, 230; his plans, 230, 231; at-

tacked by Tantia Topi, 231–5; commands bridge-head entrenchment at 2nd Battle of Cawnpore, 236 etc.
Wood, D. E., Lieut.-Col. and Brev.-Col. R.H.A., commands field artillery of Outram's Force at Siege of Lucknow, 279
Woodland, C., Capt. 1st Mad.N.I., at Kirwi Palace, 421

Yellow House, near Alam Bagh, rebel guns at, during Havelock's advance, 197, 199

Zeenut Mahal, Queen of Delhi, 12, *ibid.* note [2]
Zemindari, Oudh Yeomen, bravery of, at Newabgunge-Barabanki, 375
Zirapore, Cen. India, skirmish at, 411

THE GOVERNMENT OF INDIA IN 1857

THE GOVERNOR-GENERAL IN COUNCIL

| Governor of Madras *including* Berar Nagpore The Carnatic Circars etc. — Commissioners etc. | Governor of Bombay *including* The Presidency and the States of Sind, Baroda, Guzerat, etc. — Commissioners Residents etc. | Lieut.-Governor of Bengal *including* Bengal Behar Assam Orissa etc. — Commissioners etc. | Lieut.-Governor of The North West Provincees *including* Delhi States Rohilkhund Doab Saugor and Nerbudda Territories etc. (See Maps 8 & 10) — Agent Commissioners etc. | Chief Commissioner Oudh — Commissioners etc. | Chief Commissioner The Punjaub — Commissioners etc. | Agents-General, Agents, Residents, etc., in Protected States *including* Dominions of Nizam of Hyderabad, of Hollkar, of Sindiah etc. Smaller States in Central India and Bundelkhund and Rajputana etc. |

POPULATION (Thornton's Gazateer, 1857)

Bengal 40,871,091   Bombay, Sind and Native States 15,578,992
Oudh 2,400,000   Madras (City) 720,000
Punjaub 10,435,710
Rohilkhund (not given)
Central India and Bundelkhund } about 44,000,000

TOTAL POPULATION of Indian Peninsula 176,000,000
(Hodge's Dictionary of Dates 1862)

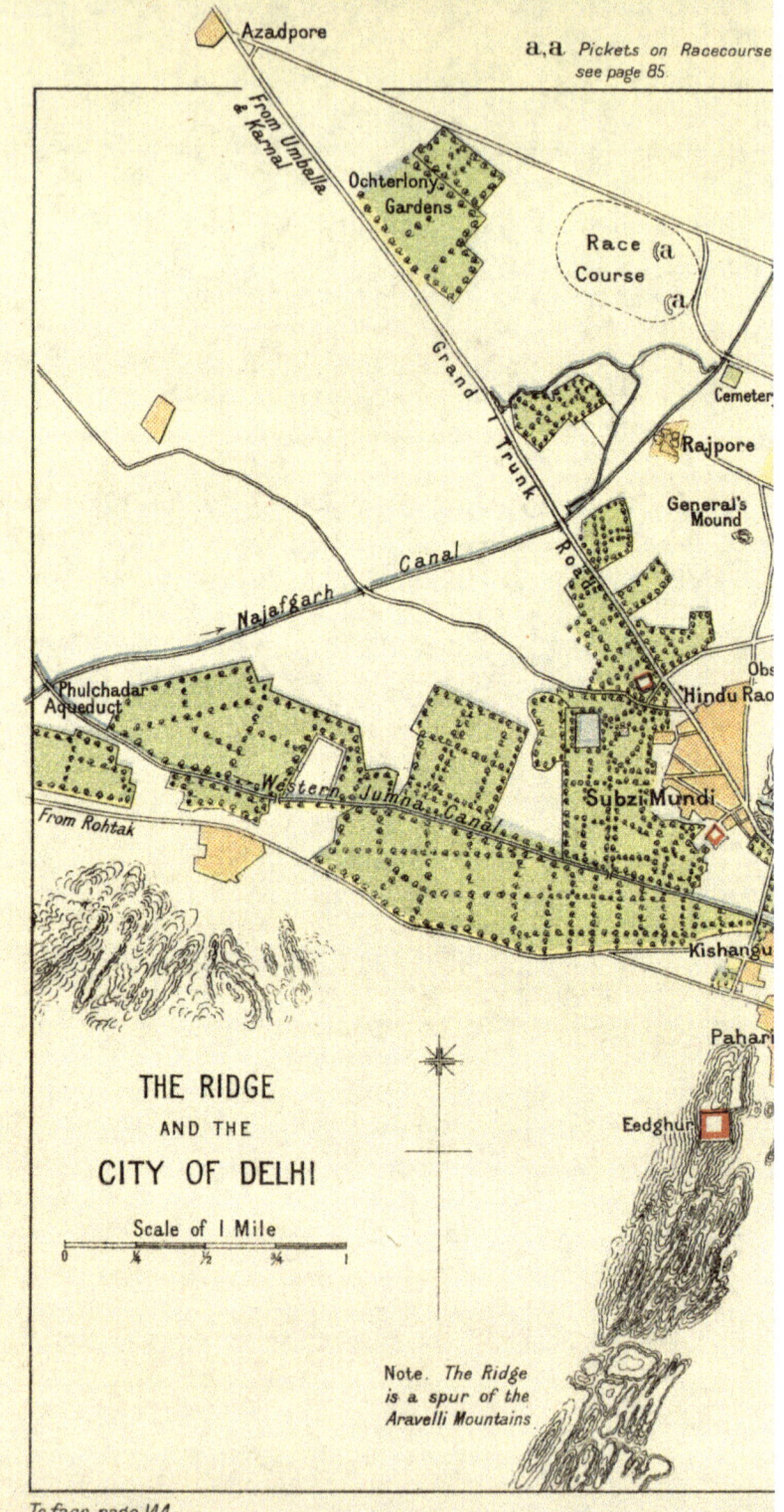

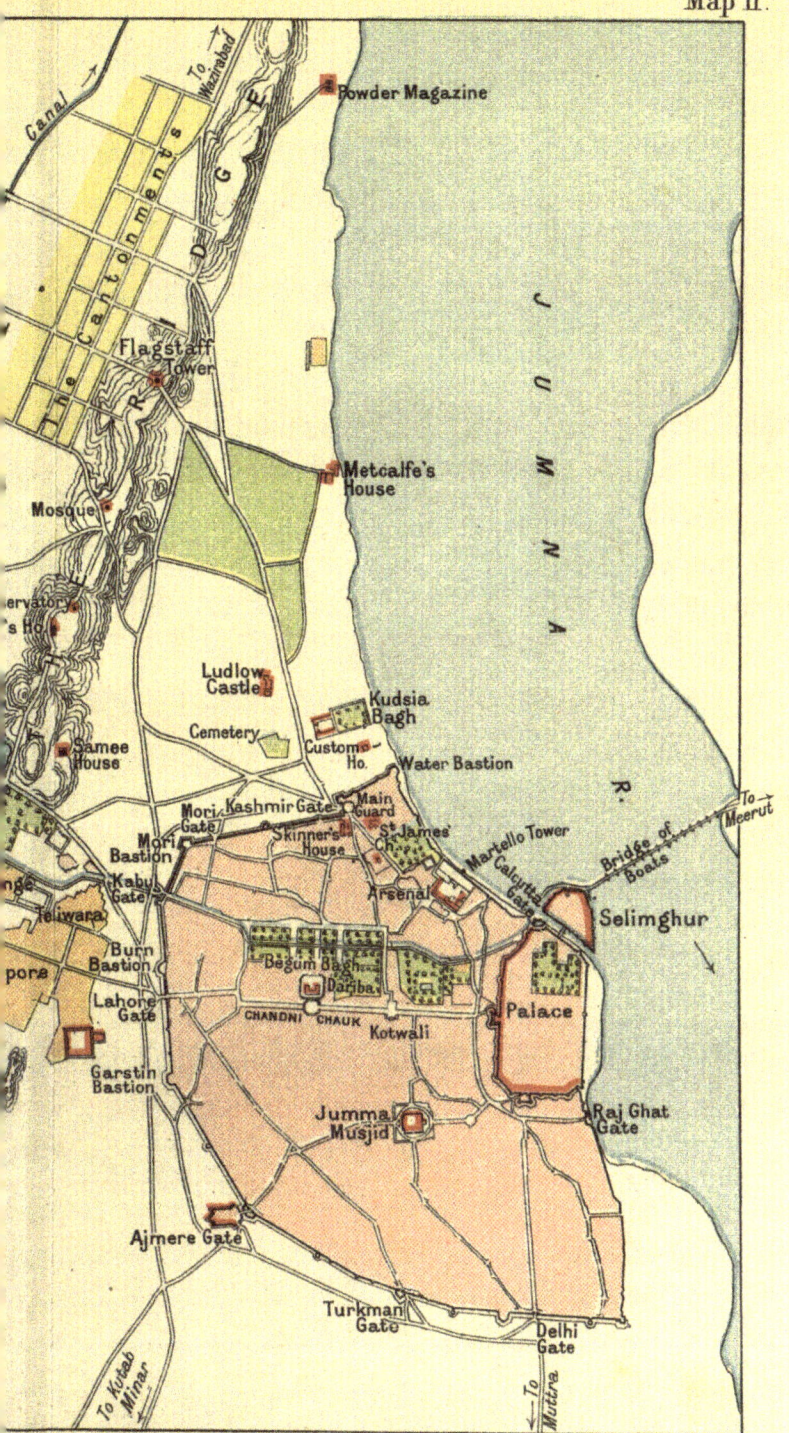

Map II.

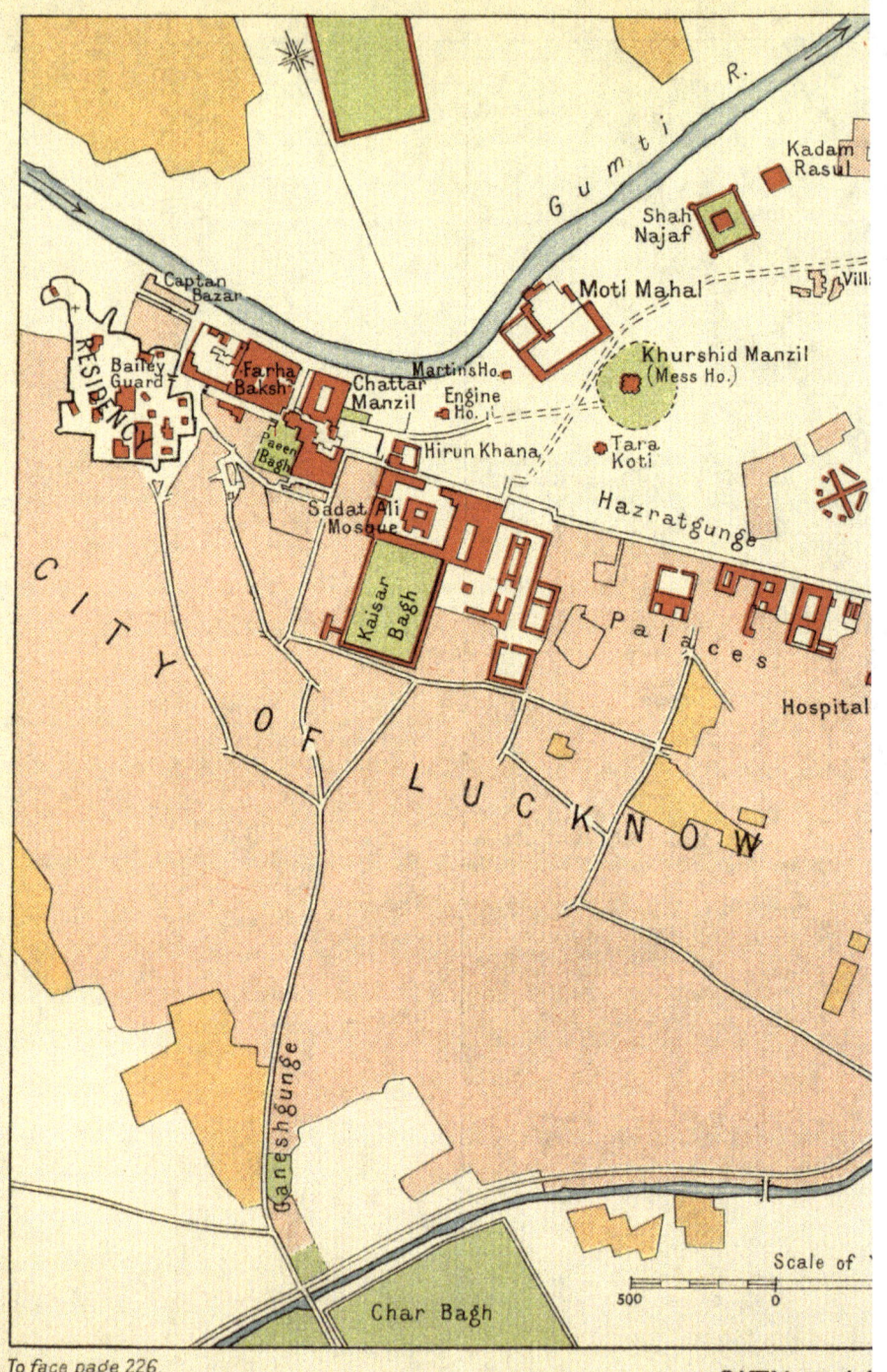

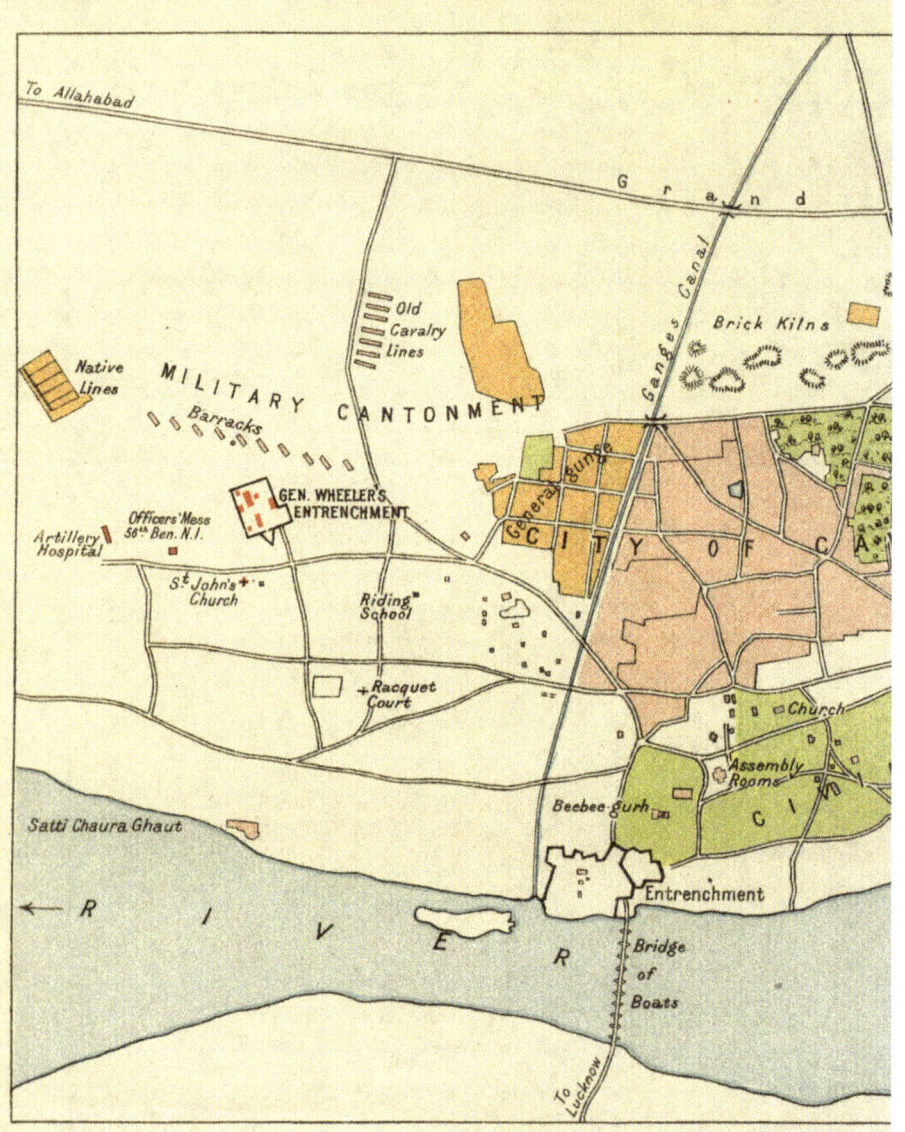

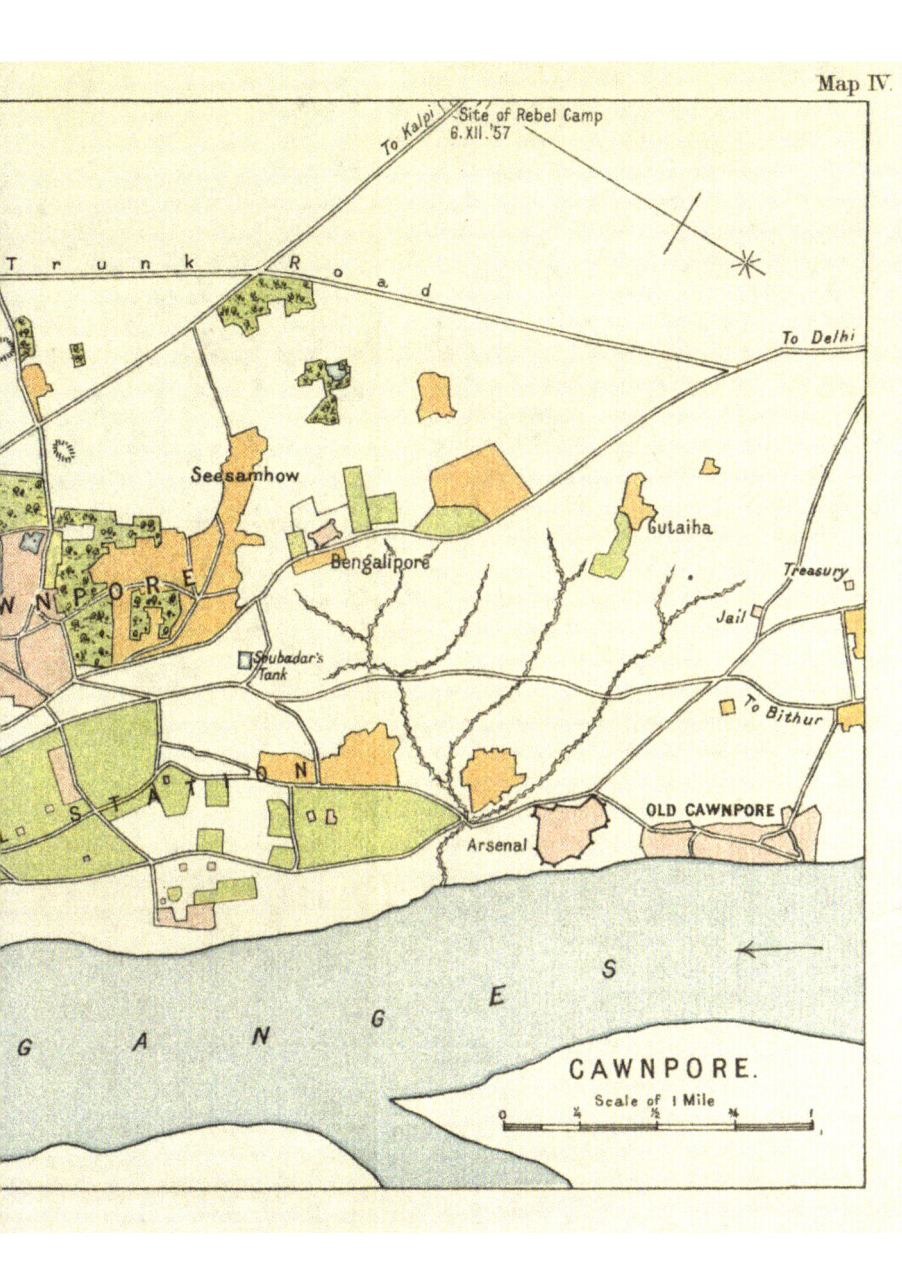

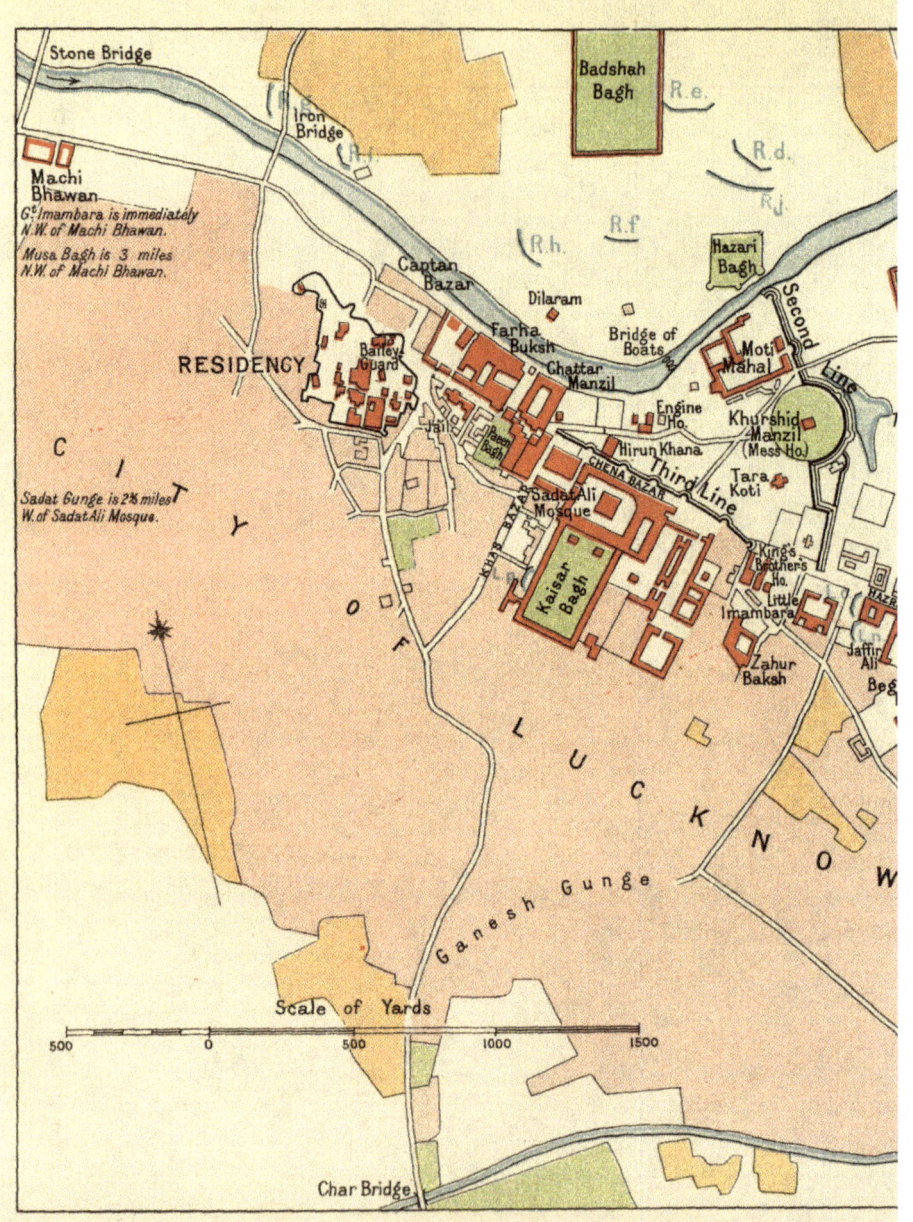

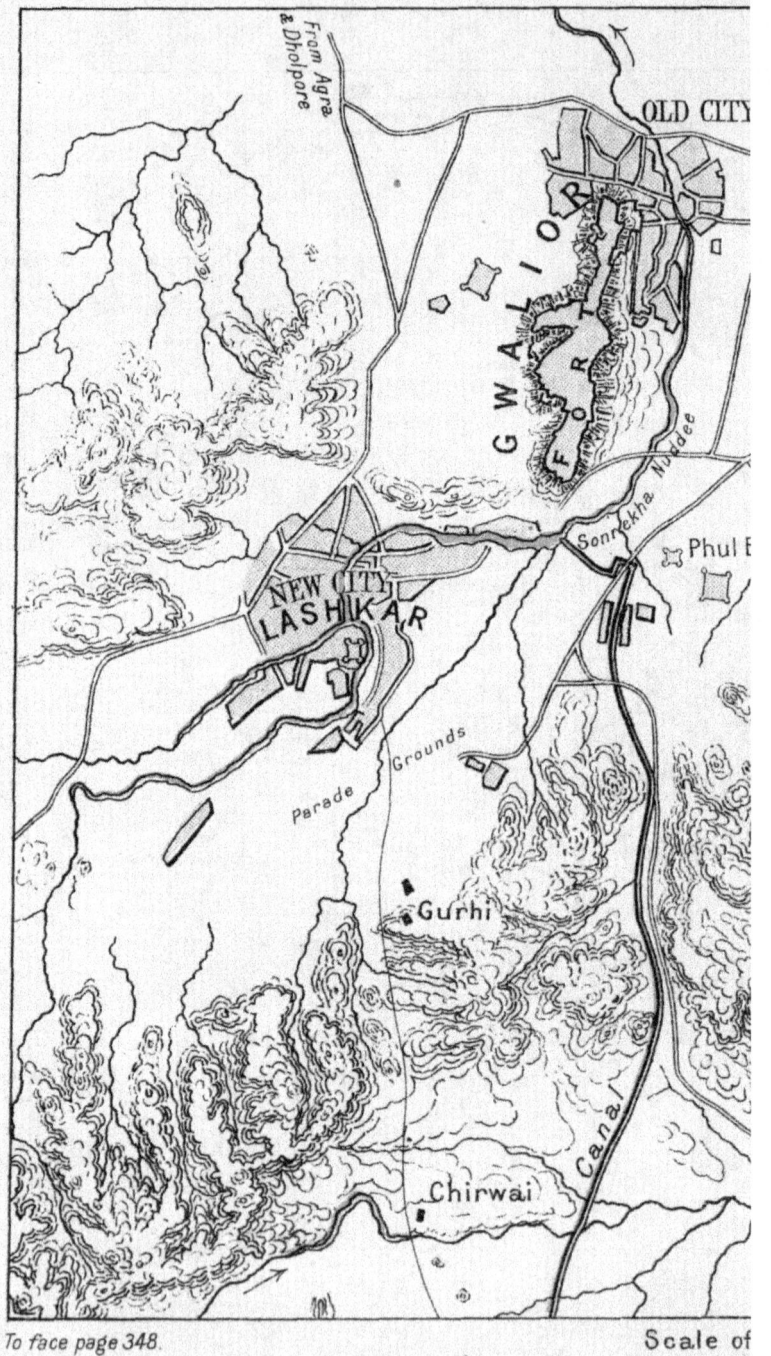

UND GWALIOR

Map VI.

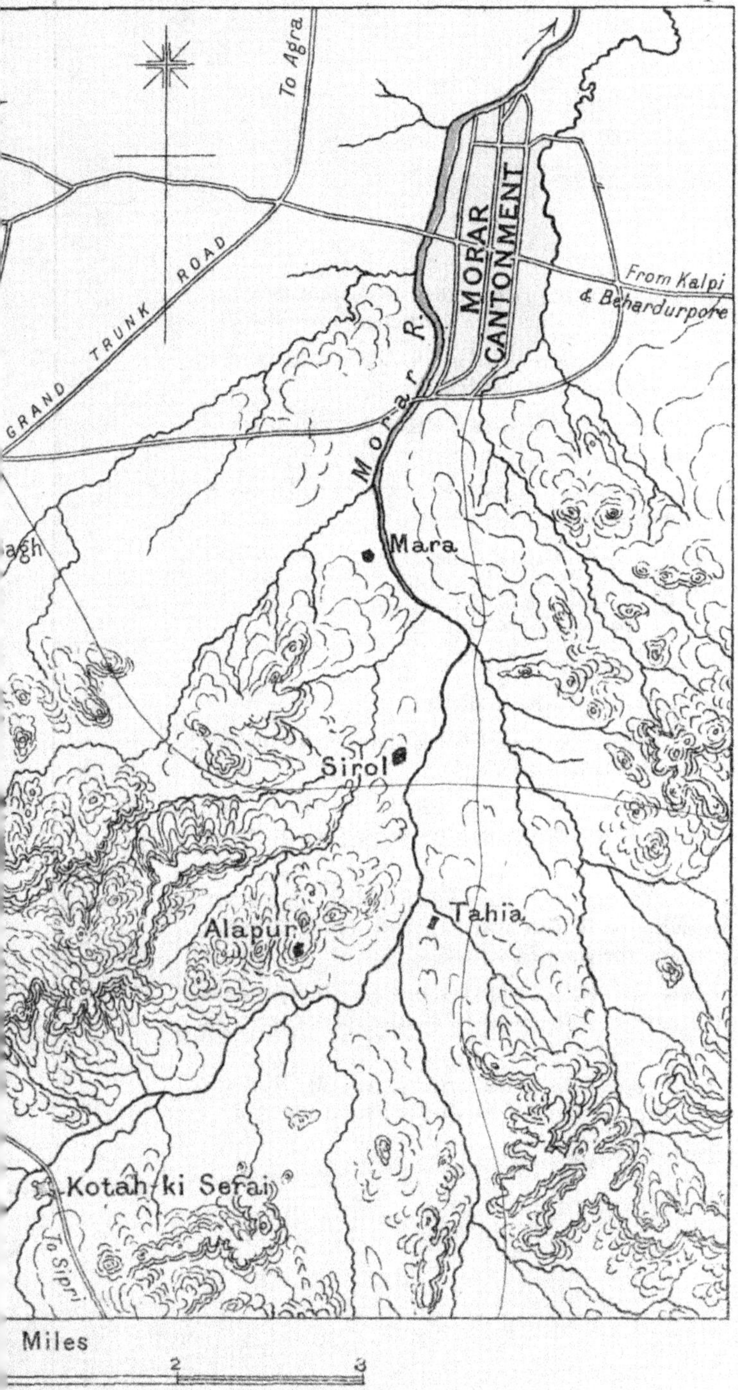

To Agra

GRAND TRUNK ROAD

MORAR CANTONMENT

From Kalpi & Behardurpore

Morar R.

Mara

Sirol

Alapur

Tahia

Kotah-ki Serai

To Sipri

Miles
2 3

There were five districts, each under a Commissioner, in the Province of Bengal (as distinct from the Bengal Presidency):

1. Nuddea, a name given to a district that included Calcutta and the neighbourhood.
2. Burdwan, the district west of the Hooghly and south of the Ganges.
3. Rajishaye, the district north of the Ganges and west of the Brahmapootra.
4 & 5. Dacca and Chittagong, the districts east of the Brahmaputra.

In Behar there were two districts:—

1. Bhagulpore, comprising Eastern Behar.
2. Patna, comprising Western Behar.

There was a Commissioner for Chota Nagpore and one for Cuttack which included Orissà.

On the western frontier of the Bengal Presidency lay two districts of the North Western Provinces, each under a Commissioner.

These were the Benares district, including Jaunpore, Ghazipore, Azimghur and Gorruckpore, and the Allahabad district lying south of the Ganges.

In Chota Nagpore was situated the small independent principality of Ramghur. The western portion of Orissà was a wild and uncivilized region.

NOTE.—*In 1858 the population of Calcutta was 413,182, of which some 7,000 were Europeans, and that of Patna 284,132.*

BEHAR AND BENGAL

Delhi and the Delhi States were uuder the Lieut.-Governor of the North West Provinces, and an Agent and Commissioner was at the Court of the Moghuls.

There were also six districts under Commissioners :—

1. Agra, including Muttra, Furruckabad, Mainpuri and Etawah.
2. Meerut, including Dehra (Map IX), Bolandshahr and Alighur.
3. Bareilly, including Shahjehanpore, Moradabad and Badaon.
4. Allahabad, including Futtehpore, Cawnpore, Banda and Hamirpore.
5. Benares, including Jaunpore, Azimghur and Gorruckpore.
6. The Saugor and Nurbudda Territories (Map X).

In addition the Jalaon, Jhansi and Chanderi districts were (in 1854) formed into a Superintendency (Map X).

POPULATION (*Thornton's Gazateer*, 1857).

Agra	125,262
Allahabad	72,093
Benares	185,984
Bareilly	111,332
Cawnpore	59,000
Delhi	152,406
Lucknow	300,000
Meerut	29,014
Philibit	26,760
Sultanpore	1,500

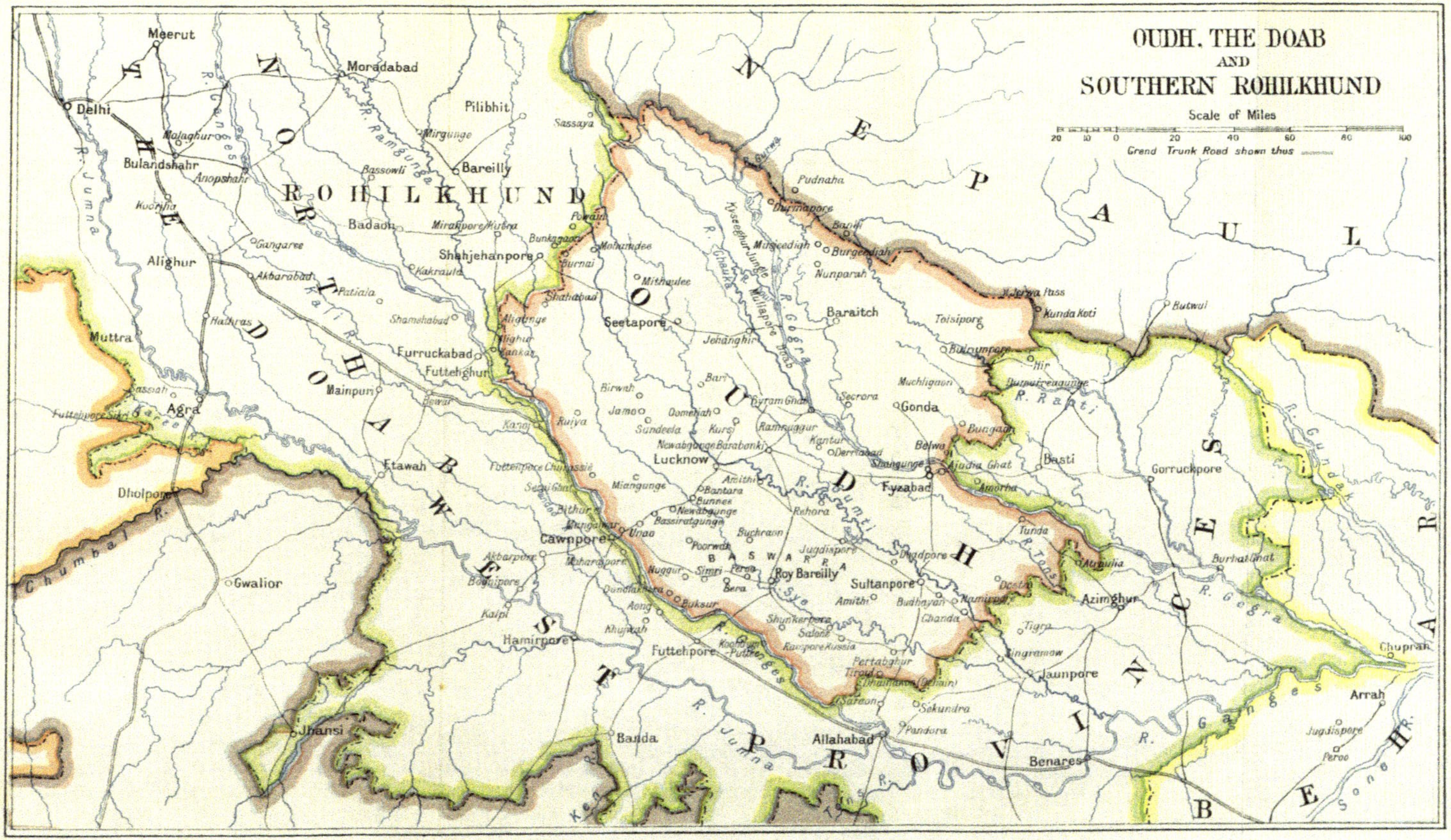

Map VIII

OUDH, THE DOAB
AND
SOUTHERN ROHILKHUND

Scale of Miles

Grand Trunk Road shown thus

At end of volume.

POPULATION
Thornton's Gazateer (1857)

Delhi	152,406
Ferozepore and District	16,890
Lahore	95,000
Meerut and District	1,135,072
Mooltan	80,966
Moradabad	57,414
Peshawar	53,295
Shahjehanpore and District	812,588
Umballa	21,962

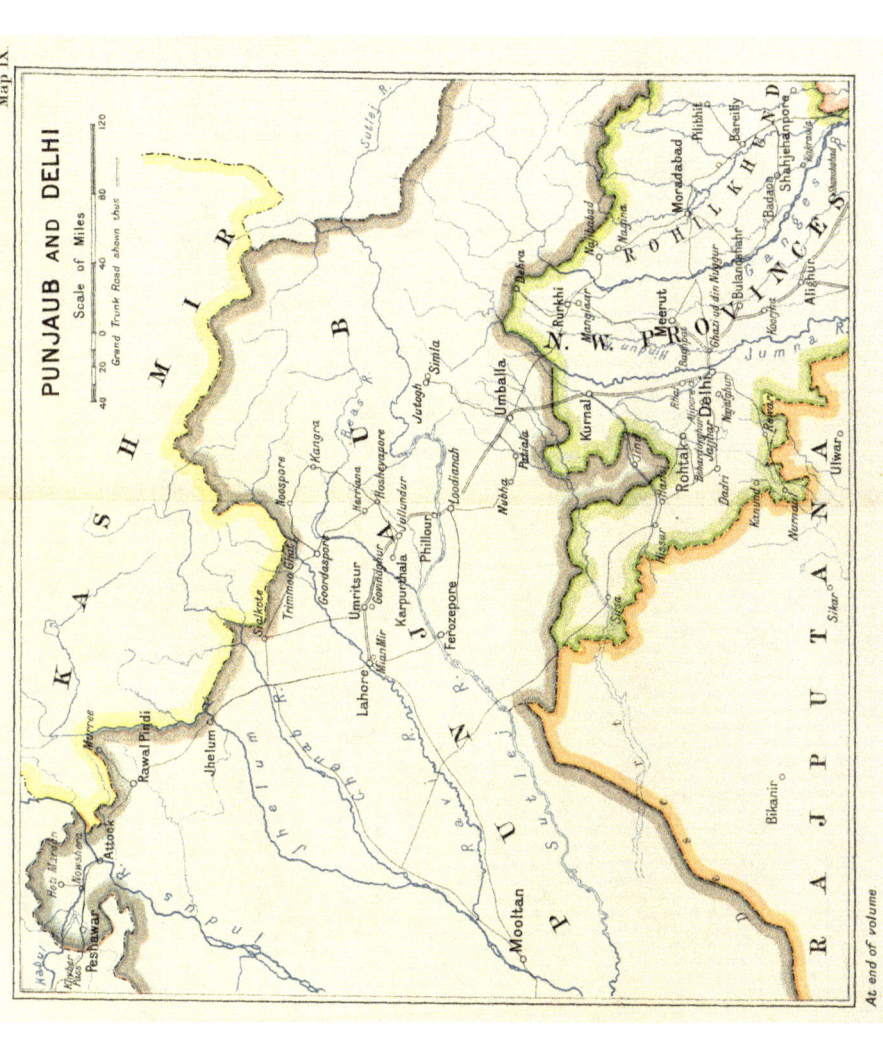

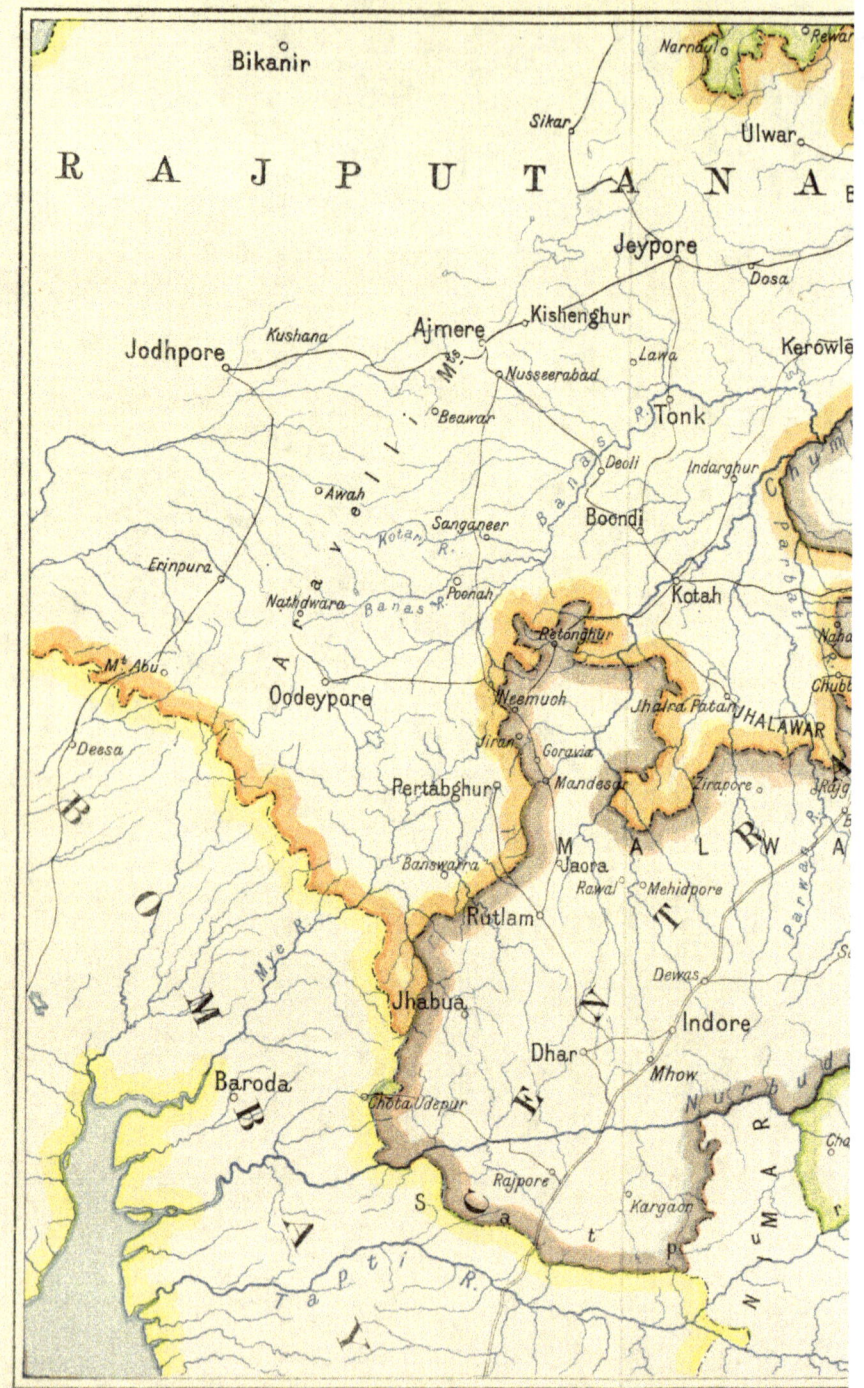

At end of volume.

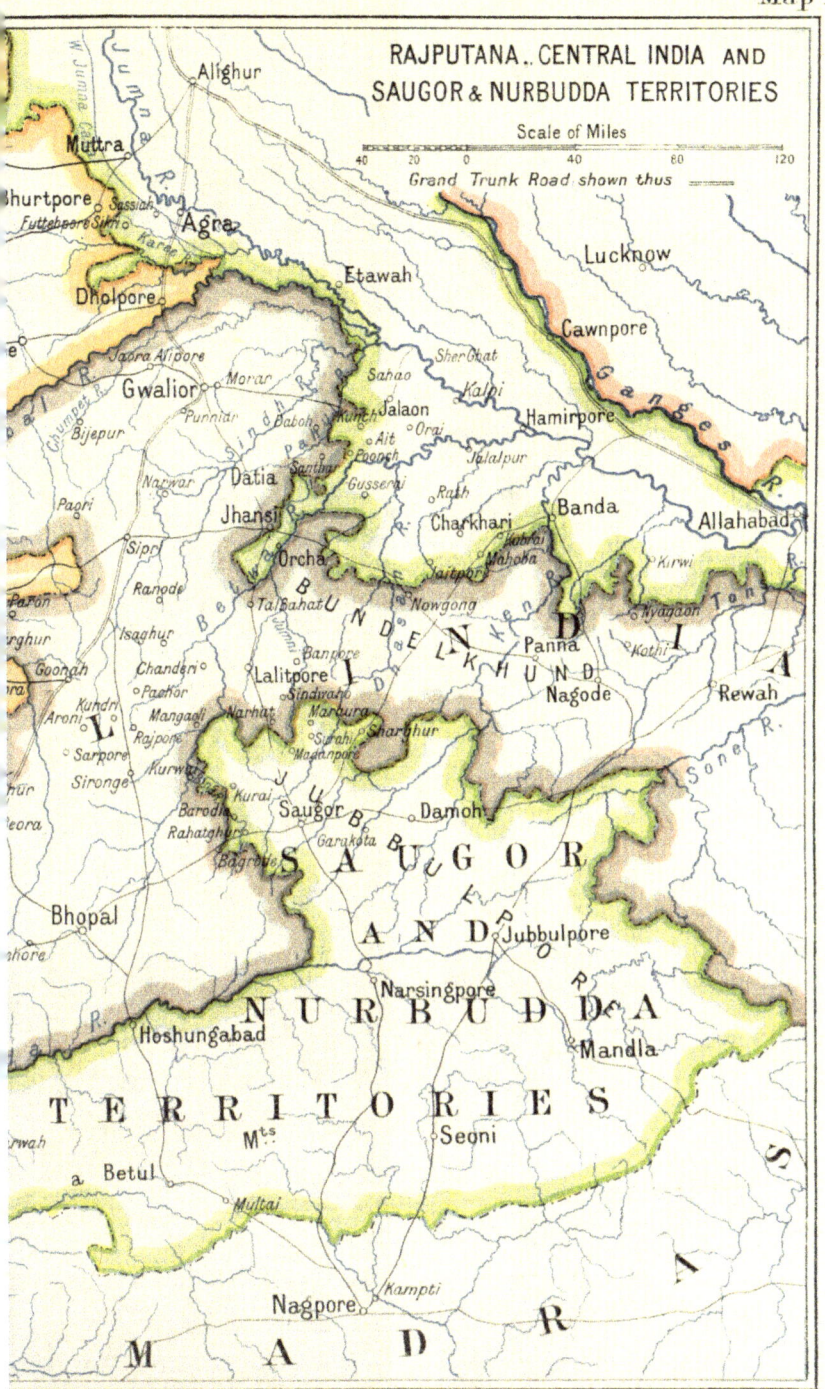

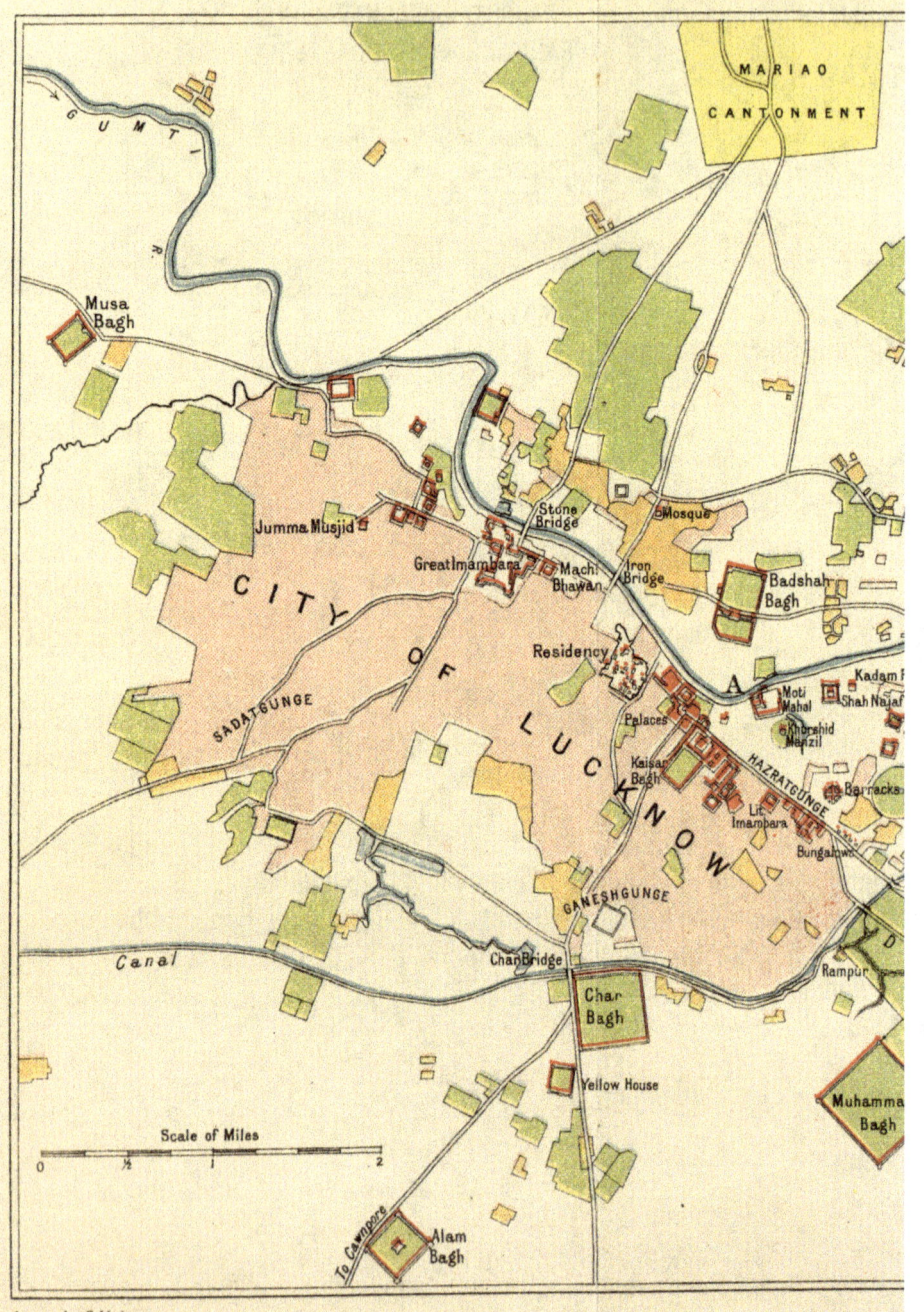

Map XI.

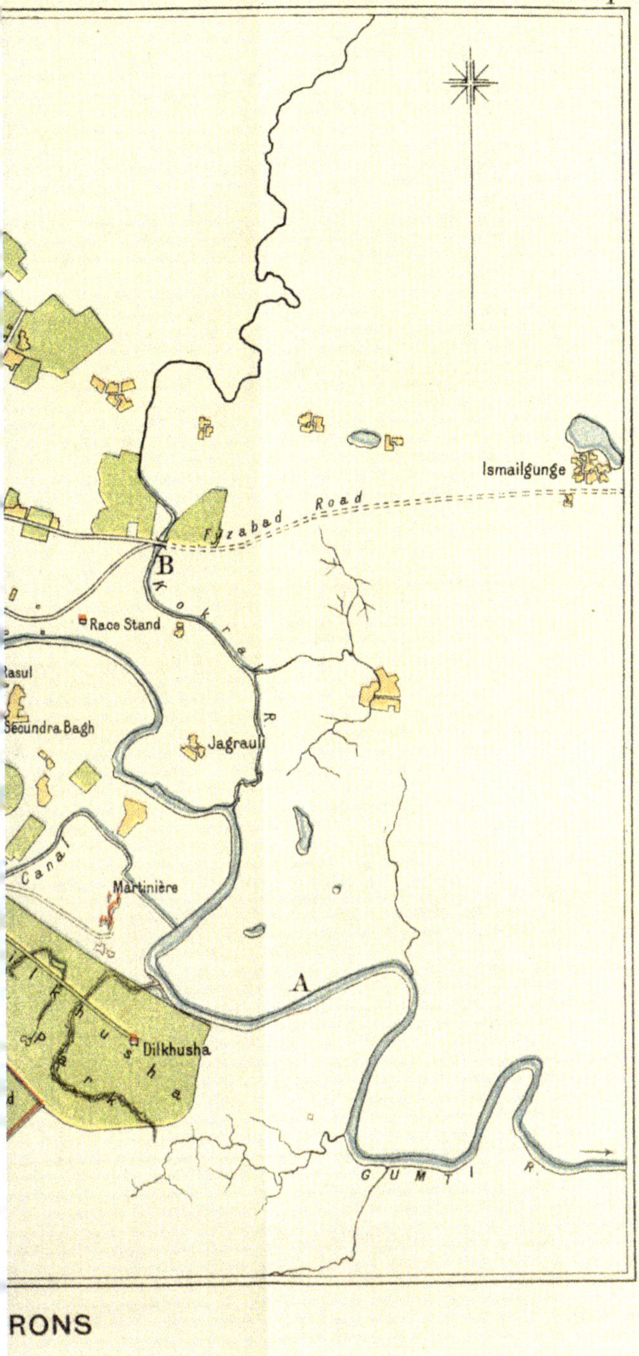
Ismailgunge
Fyzabad Road
B
Kokrail R
Race Stand
Rasul
Secundra Bagh
Jagrauli
Canal
Martinière
A
Dilkhusha
Dilkhusha Park
GUMTI R.
RONS

A Positions where bridges of boats were made for the passage of Outram's Force, 6th and 16th March.

B Position where the first emplacement of the Right Attack (R.a.) was made. See page 281.

www.ingramcontent.com/pod-product-compliance
Lightning Source LLC
Chambersburg PA
CBHW070754300426
44111CB00014B/2398